FORESTRY SCIENCES

Volume 62

The titles published in this series are listed at the end of this volume.

J. Buongiorno

Global Concerns for Forest Resource Utilization

Sustainable Use and Management

Selected Papers from
the International Symposium of
the FORESEA MIYAZAKI 1998

Edited by

ATSUSHI YOSHIMOTO

and

KIYOSHI YUKUTAKE

KLUWER ACADEMIC PUBLISHERS

DORDRECHT / BOSTON / LONDON

A C.I.P. Catalogue record for this book is available from the Library of Congress.

ISBN 0-7923-5968-2

Published by Kluwer Academic Publishers,
P.O. Box 17, 3300 AA Dordrecht, The Netherlands.

Sold and distributed in North, Central and South America
by Kluwer Academic Publishers,
101 Philip Drive, Norwell, MA 02061, U.S.A.

In all other countries, sold and distributed
by Kluwer Academic Publishers,
P.O. Box 322, 3300 AH Dordrecht, The Netherlands.

Printed on acid-free paper

Printed in the Netherlands.

Contents

Contributors

Martin, R.M.
Forestry Planning and Statistics Branch, Forestry Dept.
Food and Agriculture Organization of the United Nations
Viale delle Terme di Caracalla, Roma 00100, Italy
e-mail: michael.martin@fao.org

Lu, W.D.
College of Forest Products, Northeast Forestry University
Harbin, 150040, P.R.China
e-mail: ljy@heb.col.com.cn

Rojas, M.
Dept. of Economics, Universidad de las Américas - Puebla
Puebla, 72820 Mexico
e-mail: mrojas@mail.udlap.mx

Wilson, B., and Wang, S.
Industry, Trade & Economics Program, Canadian Forest Service
506 West Burnside Road, Victoria, B.C., Canada V8Z 1M5
e-mails: bwilson@pfc.forestry.ca

Paredes V., G.
Instituto Forestal
Huerfanos 555, Santiago, Chile
e-mail: gparedes@infor.cl

Yoo, B.I.
Forest Land Use Div., Forestry Research Institute
Seoul, 130-012, Korea
e-mail:frirok@chollian.net

Khattabi, A.
Dept. of Forest Economics, Ecole Nationale Forestière d'Ingénieurs
B.P. 511, Tabrikt SALE – Morocco
e-mail: abdelk @ bigfoot.com

Eba'a Atyi, R.
The University of Dschang, Tropenbos Cameroon Programme
B.P. 219 Kribi, Cameroon
e-mail: 105405.1222@compuserve.com

Brännlund, R.
Dept. of Economics, Umeå University
S-901 87 Umeå, Sweden
e-mail: Runar.Brannlund@natek.umu.se

Sondell, J.
Project 'Market-driven saw log production', The Forestry Research Institute of Sweden, Science Park, S-751 83 Uppsala, Sweden
e-mail: jan.sondell@skogforsk.se

Robertson, G.
USDA Forest Service, PNW Research Station
3200 SW Jefferson Way, Corvallis, Oregon 97331 USA
e-mail:robertsong@fsl.orst.edu

Howard, T.E.
Dept. of Natural Resources, University of New Hampshire
215 James Hall, Durham, NH 03824 USA
e-mail: tehoward@christa.unh.edu

Adams, D.M.
Dept. of Forest Resources, Oregon State University
Corvallis, OR 97331 USA
e-mail:adams@ccmail.orst.edu

Haynes, R.W.
USDA, Forest Service, Pacific Northwest Research Station
P.O. Box 3890, Portland, OR, 97208 USA
e-mail:rhaynes/r6pnw_portland@fs.fed.us

Buongiorno, J., Zhu, S., and Tomberlin, D.
Dept. of Forest Ecology and Management, University of Wisconsin,
1630 Linden Drive Madison, WI 53706 USA
e-mail: jbuongio@facstaff.wisc.edu

Yoshimoto, A.
Dept. of Agricultural & Forest Economics, Miyazaki University
Miyazaki 889-2192 Japan
e-mail: yoshimoa@fec.miyazaki-u.ac.jp

Phelps, S.
Canadian Forest Service, Dept. of Natural Resources
7th Floor, 580 Booth Street, Ottawa, Ontario, Canada K1A 0E4
e-mail: sphelps@nrcan.gc.ca

Obersteiner, M.
Forest Resources Project, IIASA
2361 Laxenburg, Austria
e-mail: oberstei@iiasa.ac.at

Perez-Garcia, J., Wang, Y. and Xu, W.
Center for International Trade in Forest Products
College of Forest Resources, University of Washington
Box 352100, Seattle, WA 98195-2100 USA
e-mail: perjohm@u.washington.edu

Ahmad, I.
Forest Research Institute Malaysia
Kepong, 52109, Kuala Lumpur, Malaysia
e-mail: iahmad@u.washington.edu

Ankarhem, M., Brännlund, R., and Sjöström, M.
Dept. of Economics, Umeå University
S-901 87 Umeå Sweden
e-mail: mattias.ankarhem@econ.umu.se

Tachibana, S., and Nagata, S.
Div. of Forest Science, University of Tokyo
Tokyo 113-8657 Japan
e-mail: binn@fr.a.u-tokyo.ac.jp

Nilsagård, H.
Dept. of Forest Economics
Swedish University of Agricultural Sciences, S–901 83 Umeå, Sweden
e-mail: Hans.Nilsagard@sekon.slu.se

Lippke, B., Bare, B.B., Xu, W., Moffett, J., and Turner, B.
Center for International Trade in Forest Products
College of Forest Resources, University of Washington
Box 352100, Seattle, WA 98195-2100 USA
e-mail: blippke@u.washington.edu

Byström, S., and Lönnstedt, L.
Group Staaf Technology and Dept. of Forest Economics
MoDo Company and Swedish University of Agricultural Sciences
891 01 Örnsköldsvik and 901 83 Umeå, Sweden
e-mail: stig.bystrom@MoDoGroup.com

Yukutake, K.
Dept. of Agricultural & Forest Economics, Miyazaki University
Miyazaki 889-2192 Japan
e-mail: yukutake@fec.miyazaki-u.ac.jp,

Kato, T.
Forestry and Forest Products Research Institute
P.O. Box 16, Tsukuba Norin Kenkyu Danchi
Tsukuba, Ibaraki, 305-8687 Japan
e-mail: kato0298@ffpri.affrc.go.jp

Iwamoto, J.
College of Agriculture, Ehime University
Matsuyama 790-8566 Japan
e-mail: junichi@agr.ehime-u.ac.jp

Yamamoto, H.
University Forests, Faculty of Agriculture, University of Tokyo
Yayoi 1-1-1, Bunkyo, Tokyo, 113-8657, Japan
e-mail: yama@uf.a.u-tokyo.ac.jp

Acknowledgements

The FORESEA (FOREst SEctor Analysis) Miyazaki Symposium was produced through the efforts of many. We appreciate the gracious support from Japanese Forestry Agency. With its great assistance, we were able to receive financial support from a wide variety of sources. Funds from Ministry of Foreign Affairs in Japan, Ministry of Education, Culture and Sports, and the Commemorative Association for the Japan World Exposition (1970) brought approximately 37 scientists from developing countries throughout the globe. Other funding sources include Japan International Cooperation Agency, Forest Policy Research Institute, the National-Land Afforestation Promotion Organization, Japan Wood Products Information and Research Center, Venichu Ltd., Jukensangyo Co. Ltd., Oji Paper Co. Ltd., Japan Forestry Association, Japan Forest Technical Association, Japan Housing Wood Technique Center, Forest Development Technological Institute, Japan Paper Association, Japan Lumber Importers' Association, Japan Plywood Manufacturers' Association, Rinya Kousai Kai (Forestry Promotion Association), All-Japan Federation of Lumber Associations, All-Japan Federation of Lumber Cooperatives, All-Japan Wood Housing Machine Precut Association, Forest Enterprisers' Association of Japan, The Forestry Mechanization Society, Japan Greenery Research and Development Center, National Forestry Extension Association, the Japan Forestry Association, Forest Development Corporation, Wooden Home Builders Association of Japan, Central Cooperative Bank for Agriculture and Forestry, National Federation of Forest Owners' Cooperation Associations, Miyazaki Prefectural Forestry Association, Miyazaki Prefecture Forest Conservation and Forest Road Association, Miyazaki Prefectural Civil Engineering Forestry Association, Miyazaki Prefectural Forestry Working

Machinery Center, Miyazaki Prefectural Forest Owners' Cooperative Association, Miyazaki Prefectural Lumber Cooperative Association, Miyazaki Prefectural Logging Cooperative Association, and Kyushu Land Afforestation Promotion Organization. We thank all for their sponsorship of the symposium.

We are grateful to Dr. Michael Martin, Dr. Richard Haynes, and Dr. Kiyoshi Yukutake for serving as invited lecturers to start the discussions on global concern for forest resources utilization. With their introductory presentations, participants deepened discussion, and shared scientific findings and practical issues during the symposium.

As is always the case, participants play a key role in the success of the symposium. The energetic group of scientists and practitioners who work in the forest sector came to participate in the symposium. They found time to get acquainted, to discuss scientific problems and findings of mutual interest, and to appreciate the academic and practical needs and situations of their colleagues. We are grateful to all for their involvement in the symposium and their success in the symposium.

We wish to thank our students for putting up with a very preliminary text, and for finding errors and misprints. Last but not least, we owe sincere thanks to Richard Haynes, Judy Mikowski and Lynn Starr, who have read the manuscript very carefully and fixed not only linguistic bugs but prevents us from quite a number of crucial mistakes.

Preface

During the last several decades, forest economists have sought to balance environmental protection and economic development through forestry activities, e.g., forest resource management, resource utilization and timber trading. We have seen a persistent interest in the use of quantitative and qualitative analysis for investigating the processes of and policies for forest sector activities as well as providing an objective framework for planning and evaluation of forest resources. These analyses are not only regional, but also national and international.

Having started in the 1980s, various forest sector analysis symposiums have created a scientific network among modelers and practitioners. During the last several years, sufficient progress had been made in theoretical and practical aspects. An initial idea to have a forest sector analysis (FORESEA) symposium in Japan was formed through the executive committee meeting in Tampere, Finland, 1996, when the last IUFRO World Congress was held. During the meeting, this idea was suggested by the chairman of the forest sector analysis working group, Dr. Richard Haynes, and was carried out by Dr. Kiyoshi Yukutake in Japan. This was mostly due to an emerging issue for Japan, as one of the largest timber import countries to deal with various international forestry issues--especially within the Pacific Rim nations. Just after the World Congress, the executive committee was formed by a group of forest economists in Kyushu, Japan. Then the symposium was named FORESEA MIYAZAKI to be held in Miyazaki, and to be co-sponsored by Miyazaki University, IUFRO Working Party 6.16, and CINTRAFOR at the University of Washington. The symposium, while slightly different from previous ones, focused upon gathering modeling practitioners and users from a wide variety of nations, including both timber exporting and importing countries. This symposium has revealed a large, energetic, and productive group of scientists and practitioners.

The symposium had three themes: understanding global forest sector issues, discussing the contribution of modeling efforts to forest sector analysis, and discussing the role of Japanese forest policy in a global sense. A wide variety of papers expressed a great diversity of ideas, approaches, and results. These papers display the broadening and deepening of the field of forest sector analysis. The topics for this book were chosen because of these themes.

This book is structured with the above three themes. Part I consists of papers expressing global forest sector issues. Part II is devoted to papers on modeling efforts for forest sector analysis. Forest sector analysis draws from the intellectual tradition of forestry, economics, and natural resource management. Forest sector modeling has evolved to meet the needs of contemporary policy analysis conducted in support of environmental issues broader than just commodity production. These models are evolving to consider a broader array of values of and outputs from the forest. They are beginning to deal with policy mechanisms broader than just market-based regulations. From a modeling perspective, changes now include more detail of spatial and multiple market levels, intergenerational welfare concerns, non-market valuation issues, and explicit treatment of the uncertainty inherent in both the policy process and in the biophysical systems. The last part, Part III, reflects the role of Japanese forest policy in a global sense. The economic, social and environmental challenges facing Japanese forestry are significant. In this part, papers expressing Japanese forestry issues were included.

There were other papers presented in the symposium, but not included here. Dr. Michael Martin, Dr. Patric Durst, and Ms. Ma Qiang organized the Food and Agriculture Organization of the United Nations session. They described their efforts to develop a broad-scale assessment of multiple issues across different forested regions. Their assessment embodies many of the techniques discussed at the symposium and their findings represent how better information can be derived from the underlying data and assumptions. They also pointed out some of the data limitations that still challenge us. Dr. David Brooks directed the two-night seminar on forest sector analysis, where an introductory presentation on forest sector modeling and examples of forest sector modeling were provided. We are pleased to acknowledge, with thanks, their efforts during the symposium.

Atsushi Yoshimoto
Chief Editor

GLOBAL CONCERNS FOR FOREST RESOURCE UTILIZATION

PART 1

UNDERSTANDING GLOBAL FOREST SECTOR ISSUE

Global Forest Products and Forest Products Trade

R. Michael Martin

Key words: Forest products trade, forest industry, sustainability

Abstract: The forest products industry has changed and evolved in response to feast and famine in the forest products markets. Our societies have adapted to meet new challenges in managing our heritage of rich and productive forests. In less time than most species mature, the nature of forest products demand and forest management has completely changed. Our ability to manage forests "sustainably" in the face of technological change, population expansion and income growth is more constrained by our perception of the "problem" than, perhaps, by biological or ecological considerations. Based on the Food and Agriculture Organization of the United Nations (FAO) most recent outlook studies for the forestry sector, six possible conclusions about the future are presented. Taken together, they have significant implications for sustainable forest management.

1. Introduction

A dear friend of mine recently described this quote from a Swedish newspaper,

"The future is not what it used to be."

This statement caused me to stop and think. In a real sense it is true. But also, in many ways, it is not true at all. Many of the forces shaping our future are the same that we have faced successfully time after time in our past. In this sense, I believe that even a cursory examination of our past successes gives us insight into how we can create choices and opportunities for the future. I am convinced that we can, within bounds, shape and define our destiny.

Here is the organization of my paper. First, I work to confirm this statement about the future. I will describe some of the market forces that compel forest products enterprises to change. Second, I will give some examples of how the profile of wood products consumers is changing. Together, market forces and evolving customers has given us wood products that are different. Alas, we all know that forestry will never be the same. I want to touch upon some hot issues in the forest policy debate and say a few words about a forever discussed topic – "sustainable forest management."

Drawing then from FAO's current work in studying the outlook for the sector, I would like to give you a preview of possible findings and their implications for choice and opportunity in the future. At the end, we will have to come back to the question – will the future be different and in what ways will it be the same. So, let's start.

A history of adaptation and transformation …

I claim that our countries have effectively responded to feast and famine in the forest products markets and that our societies have adapted to meet new challenges in managing our heritage of rich and productive forests.

The forest products industry is a wonderful example of change, evolution and adaptation. For years, the challenge in the forest products industry has been to find markets. Developing new products and new end-uses remains the basis for profitability and survival in a very competitive sector. Product differentiation and market segmentation allow producers an opportunity to take advantage of price elasticity to boost margins.

Sadly, as each producer knows so well, this moment of opportunity is short-lived. Competitors abound, each constantly searching for profitable market niches. As substitutes and new producers emerge, each "new" product faces an ever more crowded field of competitors.

Within mature markets, each producer is under constant pressure to cut costs. Labor must be deployed effectively, fiber supplies closely scrutinized and new technologies ruthlessly deployed to contain costs. Price pressures on commodity product lines are always downward, as capacity seems to be expanding everywhere. And yet, in the face of these intense pressures, new investments are made, new firms are born and publicly listed corporations report growth in earnings and margins.

Wood products consumers are not who they used to be…

In a sense, market demand is evolving as fast as competitors seem to multiply. Information technologies have created an exploding demand for the printing and writing paper that feeds our fax machines, laser printers and photocopiers. Glossy magazine inserts help the weekend edition of your favourite newspaper tip the scale at more than a kilo. Knockdown furniture has components from enough countries to teach your children a good lesson in world geography.

While increasing wealth supports many of the habits I have just described, we must not overlook the fact that there are many more of us today than we can even imagine. It is not only the sheer number of people that count, equally important is the form we take. As average family size shrinks around the world, the same level of population translates into many, many more households – each creating its own demand for housing, books, newspapers, furniture and the like. The increase in household formation is occurring in most parts of the world. Rich economies have populations that are ageing, creating one set of forces for new household formation while many developing economies have demographics that reveal a wave of young consumers rushing into the marketplace.

Wood products are not just wood anymore…

In almost every end-use application from doors to liquid packaging, wood fibre is being combined with plastics, laminates, metal fixtures and the like to yield a product that competes head-on with substitute materials. The driving force that determines price, contains costs and enhances performance in the sector is the ready willingness of plastic manufactures, aluminium smelters and concrete moulders to push wood products to the side with competitively produced substitutes. The forest products sector responds appropriately, fibre based liquid packaging is coated or lined to compete with glass – wooden window manufactures have redesigned an old standard to fight back against market share lost to aluminium and PVC.

And forestry is not what it used to be….

Against this confusing background of evolving and shifting market demand, new uses, and rapidly changing technologies, we ask ourselves – what is the connection to forests and forest management? Is there any connection or should there be? We've been confronted with newspaper headlines screaming deforestation, climate change, and collapse of biological diversity. Do growing populations, increasing incomes and new wood products condemn us to run out of forest? We will face a shortage of forests, that marvellous natural resource that both meets our needs for wood products and performs so many other important natural functions. Some say yes and some say no. What's the problem here? Can the experts not make up their minds?

Forest policy debate taken hostage…

I will argue that, sadly enough, the forest policy debate has been hostage for almost the past 30 years to two rather simplistic and unsophisticated assertions –

1. Access to all forest resources is necessary to meet demands for forest products. If access is denied then timber famine will follow.
2. Access to forest resources must lead to deforestation, forest fragmentation, loss of biological diversity and a host of other bad things.

I assert that this dichotomy has not served us well. It is not well founded in fact and distracts us from appreciating the rich history of change, evolution and adaptation demonstrated by the forestry sector. Most importantly, it prevents us from actively working together to understand both that

- We cannot recycle our way out of a briskly growing demand for forest products, and
- We can do a lot to improve forest management.

Equally, I advocate that a narrow focus in the policy debate can lead us to overlook possible choices that could increase our opportunities in the future. Let's be honest with ourselves, the general public is concerned about the loss of forests and recognises forests to be the storehouses of biological diversity, the habitat of wild animals, and stores of pooled carbon. More generally, forests have an intrinsic value - people want nature, vistas and a patrimony to share with their children.

Sustainable forest management ... public choice about private action

A major force driving change in forestry is the pursuit of a new vision of forest management called by many "sustainable forest management." This is not merely the development of a set of techniques to maintain the physical attributes of forests, but a far broader philosophy for forest management. At its core, sustainable forest management (SFM) is fundamentally about choice: sustaining what, and for whom. Because it will be the product of public choice, SFM must – and will – reflect social values. With this in mind, we face a great challenge in developing effective processes and institutions to decide among choices that countries face with respect to their forests. While SFM may be debated and discussed publicly, it is individuals that take action in the forests. It is the logger, subsistence farmer, miner or explorer who ultimately decides what to leave and what to cut, where to open a path and where to leave the forest untouched. Public choices about SFM are only realised if there is a complementary and supportive set of economic rewards and regulatory incentives.

2. Outlook for Forestry, Forest Products and International Trade

The Food and Agriculture Organization of the United Nations, the organization that I represent, supports the efforts of its 176 member countries to improve forest management. With this in mind, substantial components of FAO's current forestry work focuses on the medium term outlook -- the major changes we expect to see during the next decade. Equally as important is our work to support countries as they frame policy options and implementation programs. I will mention a few highlights from a global overview of the sector that we are currently undertaking. What I mention now is a preview of some of the emerging highlights.

Consider these possible conclusions about the future:

- More wood will be cooked than sawn.
- Surplus production capacity in the forest products industry may well lead to stable or declining real prices for forest products.
- Trade is likely to increase and remain controversial.
- The private sector is likely to play a much larger role in forest management.
- Financing sustainable forest management may become the main public policy issue for the sector in the next decade.
- Difficulties in financing sustainable forest management may cause some countries to re-examine basic issues of tenure and the role of the public administration in the sector.

Each of these possible findings has potentially significant implications for the forestry sector and the management of forests.

In the future, more wood will be cooked than sawn.

I learned this from Richard Haynes, another keynote speaker, about 20 years ago and the trend remains relevant today. As I mentioned earlier, the drive for profit condemns successful firms to endless pursuit of new products – continuously adapting, developing and evolving to respond to marketplace opportunities. The road to profit is in product differentiation through speciality packaging, sizing, forming, branding, whatever.

Businesses are willing to pay just a tiny bit more for something that helps them serve their customers a lot better. Recall that forest products manufacturers are typically selling to other converters who are producing envelopes, windows, furniture or printing magazines. A tiny premium can be paid for guarantees of quality and consistency. The finer the market segmentation, the more rigorous the products specifications.

For the forest products manufacturer, the sourcing of wood fiber quickly becomes important. On the pulp and paper side, a mix of species, sizes, age classes and chemical characteristics becomes a nightmare in meeting a narrowly banded performance standard for the end product. In part, the drive for clonal, plantation-grown wood is an outgrowth of these pressures -- control specific gravity, fiber length and wood chemistry.

In the production of structural products, traditionally lumber and plywood, technological innovation allows us to use species that only ten years ago we called weeds. Further, we are now able to transform what was waste and end cuts from the sawing lines, and through techniques such as finger-jointing, produce a very desirable product.

In short, the more the product is cooked, digested, bleached, engineered and reconstituted, the more exacting product standards you can maintain and the more you can segment (and therefore profit from) your market. The emergence of these technologies gives countries and wood product firms new options in sourcing wood fiber. Vast areas of forest that were considered non-commercial now represent supply. These techniques allow more use of recovered fiber and they allow us to make much more extensive use of every log taken from the forest. Each of these advances, help expand our fiber supplies and give us new options in forest management.

Surplus production capacity in the forest products industry may well lead to stable or declining real prices for forest products…

In spite of some real price increases in forest products in this decade, financial returns on existing investments in forest products manufacturing have been generally disappointing. For example, studies show that few, if any, of the major pulp or paper mills built in Europe or North America in the past 10 years have made a competitive rate of return[1]. Profitability has been driven downwards by a combination of constant real prices and undisciplined capacity growth.

Still, substantial changes in markets and fiber sources are anticipated and this will require new capital flows to finance those opportunities. Restructuring and consolidation of existing capacity are inevitable if new private capital flows are to be attracted into the industry. The Asian financial crisis may help to speed up the process.

For the changes to be sustainable, however, they have to be accompanied by structural reforms that favour market-based pricing of raw materials. To a significant extent, the overcapacity is a by-product of previous government policy over-emphasising forestry as a source of rural employment creation. This has often resulted in timber being offered at low rates to attract investment and infrastructure. Accompanying concessions on resource pricing are often tax holidays, offsets or below-market capital infusions.

These market distortions have had at least two consequences. First, the industry faces heavy overcapacity in some regions and product lines. Low returns make it hard to attract capital needed to improve product and harvesting efficiency. Second, critics of timber harvesting point to real and glaring examples of forest waste and destruction and then, unfortunately, generalize from the specific examples, to all forest sector operations. It has given us a bad name.

Any move to increase the market basis for timber pricing will be regarded by industry as a further squeeze on profit margins. However market based pricing for timber is an important public choice which individual countries have to consider and should be an element of the discussion. It may seem to each operating firm that wood is becoming increasingly scarce. It may also be that a region or product line has excess production capacity. Experience suggests that capacity adjustments, employment concerns and impacts on regional development are difficult choices for nations. For this reason, the outlook for the future is for continuing overcapacity. This has important implications for the outlook of future product prices, investment and opportunities to amend current forest practices. In general, the outlook is for stable or declining real prices for forest products.

Trade is likely to increase and remain controversial…

[1] Dargan, A. (1996). PPP Posts Poor Performance; Banking and Finance. PPI, September 1996, p.20-21. Many mill projects have been earning a return on capital of 5-10% per annum whereas the weighted average cost of capital (WACC) has been 10-15% depending on the nature of the installation and location.

International trade in manufactured forest products has been expanding for more than twenty years. The outlook is for more trade both in gross volumes and as a proportion of total production. Several forces are at work and the picture is likely to become foggy before it becomes clear. Historically, trade increases have largely been intra-regional rather than interregional. For example, lots of trade within Europe, between Canada and the US and within Asia. Trans-oceanic trade, for example, eucalyptus pulp from Brazil to Europe, has had to overcome rather daunting barriers such as escalating tariffs, phyto-sanitary regulations, export-import financing, shipping bottlenecks and currency movements. The further the boat had to go, the greater the risk. Today, increased product differentiation along with tariff reductions and forward currency hedging has made trade less risky and more profitable. Competitive forces of comparative advantage are hard at work and are manifesting themselves in international trade.

Recall that competitive advantage is not characterised by where the net annual increment for timber growth is the greatest. The forestry and forest economics professions have been plagued by this misconception for years. Comparative advantage and success in international trade manifests itself in the ability to deliver the most attractive end-use product to the consumers' door. This is a combination of product design, manufacturing efficiency and the cost of all the factor inputs such as labor, capital and raw materials. Assembly and distribution networks are vital. Wood may grow very slowly in Finland, but the composite picture of that country's secure capital base, attractive designs, efficiency in manufacturing and distribution and proximity to key European markets, makes it a competitor to beat. When we contrast its performance in European markets with distant countries that may have much lower labor costs and lots of cheap wood, we quickly realise the importance of looking at the full picture of competitiveness in the forest products market and not dwelling only on forest volumes, increments or labor costs.

The areas of trade that are growing significantly are manufactured wood products, furniture components, distinct paper and pulp grades, and interestingly, waste paper. Three things make the future of trade cloudy. The first is purely statistical. The integration within the European Union means that much trade that had been recorded as "international" will no longer be reported. In a sense, it was always intra-regional trade but the net result will be that statistics will show a drop in trade.

Second, the economic crisis that began in Asia has increased trade tensions. A number of countries have felt forced to take extreme measures to respond to adverse macroeconomic conditions. As inferred above, stable currency exchange rates and access to financing for trade are vital lubricants to trading abroad. Until the global financial system stabilises, trade is likely to suffer.

The third and most controversial factor affecting trade in forest products is the vibrant discussion on ecolabelling and certification of forest management. Since sharing opinions on this topic has the tendency to highjack any conference discussion, let me suffice to say that it would appear that we can look forward to a long but perhaps, we fear, fruitless discussion on the pros and cons of these tools and their relative effectiveness. At FAO, we tend to think in small steps and encourage countries to consider concrete and visible measures to "improve" forest management now, even if we do not yet know all the elements of "sustainable forest management." The public quickly forms opinions on what is not pleasing forest management, even if we cannot agree on what is the right balance in forestry. In any case, the outlook for trade in forest products is that it will continue to increase, trade will remain controversial, and it will stay high on the agenda for international discussions.

The private sector is likely to play a much larger role in forest management

Privately owned forests, which constitute a fraction of the globe's total forest area, already account for nearly half of world timber production. This share is likely to increase. Private ownership and the private sector will play an increasingly important role in sustaining production of industrial wood products. In part, this will be a consequence of the trend toward a greater role for the private sector in political and economic policies in many countries. As more wood comes from private forests, countries can choose to increase their investment in forestry extension.

Most research suggests that private owners have multiple objectives in owning forestland – some owners have economic objectives but many hold these lands for other reasons. Solid forestry extension programs help the landowner articulate and act upon their stewardship objectives. Extension programs assist the landowner to market their timber at competitive rates. On private forests, the landowner is less motivated by broad social objectives than by the benefits they can enjoy from the forest. The importance of private owners and the trends toward reliance on market processes will present new and different challenges for choosing and implementing forest policies. In this sense, the future of sustainable forest management will be shaped more by landowner perceptions and wishes than government fiat. If the returns to forestry remain low, the private landowner is less likely to invest in intensive management and protection from fire and disease.

Financing sustainable forest management may become the main public policy issue for the sector in the next decade

Improving forest management may not increase costs. Much can be done through training and contract administration that merely enforces the set of rules and plans now on the books. Solid application of what we already know about good forest management will do much to protect vital ecosystem functions. In fact, advocates of reduced impact logging assert that well-planned harvesting operations reduce total extraction costs to the logger.

Implementing "sustainable forest management" may be a whole new kettle of fish. In large part, widely varying cost estimates exist for implementing sustainable forest management because we cannot agree on a clear operational definition. Our market analysis has yet to reveal a "premium" for wood flowing from "well-managed" or sustainable managed forests. Whereas consumers may make good forest management a necessary condition for market access, there is little evidence yet that sustainable forest management will be a sufficient condition to extract enough of a market premium to cover increased costs borne widely across the forest landscape. While the public may insist on new forms of forest management, it is unclear how more rigorous protection of the environment would be financed. At least two operational options come to mind.

First, simply impose the costs on the operator and/or landowner through strict regulation. Regulating timber harvest would tend to focus silviculture on achieving desirable stand characteristics and ecosystem functions through careful and limited tree removal.

Another approach is to make public payments to landowners when they meet or exceed agreed upon stand management objectives. Notice however that choosing either more rules and strict regulation or public support will require more public expenditures for forest management than are being made today. The deeper you explore this issue, the more you find it a ripe and rich topic for future research.

To some extent, the current controversy over sustainable forest management is rooted in our inability to define what it means but equally, there is a vibrant debate over who will or should bear the imagined costs. Now, please imagine a much more controversial option. What would it take to increase the general returns to good forest management? What about a bold action to increase the price of stumpage. The world price of market products is set by producers accessing non-competitively priced timber that may or may not meet a market test. Examples include forest concessions that are established largely to meet development objectives or large scale land clearing for agriculture and transmigration. Each of these actions dumps large quantities of wood on the market. Converters of this wood strongly influence the world price for commodity grade lumber, panels and pulp. If these actions consistently spoil economic incentives for good forest management, then we may wish to consider dramatically reducing the wood availability from public forests to boost the outlook for decent returns to good forest stewardship on private forestland. While radical, it has the potential of a win-win situation. Increase broadly the closed natural forest areas held in reserve pending future needs, while favouring the price outlook for private sector investments in plantations, management intensification, waste recovery and new technology.

Difficulties in financing sustainable forest management may cause some countries to re-examine basic issues of tenure and the role of the public administration in the sector.

Let me assume for the moment that sustainable forest management will require more investment in silviculture, harvesting equipment, new processing technologies and product marketing. Let me also assume that there are serious constraints on new commitments of public finance. Jointly, these two conditions suggest a lead role for the private sector in financing public choices about sustainable forest management.

As we know, the private sector will pursue improved forest management most aggressively when there is a clear economic incentive. Considering these facts, countries may wish to re-examine their current tenure arrangements for trees, forests and forestland. Experience shows that even small changes in tenure and rights to the benefits of forest management can show remarkable new commitments from the private sector.

3. Conclusion

The implications for sustainable forest management are significant...

Assuming that our outlook findings are upheld, then the implications for sustainable forest management are significant. The outlook foresees a future that is relatively rich in fiber supplies with more than adequate industrial capacity to meet wood products demand. I have suggested that on the product side, this is likely to manifest itself in stagnate prices (some say a symptom of overcapacity, others would say simply a reflection of a truly competitive market). On the resource side, the price for logs would remain far below the cost of growing timber. As long as this prevails, few private investors will commit to growing fiber without subsidies or hidden price transfers. I assure you that our outlook findings will be contested because they fail to support the false dichotomy in the forest policy debate of gloom and doom. I characterise the gloom as the assertion that not keeping all forestland open for logging means timber famine and I characterise the doom as the assertion than any forest entered for timber extraction is subsequently worthless in the future for other forest functions.

The future is going to be what we want it to be...

Some people characterise the forest products industry as ruthless, exploitative or destructive. In some circles it is described as "predatory." In reality, it is a construct of our demands for wood products. It is a means to an end that we, as consumers, have set in motion. We seek and it produces. In practice, the forest products industry in particular, and the forestry sector in general, has a rich history of change, evolution and adaptation. In a particular sense, it is a model industry for sustainable development. The forest products industry commits vast pools of private capital to produce products from a renewable natural resource. It adjusts and adapts to changing markets and resource availability. With skill and determination the industry introduces technologies that extend the resource, protect the environment and give us durable products with a long useful life. Broadly, the FAO global outlook study suggests that the forest products industry is operating within the carrying capacity of our tree and forest resources.

Still, the industry will only give us the kind of future forest management that we demand of it. If we support extension activities and research into engineered wood products, then the industry will give us an exciting array of consumer products based on fiber from lands outside the closed natural forest. If instead, we promote and offer timber from closed natural forests at bargain prices then the industry will dutifully pursue those sources of fiber.

The role that forests will play in the future is in our hands. Each country will choose either explicitly or implicitly the role it wants the nation's forests to play in supporting economic, social and environmental goals. Not establishing a national direction is also a choice. Individual action will produce a result.

At present, our institutions for making public policy decisions are struggling to cope with issues of natural resource management. We have much work to do in this area. Our challenge is not necessarily to have more participation and more discussion. Participation and discussion are important but dialogue without conclusion has led to paralysis.

In part, we are unable to make incremental gains toward better stewardships of our forest because so much of our energy is directed at debating the size and shape of the "sustainable forest management" goal posts. By reflecting on our history and experience, we should calmly acknowledge, however discomforting the realisation, that the goal posts will change shape, size and location every time we learn something new about the functioning of the forest ecosystem.

In summary, the future for the forestry sector is not what it used to be. Consumer demands are expanding rapidly. The types of products sought by customers are becoming ever more diverse. Traditional concepts of forestry have been abandoned but a new paradigm to replace these foundations is not yet operational. Funny, but today, everyone considers himself a "stakeholder" in sustainable forest management. Some crazy people are trying to cook wood instead of cook with wood.

In spite of these obvious examples of a very different future, some things remain forever the same. Clever and useful products continue to flow from the forest products industry. The industry's continuing problem remains markets and market development. Returns to forest stewardship are still very low. And finally, the central problem in forestry continues to be getting people to agree on what defines proper management of the forest, becoming responsible for making choices and then acting on these decisions.

Let me close by encouraging us to embrace change, to seek out choices, make decisions and be responsible for our actions. Our future depends on it.

Challenges for China's Forest Resource Management and Utilization in the 21st Century

Wen Da Lu, Dong Xiang Wang and Nan Ling

Key words: Forest resource, timber utilization, fast-growing trees, forestry ecology project, Forest Products Industry (FPI)

Abstract: This paper is divided into three main sections: 1.General situation of forestry in China--This includes basic data on Chinese forestry and forest resources. In general, China's forest resources are limited, with low forest coverage and per capita volumes. The forest resources have a heterogeneous distribution, unusual forest structure and age distribution, and are generally low in productivity. Bamboo resources are relatively rich. 2.Strategy for sustainable forestry--Major efforts to develop sustainable forestry are described, including efforts to plant fast-growing trees, the establishment of five large forest ecology projects, use of advanced western intensive forest management methods, and protection of biological diversity. In addition, the enhancement of timber utilization capabilities, efforts to decrease consumption of forest resources, promotion of the forest products industry (FPI), use of materials as substitute for timber, and the moderate importation of timber and wood products are described. 3.Existing problems and countermeasures--With growth of the domestic economy and the transition from a centrally planned economy to a market economy, a series of social and economic problems have developed. These include: pollution; overuse of natural resources, e.g., water, soil, forest; disruption of ecological balance; and management and operational problems of mid- to large-scale state-owned forest enterprises. The author describes a series of corresponding countermeasures.

1. Introduction

People's Republic of China is the biggest developing country in the world. It is located in the east of Asia and its area is about 9.6 million km^2 and equal to 7% of the total land area of the world. It's a little larger than the United States, but its population is about 1.2 billion and amounts to 21.5% of the total population of the world. The general forestry situation in China is as follows.

1.1 Basic data of Chinese forestry

Area of forests: 128.5 million ha

Coverage of forest: 13.92%

Volume of living trees: 10.7 billion m^3

Area of plantation forest: 41.39 million ha

Annual growth amount of forest: 0.419 billion m^3

Annual consumption of forest: 0.320 billion m^3

Area of desert: 2.62 million km^2, equal 27.3% of total land area (Zhou 1990).

1.2 Characteristics of Chinese forestry

1.2.1 Forest resources are not rich, the coverage of forest is low, and the average per capita of forest resources is very low. Table 1 is used to explain this situation (Li et al. 1994).

Table 1. Comparison of the forest resource of China with other areas of the world

Item	Asia	SU(F)	USA	Sweden	Finland	World	China	*%
Area of forest (million ha)	463.22	765	308	23	19	3410.9	12.85	3.77
Coverage of forest(%)	18	36	34	57	61	26	13.92	53.54
Average per capita of forest (ha)	0.149	3.2	1.5	3.0	4	0.644	0.113	17.7

* Chinese forest resource share of the whole world.

1.2.2 Heterogeneous distribution of forest resources.

Because of natural conditions and social economic development, the Chinese forest resources are mainly distributed in the northeast, southwest, southeast, and south where annual precipitation is >=400mm. In the vast northwest area and in the economically prosperous north area and central regions there are very few forest resources.

1.2.3 Unusual structure of forest: see Table 2 (Li et al. 1994)

From Table 2, we find that the proportion of timber forest is high, but the share of other kinds of forest is very low. Without doubt this situation affects the multiple functions of the forest.

Table 2. The area and volume of different kinds of forest in China

Type of forest	Area, million ha.	Area, %	Volume, million m^3	Volume, %
Timber forest	83.84	74	7704.6	87.6
Protect forest	10.0	8.8	883.6	10.0
Fuelwood forest	3.69	3.3	69.6	0.8
Economic forest	12.58	11.1	143.2	1.6
Bamboo	3.2	2.8	80	0.9
Total	113.31	100	8881.0	100

1.2.4 Irrational structure of forest age.

The unusual structure of forest age is shown in Table 3 (Li et al. 1994).

Table 3. The structure of forest age in China

Forest age	Area, million ha	Area, %	Volume, million m^3	Volume, %
Juvenile	41.33	38.05	1023.18	11.26
Middle	36.13	33.26	2660.34	29.28
Mature	31.17	29.69	5403.65	59.46

The juvenile forest and middle forest are higher in forest area (Table 3). The available forest resource is not enough for log harvesting. This situation will be sustained a long time.

1.2.5 The low productivity of forestry.

As compared with western forests, the productivity of Chinese forestry is not high, as shown by the utilization rate of forest land, the growth rate, and growth amount (Table 4).

Table 4. The productivity of forestry in China compared with other areas of the world (Li et al. 1994).

	S.U.(F)*	USA	Japan	Sweden	Germany	Finland	China
Utilization rate of forest land %	90.4	95	76.2	89	97	100	43.2
Volume m^3/ha	106	88	76	92	149	97	90
Growth rate %	1.18	3.33	3.4	3.1	3.02	3.04	2.88
Growth m^3/ha.y	1.2	3.1	3.1	3.3	6.0	3.0	2.4

* Former Soviet Union

1.2.6 The relative rich resource of bamboo.

The bamboo resource is relatively abundant in China. It is mainly distributed in the south and southeast. The total area of bamboo is about 3.2 million ha, and the total volume is about 80 million m^3.

2. Strategy for Sustainable Forestry

Chinese forestry is faced with formidable challenges. On one hand, the available forest resources are low. On the other hand, the demand for timber will increase because of the population explosion and economic development. The double pressure may cause the deterioration of our environment and threaten our biosphere directly. So our state should pay much attention to developing sustainable forestry. The main strategies are as follows:

2.1 Devote major efforts to planting fast-growing trees

In China on March 12th of every year is a planting festival. Many people volunteer to plant trees in order to accelerate the construction of the green belt. We give precedence to fast-growing tree species such as larch (*Larix dahurica, Larix gmelinii*) and poplar in the northern area; China fir, kiri (*paulownia*), and poplar in the central area; slash pine (*Pinus elliottii Engelm*), loblolly pine (*Pinus teada L*), China fir, Chinese red pine (*Pinus massoniana Lamb*) and gum tree (*Eucalyptus*) in the southern area. We introduced the seeds of slash pine from the United States and set up our asexual seed station. We planted slash pine in the broad area of south China, especially in GuangDong province. So far, the juvenile and middle forests of slash pine have reached about 1.5 million ha (Pan and You 1991).

2.2 Construct 5 large forest ecology projects

We received a loan from the World Bank of about 300 million US$, and our government invested corresponding funds. We are constructing five large forest ecology projects. They are:

(A) Coastal protection forest: about 6.6 million ha.

(B) SanBei (north, northeast, and northwest) protection forest: about [illegible]3 million ha.
(C) The protection forest of upper and middle Yangtze River for conservation of water and soil: about 5.46 million ha.
(D) The ecological plantation of TaiHang Mountain (in ShanXi province): about 1.88 million ha.
(E) The ecological plantation of the central plain: about 5.5 million ha.

So far, the yield of timber from plantation forests have reached 25 million m^3 per year equal to 1/8 of the total yield in China.

2.3. Use western advanced, intensive forest management

Example are: introducing excellent seed sources, thinning for tree growth, and for enhancing pest control, research for wood quality and utilization of plantation forests, etc.

2.4 Protect biological diversity

China has had a very quick decrease in natural forests and a corresponding increase in area and volume of plantation forests. The latter has the disadvantage of a single species that is easy to infect by disease and insect pests. This situation will cause the decline of biological diversity. Therefore our state is taking a series of effective protecting steps as follows:

(A) Setting up 165 Natural Protection Regions (NPR): a total area of 0.57 million km^2--about 5.54% of the total land. Among them are the Chang Bai Mountain, Xi Shuang Ban Na, and others. Ten NPRs have been listed in the International Biological Circle by UNESCO (Zhou 1990).
(B) Preservation and propagation of species that are precious and rare as well as in imminent danger of extinction, such as panda, DongBei Tiger, YangTze Crocodile, etc. So far we have set up the cultivation center of 277 wild animals and a national resource storehouse of species preserving 350,000 species (Zhou 1990).

2.5 Enhance utilization of timber; decreasing the consumption of forest resources

(A) Promote the development of the forest products industry (FPI).

Our state is accelerating development of composite production, especially particleboard, medium-density fiberboard (MDF) and chips, thereby obtaining better utilization of the residues of logging timber and lumber processing than before. Table 5 shows a survey of FPI development.

Table 5. Summary of the enterprises, products and yield of FPI (1994)

Products	No. of Enterprise	Yield (million m^3/yr.)
Plywood	515	2.6
Fiberboard	391	1.93
(including MDF)	17	0.289
Particle board	216	1.68
Other composite	78	0.5
Total	1200	6.999

The yield of chips is 0.6 million m^3 per year. The total yield of wood-based composites will reach 10 million m^3 per year and the yield of chips 5 million m^3 per year in the 2000s. Production line equipment for composites is partially introduced from western countries such as Sweden, Germany, and the United States, and partially manufactured by ourselves.

(B) Use substitutes for forest resources.

The state has taken a series of effective measures to develop non-wood material industry in order to save forest resources. So far, we have substituted concrete crossties for about 90% of the traditional wooden crossties. Currently the total yield of non-wood-based composites (including bamboo and other agriculture residues) is about 0.5 million m^3 per year; it will reach 1.0 million m^3 per year in the 2000s. Meanwhile the state makes use of recycled wood fiber such as waste paper to produce paper products.

(C) Import timber and wood products.

In order to decrease the consumption of forest resources, the state imports timber and wood products from abroad on a large scale (average 5 million m^3 per year in the 1980s, 4 million m^3 per year in the 1990s). In the 1980s our state became the second largest importer of timber in the world (Zhang et al. 1997).

3. Existing Problems and Countermeasures

3.1 Existing problems

3.1.1 Overuse of natural resources and aggravation of environmental pollution.

Nation-wide industrialization and the population explosion produced double pressure on natural resources (water, soil, and forest). The direct result of overuse of natural resources is that the groundwater level of urban areas has declined quickly, the total days of cut-off of water in the Yellow River increase year by year, in some areas the water of Huai River is undrinkable, and recently the natural disasters (flood, drought, etc.) have happened frequently. The whole environment is faced with serious disruption of ecological balance. In addition, mid-sized to small locally owned enterprises such as pulp and paper factories, chemical plants, farm chemical mills, etc. produce a lot of waste gas and waste water and cause serious environmental pollution.

3.1.2 Keeping the mid- to large–scale state-owned forest enterprises going.

China's forest resources are gradually being exhausted by exploitation excess. A lot of mid-sized to large state–owned forest enterprises based on timber production are in the midst of a forest resource crisis and an economic crisis. During the transition of enterprise from a centrally-planned economy to a market economy, a series of problems have developed such as the out-of-date facilities, arranging for off-site workers, and financial difficulty.

3.2 Corresponding countermeasures

3.2.1 Put the natural forest protection project into effect.

Our government decided to sharply decrease the volume of natural forest harvest and invest a lot in forestry, promote the development and utilization of timber from plantation forests, and use substitute materials more in order to maintain sustainable forestry and improve the environment of human beings.

3.2.2 Close a lot of small- to mid-scale enterprises such as pulp and paper factories, chemical plants, farm chemical mills, etc. where the technology is backward and the pollution problems are serious. The amount of waste gas and waste water can be decreased sharply, saving water and improving the quality of water.

3.2.3 Promote forest products industry (FPI) and multi-management enterprises.

Introduce composite production line from western countries, enhancing competitiveness of forest products. Develop forest plantations and cultivation and a diversified economy, in order to change the single timber production pattern and strengthen the vigor of the forest-based economy.

3.2.4 Restructure the economic system of forest enterprises.

According to different situations, engage in different policies: in the small- or middle-scale enterprises, we use the contract system of managerial responsibility or lease contract. In the large-scale enterprises, we engage in the share system. Meanwhile, in order to prosper these forest enterprises we must enhance the consciousness of products and increase the sales volume.

4. References

Li, J., Lu, W., Liu, Y. et al. 1994. Wood science. The Press of Northeast Forest University, Beijing.

Pan, Z., and You, Y. 1991. Introduction and cultivation of slash pine (*Pinus elliotti Engelm.*, loblolly pine (*Pinus taeda L*) and Caribbean pine (*Pinus caribaea Morelet*). The Press of Beijing's Science & Techonology, Beijing.

Zhang, Y., Buongiorno, J., and Zhang, D. 1997. China's economic and demographic growth, forest products consumption and wood requirements: 1949 to 2010. For. Products Journal. 47: 27-35.

Zhang, D., Liu, J., Granskog, J., and Gan, J. 1998. China: Changing wood products markets. Forest Products Journal. 48: 14-20.

Zhou, Y. 1990. Chinese forest. The Scientific Press, Beijing.

Selling Clean Air

- The Costa Rican experience with carbon dioxide bonds -

Mariano Rojas

Key words: Environmental services, greenhouse gas emission mitigation, Costa Rica

Abstract: In 1997 Costa Rica created a new kind of environmental bond to be introduced in the international financial markets, becoming the first country to issue environmental bonds. The bonds, called Greenhouse Gas Emission Mitigation Certificate, aim to sell the world the clean air generated by Costa Rican forests. The scheme is based on the air purifying process of trees through their fixation of carbon dioxide. The bonds' value depends on the amount of carbon dioxide fixed, using cubic meters of wood as a measurement unit. Several difficulties have been faced: setting a price basis (measurement of carbon dioxide fixed), creating market value (property rights, global agreements, value-creating schemes), and avoiding moral hazard problems (verification process). The objective of the project is to channel its revenue to the territorial consolidation and funding of national parks. In addition, the scheme recognizes the market value of a positive externality, increasing the likelihood of applying market solutions to the world's environmental problems. This paper studies: (a) the theoretical background behind the carbon bonds; (b) the original scheme proposed by Costa Rica and the global and local requirements of the scheme; (c) the implementation problems faced by the program and; (d) the preliminary results of the project. It also elaborates on the future of the carbon dioxide bonds and the modifications needed to ensure their success.

1. Introduction

Costa Rica enjoys a privileged location; in an area of 50 000 km^2, the country holds about 5 to 7% of the world's biodiversity. During the decades of the sixties and seventies Costa Rica experienced one of the highest deforestation rates in the world. The area covered by forests went down from about 80% of the national territory in the forties to around 20% in 1985. However, important changes happened at the end of the seventies: a strong environmental concern built up in Costa Rica, awareness about the need for a responsible use of the country's natural resources grew. The government decided to strengthen the national park and forest reserve system. New national parks were created, and some forest areas were declared as protected, even though the government did not have the financial resources to purchase the land. The country enhanced the institutional and legal framework related to the protection of the environment; in addition, it created the Ministry of Environment and Energy (MINAE). As a consequence from these efforts, national parks and protected areas represent approximately one quarter of the country's territory at the present time.

The environmental sentiment also produced some economic rewards to the country. During the eighties Costa Rica became a prime ecotourism destination, and an important lesson was learned by Costa Ricans: protection of the environment is also an economic activity and it can be profitable. Tourism became the main source of hard currency on top of traditional products such as coffee and bananas. In addition, the country cultivated a good reputation as a pioneer in the protection of the environment. Different kinds of incentives provided by the government to protect the environment, such as preferential credit, were typical in the early eighties. Later, the government introduced a *forestry certificate* to promote reforestation programs. The certificate was accepted by the government as a tax payment instrument and it was tradable in the market. This seems to be the beginning of Costa Rica's market approach in the protection of the environment. The certificate was criticized because of its bias in favor of large landowners. Thus, in 1988 the *Forestry Development Fund* was created with financial resources coming from the *Debt for Nature* program. Its purpose was to provide incentives to small-scale reforestation programs. In 1996 a new law recognized the environmental service provided by forests in the mitigation of greenhouse gas emissions. Taking advantage of the new institutional setting defined by the United Nations' Framework Convention on Climate Change (UNFCCC), Costa Rica became a pioneer in the Third World in the development of projects supported by Actions Implemented Jointly (AIJ) and by Joint Implementation (JI) programs. More recently, in 1997, Costa Rica created and issued *Certifiable Tradable Offsets* certificates (CTOs), also known as *Greenhouse Gas Emission Mitigation* certificates. Following a commodity approach, the country aims to facilitate the selling of the environmental services from its forests and from other related programs.

This paper studies the experience of Costa Rica in the environmental-services business, with an emphasis in the development of the CTOs. There is a review of the recent history, starting with the United Nations' FCCC and explaining how Costa Rica took advantage of the new institutional framework. Implementation problems associated with AIJ and JI programs, as well as the high transaction costs involved in the original scheme, led the country to search for and innovate with new schemes, the CTOs among them. The preliminary results from the new schemes are mixed.

2. The United Nations Framework Convention on Climate Change

Increasing concern for global environment problems led the United Nations to organize the first world conference on the environment (held in Rio de Janeiro, Brazil in 1992), followed up by many other conferences. Since then, a global perspective on environmental problems has been gaining momentum. It is now recognized that the solution to many environmental problems is beyond the capabilities of any single country. Environmental problems do not discriminate among country borders, and actions from any country could damage or benefit other countries. Thus, world-level externalities have been recognized, for example: acid rain, ozone depletion, pollution, carbon dioxide emission and mitigation, etc. (Tietenberg 1992). Accepting the existence of these externalities is a necessary step towards a solution to global environmental problems. However, the problem is that no single authority with the coercive power to deal with these global externalities exists. In consequence, negotiations at a global level need to take place to internalize the environmental problems and benefits generated from the activities of any country. There are two main problems faced in these negotiations: the existence of enormous asymmetries across countries, which make it difficult to reach a satisfying agreement for each country; and the free-riding problem typical in collective-action activities, where countries would like to enjoy the benefits from any agreement without carrying out with the costs (Carraro and Siniscalco 1993, 1996). Thus, global agreements are difficult to reach and even more difficult to sustain in the long run.

Unilateral action at the country or regional level is likely to take place when the benefits for the region are greater than the cost of the action (Olson 1994). Hence, many developed countries committed themselves to reducing the levels of greenhouse gas (GHG) emissions, as well as undertaking other measures to protect the environment. The agreement required the creation of a coordinating mechanism across countries (market or non-market mechanism), as well as the clear definition of property rights, meaning the responsibility of each country for polluting and protecting the environment. Joint implementation was one of the initial country coordination schemes approved at the United Nations conference. The Costa Rican experience is related to the continuous search for a more cost-efficient mechanism to facilitate the coordination across countries and to clearly define global property rights.

It is obvious that the value of any measure to mitigate GHG emissions depends on the definition and enforcement of a global system of property rights. A vague definition and enforcement of property rights would make it almost impossible for any country to internalize the benefits from its environmental protection efforts and to charge the rest of the world for that. Therefore, incentives have to be large and certain for any country to unilaterally undertake mitigation efforts.

3. Early Conditions in Costa Rica

As it was stated before, a strong environmental concern developed in Costa Rica during the eighties. The economic benefits from ecotourism activities proved that protecting the environment was not only a matter of concern but also a matter of enhancing economic welfare for the population. Thus, Costa Rica gave priority to the sustainability of its development. The country set the following goals in the sustainable development area: increasing the generation of energy from low-emission sources, such as hydroelectricity and wind energy; increasing energy efficiency; increasing the country's capacity to fix carbon dioxide; and generating revenue from gas emission reduction programs to finance its sustainable development, in specific, to finance the consolidation of its national parks and protected areas.

Costa Rica began protecting the environment through costly fiscal incentives, such as tax exemptions and preferential credit. In a developing country, sustaining a large national park system was considered a luxury, even after considering the revenues from ecotourism. In the Costa Rican case the intentions went farther than the real possibilities of the government to purchase the protected areas. Therefore, there was a necessity to find alternative ways to protect the environment and to initiate a reforestation campaign. The agreements on Joint Implementation and Actions Implemented Jointly at the United Nations conference opened an opportunity that the country not only needed but was also well prepared to take advantage of. Costa Rica planned to finance its environmental program (sustainable development) by selling the environmental services from its protected areas, national parks, and related greenhouse gas emission mitigation programs to the developed countries, which are responsible for polluting the environment and, on the basis of their commitments at the UNFCCC, were in need of reducing or compensating their greenhouse gas emissions.

The fundamental point is the air purifying service provided by trees through their fixation of carbon dioxide (Sedjo et al. 1995). It is estimated that, on average, a cubic meter of wood absorbs one-fourth ton of carbon dioxide. By considering this sequestering process as an environmental service provided by the country, and by setting GHG emission reduction commitments to the developed countries, the UNFCCC created a market value for this positive externality. Therefore, property rights were clearly defined: the GHG mitigation services belong to the country where the project is located; and the recognition that these services have cross-border effects (at a global level, the mitigation of greenhouse gases does not require the location of the air-polluting source and the air-purifying source to be the same) created the opportunity to cash in on these services from forests.

4. The Initial Challenge

In early 1995 Costa Rica realized that to take full advantage of the UNFCCC it was necessary to undertake an effort at the national level, instead of the isolated efforts that were being undertaken by private agents at the time. Thus, in August 1995, the Costa Rican Office of Joint Implementation (CROJI) was created, with the financial and institutional support of the private initiative, the government (MINAE), and private and public foundations. CROJI was conceived as a decentralized public organism in charge of all the activities related to joint implementation: preparation, evaluation, and approval of all joint implementation projects. In addition, CROJI was in charge of developing the financial, legal, and administrative mechanisms to market and finance the joint implementation projects (Joint Implementation Quarterly 1996).

CROJI faced many challenges:

a. It is difficult to estimate the market value of any project because the demand is unknown and depends on the fulfillment of the commitments made by developed countries. There is a technical problem in measuring the production from the project, but there is also an economic problem in estimating the market value of the production because of undefined property rights and high enforcement costs. Thus, there was great uncertainty about the profitability of any investment.

b. There were initial costs (sunk costs) related to starting the program. Some institutions had to be created and developed, and programs had to be designed and promoted. These sunk costs represented the non-recoverable investment that the country was risking in case the program was unsuccessful.

c. The creation of a market to join suppliers and demanders at a low cost was the main challenge. It was necessary to create a standard product (to develop national and international standards for measuring and monitoring GHG emissions in the different activities and countries); to introduce confidence in the market; to find or create a mechanism linking potential investors with potential investment projects; to provide adequate information about the projects; to reduce the risk to investors; and so on. Because Costa Rica is a pioneer country in the development of the market, a free-ride issue emerged: the benefits from the creation of a market become a public good, while the costs from its creation are private. The use of stock markets associated with the invention of the CTOs, partially solved many of these tasks, as will be explained below.

d. Costs are associated with the initial searching and financing of implementation programs. There were no databases about potential investors; the financial support committed by the governments is always subject to political good will; and a methodology to prove that the projects were beyond the so-called base line and, hence, were really mitigating GHG emissions needed to be established.

e. It was necessary to locate potential investors: national or international companies interested in investing in GHG emission reduction projects. These companies could be firms with mitigation compromises willing to finance a project in Costa Rica and companies investing in the projects as a business opportunity. The matching of a project to a group of specific investors created additional transaction costs.

5. The Costa Rican Scheme

At the Kyoto Conference (in Kyoto, Japan in December 1997) Costa Rica and Brazil promoted an agenda to benefit developing countries in the selling of GHG emission reductions (Tattenbach and Villalobos 1998). The Kyoto Protocol states that by 2008 to 2012 total emissions from developed countries must decline to levels 5% below the 1990 emission levels. To reach this goal, the protocol set emission reduction compromises for every developed country. Besides, the developed countries could satisfy their compromises either individually or jointly among them. This agreement excluded the possibility of Joint Implementation programs between developed and developing countries, practically killing the aspirations of the developing countries of using the Joint Implementation mechanism (JIM) to sell their environmental services. However, the Protocol allowed for the transferring and acquisition of emission reduction units (ERU) among all kind of countries, meaning that developing countries could sell ERU to developed countries through a new mechanism created by the Protocol: the Clean Development Mechanism (CDM). To use this mechanism, the ERU had to be certified by a trusted entity, had to provide real and measurable long-run-sustained mitigation benefits, and had to represent additional mitigation benefits (meaning that the mitigation benefits had not taken place without the CDM). The certified ERU purchased by the developed countries between 2000 and 2008 could be used by these countries to partially fulfill their commitments (the Protocol states that purchasing of certified ERU must not become a complete substitute for the domestic efforts of the developed countries; thus, ERU must be considered as complementing the domestic efforts).

The CDM has several advantages over the JIM: first, it facilitates the creation of a market by reducing transaction costs in the linking of buyers and sellers and by resolving some technical considerations such as homogenizing standards and measurement methodologies. It also created a real demand for certified ERU originated in developing countries. In addition, the CDM facilitated the creation of a multinational coordination mechanism, while the JIM was vague on this respect. Of particular importance to Costa Rica was the inclusion in the Kyoto Protocol of forestry as an accepted activity in the reduction of emissions.

The CROJI initiated an aggressive program to take full advantage of the new rules of the game. The strategy followed a *commodity approach* to environmental services, involving the creation of a portfolio of potential projects with a GHG emission mitigation benefit and the finding of investors for these projects. In March, 1996 the government signed into Costa Rican Law the Forestry Law 7575, which legally strengthens the *commodity approach* by incorporating the concept of *payment for environmental service*. The law also created the Carbon Fund to bank AIJ foreign investments.

The objective was to sell the environmental services provided by different GHG emission mitigation projects as a commodity in the international markets. CROJI designated the following areas where investment was possible (Tattenbach and Villalobos 1998, Dutschke and Michaelowa 1997):

a. Energy sector: renewable energy projects, mostly the building of hydroelectric installations, but also wind, geothermal and biomass projects. The baseline in this case is the GHG emissions of fossil fuels needed in the event that the renewable energy projects had not taken place.

b. GHG sequestering projects: such as the promotion of sustainable private forests and protected areas and the consolidation of the Costa Rican national park system and biological reserves. The baseline is established by the alternative use of the land in case the government could not sustain the national park system and protected areas. It is important to note that the government had declared many areas as protected or as national parks without being able to compensate the formal landowners; thus, a significant portion of these areas was at risk.

c. Energy efficiency projects.

The baseline, defined as the "without project" reference case, had to be certified by an independent third party (verifier). Determining how many tons of carbon dioxide the country would fail to fix if the land were not allocated to national parks, reserves, or protected areas, or if the renewable energy projects were not undertaken is a difficult task; it requires comparing the amount of carbon dioxide fixed to a hypothetical reference. Without an external verifier the projects could be too risky to attract investors. It is by designing high-quality projects that can be certified by a well-known international organization, that the country can add value to the project as well as create a good reputation in the international arena. Country entrance to the GHG emission mitigation market is easy; thus, it is very likely that in the long run competition across countries will be based on the quality of their projects. Costa Rica took actions to differentiate and add market value to its products. The country contacted the Societé Generalle de Survillance (SGS), a prestigious Swiss organization, to be in charge of verifying the existence and permanence of the programs as well as the use of scientific models to correctly measure the amount of carbon fixed, and to certify the baseline.

It is important to note that before the Kyoto Conference there was doubt about whether developing countries had to adhere to the emission reduction requirements set by UNFCCC. It was unclear whether the requirements were set just for developed countries or for all countries. Furthermore, the required emission reduction is set at the country level; how the country satisfies the goal is not stipulated. If Costa Rica had to commit to emission reduction requirements it would have been very difficult to distinguish (and to economically reward their environmental services) between those projects that were satisfying the country's required emission reduction and those projects that were mitigating GHG emission. Thus, the transaction costs from the whole program would have increased significantly. However, the Kyoto Protocol clearly specifies that the requirements are set only for developed countries, thus, facilitating the implementation of the Costa Rican strategy.

In 1997, CROJI innovated a promising market-oriented approach to sell environmental services internationally by issuing the first environmental bonds in the world: the *Certifiable Tradable Offsets* (CTOs), also known as *Greenhouse Gas Emission Mitigation Certificate* (Tattenbach and Villalobos 1998). CTOs represent a certification of a specific number of units of GHG emissions, expressed in carbon equivalent units, reduced or sequestered, or to be reduced or sequestered, by projects developed in the country. CTOs are certified and fully transferable, and they are issued when all project implementation phases are completed. CTOs are guaranteed by the Ministry of Environment and Energy of Costa Rica for a certain period of years, according to the lifetime of the project, and they are certified by an external verifier: the SGS. The investor simply purchases the offsets in a stock market, like any other bond. To reduce the risk to investors and add market value to the certificates, the Costa Rican government created a bank or reserve of unused carbon offsets. In case a program is declared invalid by external monitoring and verification, the government commits itself to use its reserves to replace any invalid bond. The mechanism is supervised by the SGS.

By separating the investors from the projects, CROJI has more freedom in developing those projects that have priority for the country, as well as developing portfolios of small projects (which reach a larger number of Costa Ricans) and issuing certificates over the whole portfolio. Transaction costs are significantly reduced by using established commodity markets to sell the certificates and by reducing information problems related to the direct involvement of investors in the financed projects. In addition, the Costa Rican government created a fund: the *National Specific Fund for Greenhouse Gases Sinks and Deposits*, to facilitate the financing and the development of national-scope projects with strategic value to the country.

The country hopes to gather enough financial resources by selling its environmental services to advance its own agenda of territorial consolidation and funding of national parks and state biological reserves: purchasing land, financing the demarcation of limits, reinforcement of fire prevention programs, and compliance, among others.

6. The Costa Rican Experience

Two stages can be clearly defined in the experience of Costa Rica.

6.1 The early stage

Taking advantage of the UNFCCC, in September 1994 Costa Rica signed with the United States the first AIJ/JI agreement in the Western World: the *Sustainable Development, Cooperation and Joint Implementation Agreement*. It was also the first AIJ/JI program in a developing country. In 1995, a similar agreement was signed with Norway. There are advanced negotiations with Switzerland, the Netherlands, Canada and Germany.

In 1994, AIJ projects in Costa Rica began as independent and isolated projects. Thus, this stage is based on a project-by-project basis that requires the direct participation of the investors. It can be considered as an exploratory stage, because there is no international experience, and Costa Rica becomes a pioneer in the developing world. The project-by-project approach is characterized by high transaction costs: foreign investors do not have adequate information about the potential projects and the costs of getting the information are high; investors being taken advantage of by opportunistic agencies; and investors directly face the potential risks from the investment. Because many of these costs are fixed, only large-scale projects become profitable.

To reduce the high transaction costs implied by the mechanism, the CROJI was created in August of 1995 with the participation of both the public and the private sector, with the mission of developing a coherent strategy to attract investments under the Joint Implementation framework. The CROJI decided to go beyond the promotion of specific projects to develop a concept of "national" or "umbrella" projects.

During the last 30 months the total investment of AIJ in Costa Rica has been about 140 million US$ in projects with GHG mitigation benefits. There are nine projects approved by the Secretariat of the UNFCCC; five of them in the energy private sector (three wind-farms and two hydroelectric projects) and four in the land-use sector (one implemented, one under implementation, and two partially financed).

6.2 The second stage

The CROJI developed national-scope projects, creating portfolios of projects and following a commodity approach. The new approach has the following advantages: the country designs projects according to its priorities; the benefits from the projects extend to a larger number of participants by not penalizing small-scale projects as a result of reducing the transaction costs (from development, marketing, monitoring, verification, and financing of projects) and enhancing credibility on the CTOs. There are currently two portfolios of projects: the Protected Area Project (PAP) and the Private Forestry Project (PFP) (Tattenbach and Villalobos 1998).

The PAP will provide protection for about 530,000 ha within the National Parks. Although declared protected areas, they have not been fully integrated into national property because the government has not been able to purchase the land; thus, there is a potential threat of deforestation. The PAP consolidates the protected areas and transfers the land ownership to the Ministry of Environment and Energy for protection in perpetuity. Costa Rica is issuing CTOs on this project. Approximately 15.5 million CO_2 equivalent metric tons of carbon to be generated by the project will be commercialized as a commodity. The CTOs will be traded at the Chicago Board of Trade, with the support of the Earth Council, the external monitoring of the Societé Generalle de Survillance, and the Centre Financial Products as a broker.

The PFP uses a tax on fossil fuels to compensate farmers for their forest conservation, reforestation, or sustainable management efforts. Recipients of the compensation have no claim on the environmental services from their projects to the Costa Rican government. The government will issue CTOs on these environmental services. The objective of this project is to create, with the participation of the private sector, a sustainable buffer zone around protected areas. The program is based on a *payment for environmental services* concept rather than on subsidies. The Costa Rica-Norway Reforestation and Forest Conservation AIJ Pilot Project constitutes the first application of the PFP (Joint Implementation Quarterly 1996). In October 1996, the Government of Norway and 3 Norwegian private companies purchased a total of 200,000 metric tons of carbon from Costa Rica for a total price of 2 million US$. It is the first transaction that clearly separates and values GHG offsets as a commodity. It is noteworthy that PFP constitutes a scissors scheme to attack externalities: taxing GHG emission activities while compensating GHG emission-reducing or -sequestering activities.

The country is currently consolidating the "umbrella" projects. CTOs will be offered in the international markets, with no necessity to have a predetermined investor. There are 11 joint implementation projects approved by the Secretariat of the UNFCCC: 5 energy projects, 5 forestry projects, and 1 agricultural project. The five forestry projects incorporate a total area of 542 400 ha and represent the GHG emission reduction-sequestering of the equivalent of approximately 20.6 million metric tons of carbon (Table 1).

Table 1. Forestry projects

Forestry Project	Type of Project	Area (hectares)	Total Cost (US$mill)	Duration (years)	Emission Reduction/ Sequestering ($x10^3$ metric tons of carbon)
Ecoland	Conservation	2,340	1	15	345.5
Klinki	Reforestation	6,000	3.8	40	1968.0
CNFL	Conservation Reforestation Regeneration	4,000	3.3	25	313.6
PAP	Conservation	530,000	150	25	18000.0
Earth	Reforestation	78	0.16	20	2.1
Total		542,418	158.16		20629.3

Source: Tattenbach and Villalobos (1998)

6.3 On the market for trading CTOs

Like any other commodity, environmental services are traded internationally. A well-defined and enforceable global property-rights system is required. The equilibrium price of the environmental services that Costa Rica is selling in the market will be somewhere between the costs of the offset suppliers in developing countries (supply side) and the alternative costs of emission reduction for the buyers in developed countries (demand side).

On the supply side the comparative advantage of a country depends on land vocation, hydroelectric capacity, location of the country, and alternative uses of land. A country's entry to the market is, in principle, easy; however, not all the countries have the institutional and legal framework, and the world reputation to add value to their certificates. This is the reason for Costa Rica stressing the importance of the quality of its bonds through both internal and external verification and monitoring. In addition, not all countries have the same opportunity cost in the allocation of their resources (land) to the production of environmental services; some countries will find it out more profitable to allocate these resources to the production of agricultural activities, urbanization, and so on. Factors to be considered on the demand side are: the GHG emission reduction commitments of the developed countries, and their compliance to these commitments, as well as the delaying of the agreement's implementation date. The implementation of these commitments within the country (e.g., across-activities reduction requirements, reduction requirements to specific activities, taxes on GHG emission activities); the cost of alternative emission reduction options (e.g., changing production technologies to more cost-efficient ones). The demand curve is highly unstable, with much uncertainty coming from the countries' willingness to obey their commitments.

7. Final Considerations

The recognition of the environmental problems as global, as well as the definition of responsibilities and property rights over negative and positive externalities constitutes the fundamental step to initiate their solution. United Nations conventions have contributed to the definition of these property rights. Costa Rica took advantage of it to foster its ecological programs by selling the environmental services from its national parks, protected areas, and buffer zones. The Costa Rican experience is based on following a market approach to the selling of these environmental services and innovating ways to reduce transaction costs. The CTO certificates represent the most recent innovation. The success of the CTOs is uncertain yet, it depends on the willingness of developed countries to implement their GHG emission reduction commitments.

8. References

Carraro, C., and Siniscalco, D. 1993. Strategies for the international protection of the environment. Journal of Public Economics, 52.

Carraro, C., and Siniscalco, D. 1996. International coordination of environmental policies and stability of global environmental agreements. *in* Bovenberg, L., and Cnossen, S. eds. Public economics and the environment in an imperfect world. Kluwer Academic Publishers.

Dutschke, M., and Michaelowa, A. 1997. Joint implementation as development policy: The case of Costa Rica. Discussion Paper No. 49, Institute fur Wirtschaftsforschung, Germany

Joint Implementation Quarterly. 1996. AIJ pilot project between Costa Rica and Norway, V.2 N.4. The Netherlands.

LeBlanc, A. 1996. An emerging host country joint implementation regime: The case of Costa Rica. Incentives for private sector investment in JI: Four case studies. Centre for Economic Analysis.

Olson, M. 1994. The logic of collective action. Harvard University Press.

Sedjo, R.A, Wisniewski, J., Sample, V.A., and Kinsman, J.D. 1995. The economics of managing carbon via forestry: An assessment of existing studies. P.139-165 *in* Environmental and Resource Economics, Vol.6, No.2.

Tattenbach, F., and Villalobos, V. 1998. De implementación Conjunta al Fonde de Desarrollo Limpio: Financiando el Desarrollo Sostenible en Costa Rica. MIDEPLAN, Costa Rica.

Tietenberg, T. 1992. Environmental and natural resource economics. HarperCollins Publishers.

Sustainable Forestry

- The policy prescription in British Columbia -

Bill Wilson and Sen Wang

Key words: Sustainable forestry, timber supply, economics, policy

Abstract: In pursuit of sustainable forestry, British Columbia, Canada, has undertaken a number of policy initiatives in recent years. First, the participation of stakeholder groups was formally institutionalized through round-table and public debate processes that serve as the foundation for spatial planning of forestland and as an input to forest policy development. Second, areas of conflict over forest resources utilization were identified, and a variety of policies were implemented in pursuit of resolution. These policies included land-use zoning, protected-area expansion, forest management prescription and timber supply analyses. Third, the legislative framework was consolidated with an orientation towards sustainable forest practices and enhanced accountability. The transition from a timber-harvesting dominant model to one of sustainable forestry presents the challenge of managing forest resources for increasingly diversified uses under a growing number of operational constraints. British Columbia is moving to a multiple-use strategy at the forest landscape level with dominant use zonation in an effort to minimize forestland dedicated to timber production. The experience indicates the importance of considering the effects of the trade-offs in forest resource sustainability.

1. Introduction

British Columbia (BC), the third largest province in Canada, has a total area of 95.2 million ha and a population of some 3.7 million. The abundant resource base and relatively small population have been central factors in provincial development.

In BC, the forestry sector has historically been the mainstay of economic activity and the finance source to economic development. In 1996, the province exported 28.5 million m^3 of softwood lumber, 4.9 million metric tons of wood pulp and 1.4 million metric tons of newsprint, which accounted for 58%, 48%, and 16% of Canada's total, respectively (Natural Resources Canada 1998).

Despite economic maturation with increased diversity and growth in the service sector, BC's forestry sector continues to be a major employer, a large contributor towards provincial gross domestic product (GDP) and the main source of export earnings. In 1996, forestry activities and the forest industry constituted about 17% of the total provincial employment, 25% of GDP, and 63% of total export earnings (British Columbia Council of Forest Industries 1998).

Compared with many other jurisdictions, BC forestry is somewhat unique. First, accounting for about 7% of global softwood growing stock, BC forestry operates largely within the natural timber endowment, and has one of the few remaining temperate rainforests and a high degree of public ownership of the forestland (only 4% of the forestland is private). Second, strident conflict characterizes decisions on forest resource management and use. At present, about 89 communities are dependent on the forest industry. And third, despite high per capita forest products consumption levels, BC's domestic market is small. Over 90% of production is exported and currently, major markets are the United States (50% by value of shipments), Japan (25%) and the European Union (EU [15%]). With a concentration in commodity-grade products such as lumber and wood pulp, BC is subject to cyclical price fluctuations in the highly competitive global market for forest products. BC's companies are small in comparison with major global competitors (e.g., only one BC company is among the top 25 global forest product companies).

This set of characteristics has presented additional challenges to the BC policy package intended to position forestry to meet the multi-faceted demands of sustainability. This package includes the commitment to double the land area in preserved status, a highly prescriptive approach to forest management, land zonation based on regional land-use decisions, and a rigorous review of public timber harvest levels. The package will affect, among other things, forest management practices, land use, forestland preservation, and access to and the cost of timber. The effect on timber supply will in turn affect government revenues, employment, and economic activity. As a result, new measures were introduced to promote sector employment by encouraging forest companies to invest in value-added manufacturing and silviculture (Delcourt and Wilson 1998).

This paper summarizes the highlights of BC's policy initiatives aimed at promoting sustainable forestry. In Section 2, the complexity of sustainable forestry is outlined, and an overview of the recent policy changes and their effects is provided. Section 3 provides a short prognosis on commercial forestry within the new institutional setting. A selection of trade-offs and challenges are discussed. The paper concludes with some general comments on sustainable forestry.

2. Programs and Initiatives for Sustainable Forestry

2.1 What is sustainable forestry?

Sustainability in forestry is neither static nor is it one-dimensional. Because forests are living ecosystems, inherently dynamic and subject to catastrophic events (e.g., fire, diseases, insect pests, and human activities), conservation has a limited meaning. Even a natural forest undisturbed by human intervention will not be the same as it was 100 years before, nor will it be the same today as it will be in 100 years.

We define sustainable forestry as one that comprises a host of management regimes to maintain and enhance the long-term health and integrity of forest ecosystems and forest-dependent communities, while providing ecological, economic, social, and cultural opportunities for the benefit of present and future generations. Sustainability involves a multitude of dimensions (Perry and Amaranthus 1997). At a minimum, sustainability includes biological, sociological, political, and economic dimensions in the definition. High education and living standards foster social expectations on forest management and preservation. This has been complemented by the rapid reduction in the time and distance costs associated with information acquisition. BC public ownership of forestlands links the social to the political dimension.

Sustainable forestry is a paradigm of trade-offs among the various dimensions that define sustainability. A complex task is further confounded by changes in social expectations, delays, and externalities in the effects of the sustainability policies adopted for implementation, and by the distribution of costs incurred in the pursuit of sustainability.

2.2 Forest policy and sustainability

In an effort to facilitate an efficient transition to sustainable forestry while protecting jobs, communities, and the provincial economy, BC implemented a package of policies and programs. The package includes a stakeholder process (consensus based) to develop comprehensive regional land use plans, a review of the harvest levels in major harvest licenses, a highly prescriptive code for forestland management, a stumpage-based forest investment fund, and a forest employment initiative (see Table 1).

Table 1. Summary of initiatives for sustainable forestry

Issue	Initiative/Program	Objective	Implementation Instrument
Conflict resolution	Commission on Resources and Environment (1992)	Stakeholder participation	Round-tables and public consultation
Old-growth preservation	Protected Areas Strategy (1992)	12% of BC land area	Legislation, regulation
Sustainable forest practices	Forest Practices Code (1995)	Ecosystem management	Legislation, regulation
Timber supply	Timber Supply Review (1992-1996)	Sustainable harvest	Determination of harvest levels
Forest renewal	Forest Renewal BC (1994)	Dedicated funding for forest renewal	Crown corporation
Employment	Jobs & Timber Accord (1997)	Employment opportunities	Intensive silviculture, value-added manufacturing

BC's public forestlands are managed by the provincial Forest Service (also known as the Ministry of Forests) pursuant to the Forest Act (RS 1979, Chapter 140), which provides the statutory basis for timber tenure administration and forest management.[1] Having evolved over several decades of changes in forestry institutions, the two primary tenure vehicles, which may be viewed as "contracts" between the Forest Service and forest companies, are Tree Farm Licenses (TFLs) and Forest Licenses within Timber Supply Areas (TSAs).[2] The 34 TFLs are area-based and they constitute about 23% of the province's allowable annual cut (AAC). The 37 Forest Licenses are volume-based agreements and represent some 75% of the AAC. The AAC in a TSA is assigned among the licensees (by the Minister of Forests) whereas the TFL AAC is available to the tree farm licensee (Wilson et al. 1998, Ross 1995, Haley and Luckert 1990).

Pursuant to the Forest Act, as a condition to access public timber, holders of major forest tenures are responsible for, among other things, (a) the preparation of forest operational plans, (b) road construction, (c) reforestation and stand-tending to a free-to-grow stage, and (d) stipulated timber processing. Forest regeneration is the law in BC, and amendments to the Forest Act in 1987 made licensees responsible for the financial cost of reforestation. Public funding, both federal and provincial, has been dedicated to reducing the non-satisfactorily stocked areas from pre-1987 harvesting. Currently, about two-thirds of reforestation is from planting and one-third is from natural regeneration. Research and improved planting has raised the seedling survival rate to around 85% (Bartlett 1996).

- **Land-Use Zonation**

With the establishment in 1992 of the Commission on Resources and the Environment (CORE), BC embarked on a process of stakeholder participation in defining land use assignments. The process, which has evolved with time and experience, involves a consensus-based negotiation, in the form of round-table meetings among the various stakeholder groups at a regional level. The government provides resource and facilitation support functions. The commitment has produced three land-use plans based on defined geographic areas (Vancouver Island, Cariboo-Chilcotin, and the Kooteneys). A similar process continues in other regions, and implementation plans and provisions for monitoring and amendment have

[1] The Ministry of Forests Act (1979) stipulates that the Forest Service assert the financial interest of the Crown in its forests in a systematic and equitable manner. Major government direct revenues from forestry are raised in the form of timber royalty, stumpage and land rental fees.

[2] Other vehicles used to allocate public timber harvesting rights include Timber Sale Licenses, Timber Licenses, License to Cut, Woodlot Licenses, Pulpwood Agreements, Christmas tree permits, free-use permits, and road permits.

also been formulated. Forestry operational plans must conform to the provisions in the higher-level regional land and resource management plan (Sahajananthan et al. 1998). These higher-level Land and Resource Management Plans (LRMPs) set strategic direction for regional planning areas and delineate use zones. Land-use planning is expected to be in place for the entire province by 2000.

- **Protected Areas Strategy**

Announced in 1992, the Protected Areas Strategy (PAS) committed BC to doubling its protected areas to 12% of the province by the year 2000. At the end of 1997, the area in protected status exceeded ten million ha, or 10.6% of the province. The designated protected areas are intended to be representative of BC's diverse ecosystem, and no extraction or industrial activity will be allowed on PAS lands. Ongoing efforts to identify and establish new protected areas are mainly through comprehensive LRMPs.

- **Forest Practices Code**

In 1995, in an effort to promote forest resource stewardship and to improve sustainable forest management, the BC government enacted the Forest Practices Code of British Columbia Act (SBC 1995, Chapter 4). It provides a rigorous, comprehensive, and enforceable code of conduct for forest management practices. The Forest Practices Code (FPC) has jurisdiction over all public land and private lands that have been integrated into TFLs. The code is highly prescriptive of forest planning and operational standards. The code also provides for independent audit and enforcement options such as timber withdrawal, fines, and legal action (Kremar-Nozic et al. 1996). The main objectives of the FPC are to balance the various competing uses and values pertaining to forestlands. In effect, it acts as a regulatory framework to guide forestry activities in the interests of maintaining biodiversity, soil, hydrology, wildlife habitat, and aesthetic values.

- **Forest Renewal Agency**

Forest Renewal BC (FRBC) is a Crown agency funded with stumpage and royalty fees charged on public timber. The Agency, pursuant to the British Columbia Forest Renewal Act (SBC 1994, Chapter 3), has a mandate to develop and support the delivery of forest and ecosystem renewal initiatives, worker training, community stability projects, and value-added efforts. Program development is based on broadly-based input from among various stakeholders in a structured process that sets targets and reviews proposals to achieve these targets. The Forest Renewal Act was recently amended to assign priority to displaced union labor for any employment arising through FRBC expenditures.

- **Timber Supply Review**

Under the Forest Act, the provincial Chief Forester is required to determine the AAC, and this determination is subject to review at least once every five years. A comprehensive Timber Supply Review (TSR) was initiated in 1992 as a four-year initiative to ensure that timber harvest levels in each of the TSAs and TFLs were appropriate and sustainable in the long term. Factors considered in the review included: (a) the timber resource, (b) current forestry practices and non-timber values, (c) production capabilities (d) socioeconomic objectives, and (e) salvage logging opportunities.

2.3 Impacts of sustainability initiatives

It is recognized that the efforts to improve forest sustainability will reduce the volume of timber harvested, and this will reduce direct employment and economic activity and will increase the cost of timber. On the other hand, the gains will include fostering environmentally responsible forest management, expansion and improved ecosystem representation in protected areas, reduced social discord on the management of public forestlands, broader participation in land-use planning, greater transparency in land-use decision-making, and improved accountability in public lands administration.

BC's commitment on protected areas is popular within the province and has quickly improved its relative position in comparison with many other jurisdictions. At present, BC is nearing its target of 12% protected areas, and the level of accomplishment is considerably higher than that achieved by countries like Sweden, the United States, or the rest of Canada. However, the province is paying a high economic price for the protection of nature (van Kooten and Wang 1998).

The FPC has, among other practices, defined new standards for clearcutting and for riparian zones. The new standards in clearcut harvesting include: (a) the type, size, and location of cutblocks; (b) incorporation of emerging environmental and aesthetic objectives; and (c) site-specific regeneration requirements and practices. BC is now comparable to many major forest jurisdictions as far as maximum clearcut opening size is concerned (Westland Resource Group 1995). Furthermore, some forest companies operating in environmentally sensitive areas are working to reduce or eliminate clearcutting and are restricting the cutting of old-growth forests.

The effects of the 1992-96 TSR are summarized in Table 2. Although the aggregate effect was a mere 0.5% reduction in AAC, it masked some significant regional variation. For instance, a southeastern region had a 9.3% AAC reduction, whereas, a southcentral region had a 0.4% decline. In addition, while the conventional AAC was actually reduced by 5.2% (3.4 million m^3), the reduction was offset by increased provision for deciduous, salvage, and marginal timber harvesting. The latter involves the inclusion of small-diameter interior lodgepole pine (*Pinus contorta*) stands.

Table 2. British Columbia timber supply review results (1992-1996) (million m^3)

	AAC	Conventional	Deciduous	Insect/Disease	Marginal
Pre-TSR	71.2	66.8	2.4	0.85	1.2
Post-TSR	70.9	63.4	3.2	1.14	3.2
% Change	(0.5)	(5.2)	34.7	34.1	175.0

Reduced public timber access, increased forest management prescription, and greater uncertainty in log volumes during the implementation of the policy changes have combined to increase the cost of timber in BC (KPMG 1997, Thibodeau 1995). Over the 1992 to 1996 period, logging costs increased by 75% (or \$37.17 per m^3). Table 3 provides a breakdown of the change in logging costs by FPC, non-FPC, and stumpage-royalty elements for both coastal and interior regions. The relative changes in logging costs over this period are presented in Figure 1.

Table 3. Changes in British Columbia's logging costs, 1996 versus 1992 (Canadian \$/m^3)

	Interior	Coast	BC
Forest Practices Code	8.41	19.68	12.22
Non-code	7.07	10.97	8.40
Stumpage/Royalty	16.98	15.70	16.55
Total	32.46	46.35	37.17

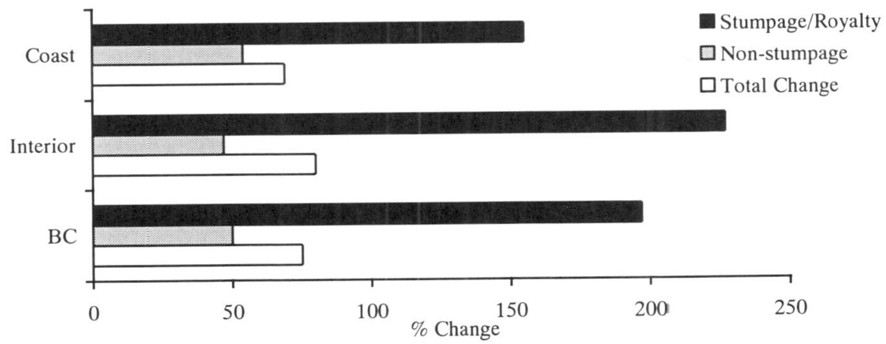

Figure 1. Percentage change in British Columbia's logging costs, 1996 versus 1992 (Source: KPMG 1997)

BC has become a high-cost timber producer and can expect to incur quick and significant demand reductions in periods of weak forest product markets in the highly competitive international markets (Price Waterhouse 1996). Recognizing that a high-cost producer of commodity products is not a financially sustainable option, a forest sector strategy committee has been working on ways to improve industry fundamentals.

3. Prognosis of Commercial Forestry

BC forestry remains largely based on harvesting from the natural timber endowment (up to 90%).[3] In contrast to international competitors harvesting second or third rotation timber or from intensively managed plantations, the extensive manner of BC forest operations reflects the geographic scale of the province and the rest of Canada (Benson 1988). BC's dependence on old-growth harvesting is the product of a traditional abundance of timber and the slow-growth character of native species (the mean annual increment for BC's coniferous forests is about 2.5 m^3 per ha), and the institutional setting for forest investment and management (Cartwright 1994). One major institutional factor influencing BC's forestry is land ownership. As a consequence of public ownership of the forestlands, provincial policy is the major driver in land use, forest management, stumpage determination, harvest levels, and industrial location and scale.

However, harvesting large, old timber creates unique problems. In the past decade, the province has witnessed waves of criticism from the public over the harvesting of old-growth timber. While the rising environmental challenge to commercial harvesting focused on coastal temperate forests in the early 1990s, there is an emerging concern about harvesting in the boreal forests; furthermore, some environmental groups are stridently opposed to plantation forests (Stanbury et al. 1995).

In response to calls for sustainability of the forest resource base within a balance of employment, timber harvest volumes, and environmental objectives, the BC government has restructured its policy to focus on two broad aspects, namely, maintaining biodiversity and safeguarding economic well-being. The expansion of protected areas (via PAS) serves to reduce the forestlands available for timber production by promoting an investment response that is less dependent on forestland access. In addition, as wood costs rise owing to a shift in logging operations to the extensive margin, higher FPC-related standards, and increased stumpage levels, opportunities are sought to replace timber with other production factors (i.e., capital and

[3] Constituting some 8% of the global harvest of softwoods, BC's current AAC is about 71 million m^3, and an additional 5-6 million m^3 is harvested from private unregulated lands. The harvest is divided roughly one-third from the coastal region and two-thirds from the interior region. The principal species of the coastal region are hemlock (*Tsuga* spp.), western red cedar (*Thuja plicata*) and Douglas-fir (*Pseudotsuga menziesii*). These stands are dominated by large old-growth timber, and the average rotation length is 70 years. The interior region forests are primarily composed of smaller-diameter spruce-pine-fir trees (*Picea* spp., *Pinus* spp., and *Abies* spp.).

labor). Meanwhile, the information technology can be expected to provide an increasing amount of knowledge into timber production and forest products (Binkley 1997).

The complexity inherent in forest resource conflicts may be dealt with by use zonation, which leads to the adoption of a dominant-use strategy. Furthermore, over time, rising wood costs will tip the economic scale from logging the old growth towards harvesting the second growth. The recent findings that BC's site index has been considerably underestimated provide new support for intensive forestry investment in higher productivity sights (Nigh 1998, Nussbaum 1998). Zonation or the dominant-use strategy contributes positively to the environmental, social, and political dimensions of sustainability (Sahajananthan et al. 1998). However, the economic implications are proving a difficult challenge.

The essence of the dominant-use strategy is to sustain forest values, both timber and non-timber, and to secure a non-declining flow of timber supply to the forest-dependent communities and the industry. Binkley (1997) estimates that the current annual harvest of about 70 million m^3 may be sustained from a much smaller land base if a timber-use zone is designated to permit intensive management of natural forests and plantations for largely commercial purposes. In addition to the operational considerations such as site selection, species choice, and organizational arrangements, such an option will require institutional reform (including forest tenure) in order to encourage private sector investment. BC has invested significantly to improve forest sustainability and has made some solid progress (Wilson 1997).

4. Conclusion

Sustainability is a complex amalgam of trade-offs among its various dimensions, and there is no easy route to achievement.[4] It is important that the policy process on sustainability be based on sound information about these trade-offs, and the selection of objectives reflect these trade-offs. BC has been successful, to a degree, in this regard. Forestry remains a key contributor to BC economic performance, and it is imperative that sector policy serve to effectively promote the sustainability of this contribution. The costs of this transition are not likely to be absorbed by the market. The market signals on sustainable forest management are mixed at best. The rhetoric has appeared in abundance but the willingness to pay the additional cost associated with sustainable forest management has not (Stanbury et al. 1995).

[4] For an extensive review of the forest policy response to changing sector demands in a selection of jurisdictions with significant forest sectors see Wilson et al. (1998).

Six years following the initiation and subsequent adoption of sustainable forestry as a nationwide guiding principle, 1998 witnessed the signing of a second Canada Forest Accord that represents a strong endorsement of the new five-year National Forest Strategy entitled *Sustainable Forests: A Canadian Commitment* (Canadian Forest Service 1998). BC has contributed significantly towards this process and has made solid progress in sustainable forestry. The sustainability policy package implemented in the province has served to greatly broaden stakeholder participation, improve the transparency in land-use planning and allocation, and to reduce the environmental footprint of commercial forestry. Social expectations on forestland management, a critical ingredient given the large proportion of public forestlands in BC, will continue to evolve but can be expected to become more demanding as the link between the gains accruing from commercial forestry become more distant to the urban dweller.

5. References

Bartlett, K.J. 1996. Just the facts: A review of silviculture and other forestry statistics. British Columbia Ministry of Forests, Victoria, B.C.

Benson, C.A. 1988. A need for extensive forest management. The Forestry Chronicle. 64(5): 421-430.

Binkley, C.S. 1997. Preserving nature through intensive plantation forestry: The case for forest land allocation with illustrations from British Columbia. The Forestry Chronicle 73(5): 553-559.

British Columbia Council of Forest Industries. 1998. British Columbia forest industry fact book 1997. Vancouver, B.C.

Canadian Forest Service. 1998. Sustainable forests: A Canadian commitment. National Forest Congress Version. Ottawa.

Cartwright, D. 1994. British Columbia forest products industry: Production, markets, competition and competitiveness. Working Paper 94.01. Canadian Forest Service, Victoria, B.C.

Delcourt, G., and Wilson, B. 1998. Forest industry employment: A jurisdictional comparison. Canadian Public Policy. 24(supp2): S11-S25.

Haley, D., and Luckert, M.K. 1990. Forest tenures in Canada: A framework for policy analysis. Information Report E-X-43. Canadian Forest Service, Ottawa.

KPMG and Perrin, Thorau & Associates Ltd. 1997. Financial state of the forest industry and delivered wood cost drivers. Report prepared for British Columbia Ministry of Forests, Victoria, B.C.

Krcmar-Nozic, E., van Kooten G.C., Vertinsky, I.B., and Brumelle, S. 1996. Managing by objectives or prescribing behavior: Achieving soil protection and economic benefits in forestry. Working Paper 96.07. Canadian Forest Service, Victoria, B.C.

Natural Resources Canada. 1998. Selected forestry statistics Canada 1997. Canadian Forest Service, Ottawa.

Nigh, G.D. 1998. Site index adjustments for old-growth stands based on veteran trees. Working Paper 36. Research Branch, Ministry of Forests. Victoria, B.C.

Nussbaum, A.F. 1998. Site index adjustment for old-growth stands based on paired plots. Working Paper 37. Research Branch, Ministry of Forests. Victoria, B.C.

Perry, D.A., and Amaranthus, M.P. 1997. Disturbance, recovery, and stability. P.31-56 *in* Kohm, K.A., and Franklin, J.F. (eds.), Creating a forest for the 21st century. Island Press, Washington, DC.

Price Waterhouse. 1996. The forest industry in British Columbia. Vancouver, B.C.

Ross, M. 1995. Forest management in Canada. Canadian Institute of Resources Law. University of Calgary, Alberta.

Sahajananthan, S., Haley, D., and Nelson, J. 1998. Planning for sustainable forests in British Columbia through land use zoning. Canadian Public Policy. 24(supp2): S73-S81.

Stanbury, W., Vertinsky, I., and Wilson, B. 1995. The Challenge to Canadian forest products in Europe: Managing a complex environmental issue. Working Paper 95.04. Canadian Forest Service. Victoria, B.C.

Thibodeau, D. 1995. Effects of environmental protection on forest management costs in the Nehalliston Creek watershed: An analysis. Working Paper 6-008. Canadian Forest Service, Victoria, B.C.

van Kooten, G.C., and Wang, S. 1998. Estimating economic costs of nature protection: British Columbia's forest regulations. Canadian Public Policy. 24(supp2): S63-S71.

Westland Resources Group. 1995. A comparative review of the forest practices code of British Columbia with Fourteen Other Jurisdictions. Prepared for British Columbia Ministry of Forests. Victoria, B.C.

Wilson, B. 1997. Commercial forestry: The road to sustainability in British Columbia. Paper presented at a conference on "Forest Management into the Next Century: What Will Make It Work?" Spokane, Washington, November.

Wilson, B., van Kooten, G.C., Vertinsky, I.B., and Arthur, L. (eds.). 1998. Forest policy: international case studies. CAB International Publishing, Wallingford, UK.

Chilean Forestry

- Wood products from an exporting country -

Gonzalo Paredes V.

Key words: Forest products exports, forest production, institutions

Abstract: Since 1990, Chilean forestry has been facing important changes in its institutional scenario. New legislation for native forests has been discussed for at least six years. The incentives program for plantation forestry is now focussed on fragile and degraded soils and on small landowners. The law for wildland protected system will be announced to include private ownership. Also changes to improve the performance of state forestry agencies are being analyzed. New policy tools are being incorporated to address social values. International trade of forest products has also experienced some important changes over this period. New manufactured products are being exported to North America, and exports to Asian countries have increased. At the same time, forest products imports have increased as the domestic economy grows. Recent timber supply projections show an increasing availability of quality logs.

1. Introduction

The forest sector in Chile has experienced an interesting development during the last two decades. Moreover, it seems that its resource base, human and technical assets, as well as favorable institutional and economic environments will allow for further development.

Along with the development of the forest sector, wood products exports have increased significantly during the same period. The expansion and fairness of international trade will also provide a good scenario for Chilean products to test their economic and social performance.

This paper's objective is to contribute to the correct knowledge, characterization and dimensioning of Chilean forest sector. The next sections describe the recent evolution and current features of forestry: production, exports, ownership, institutions, and issues.

2. The Chilean Forest Sector

2.1 Land and forest resources

The natural resources most relevant to forestry in Chile are the land, the climate, its relative isolation from other South American countries, and the natural vegetation. The latter have been under heavy antropogenic pressure for, at least, the last three centuries.

Land, climate, isolation and the characteristics of natural ecosystems, have allowed the development of fruit crops, forest plantations, and agricultural crops with a low to moderate load of pests and diseases, relative to other countries. Actually, most, if not all, of the exports of products from forest plantations, agricultural crops, and livestock are based upon vegetal or animal species not native to Chile, growing in a healthy environment provided by the natural obstacles to the entry of pests and diseases from neighboring countries and from overseas.

In this sense, the Andes' altitude, the Pacific Ocean with its cold Humboldt current, a temperate climate, and natural barriers to the entry of undesirable insects and pests contribute to a good and healthy environment. Thus crops, trees and livestock can grow productively. Obviously, these natural barriers do not guarantee absolute isolation; however they contribute to environmental conditions quite different from other countries of the Southern hemisphere.

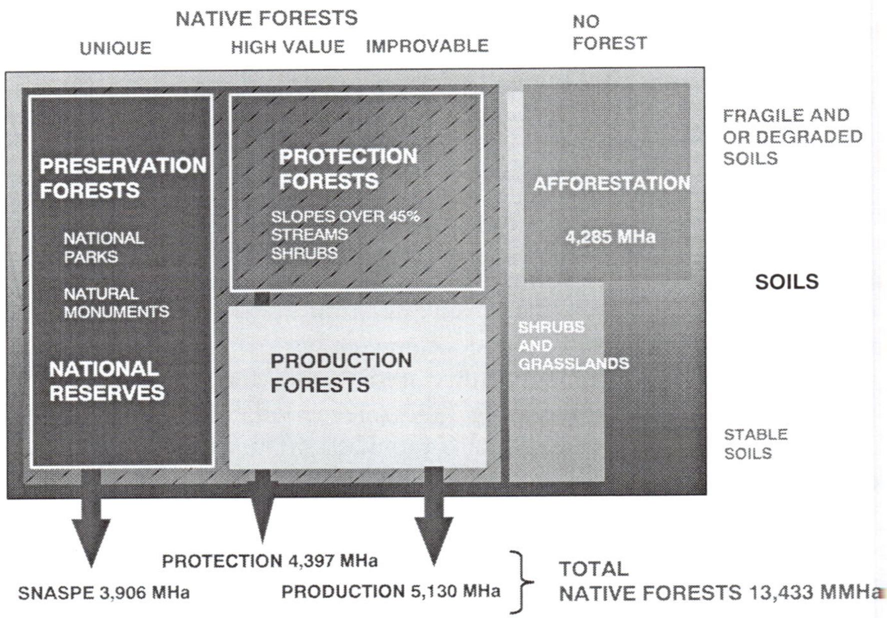

Figure 1. Categories of soil-forest situations in Chilean forestry

As observed in Figure 1, about 30% of the native forests are included in the National Wildlands Protected System (SNAPE), 33% are Protection forests and the remaining (38%) are Production forests. Even though it is a large area, poor access to the production forest constrains its economic potential. It should be also noted that the additional plantations potential accounts for more than 4 million ha.

With regard to native forest resources, Table 1 summarizes natural vegetation classified by forest types, their area in the National System of Protected Wildlands (SNASP), and their total area in the country.

The above and below figures have been recently obtained through a National Survey of Native Vegetation. This survey is providing an improved basis for policy decision-making with regard to native forests.

Table 1. Natural vegetation and wildlands protected areas

Forest Type	SNASP	Total	%
Alerce	47,395	264,993	18
Cipres de las Guaitecas	676,287	972,181	70
Araucaria	122,679	253,715	48
Ciprés de la Cordillera	2,862	45,079	6
Lenga	566,531	3,400,346	17
Coihue de Magallanes	885,222	1,801,637	49
Roble-Hualo	885	184,783	1
Roble-Raulí-Coihue	23,871	1,370,218	2
Coihue-Raulí-Tepa	43,389	456,919	10
Esclerófilo	6,810	342,631	2
Siempreverde	1,491,535	4,350,814	34
Total	3,867,543	13,443,316	29

Source: CONAF-CONAMA-BIRF (1997)

Of the total area covered with natural vegetation, 38% is adult forest, 23% is second growth (renovales), 5.5% is a mixed adult second-growth forest, and 19% is forest with tree height less than 8 meters.

The National System of Protected Wildlands is currently under revision and expansion to include the habitats of threatened species, unique landscapes, and unprotected ecosystems and forest types. Legal arrangements are being made for some tax exceptions for private landowners incorporating their forests into the System.

To understand Chilean forestry it is necessary to highlight the fact that all production activities are carried out on private ownership. As Table 1 shows, the SNASP includes 29% of the area covered with natural vegetation, whereas man-made forest plantations (2 million ha) are totally on private lands.

Another important dimension to describe Chilean forestry and policy implications, is ownership size. Cost of capital varies significantly among landowners, thus affecting their decision-making and managerial implications.

Natural forest is mostly concentrated in large farms. The 1976 national agricultural census showed that only 20% of the productive native forest was owned by 50 thousand “small” landowners (each with less than 100 ha), with an average of 8 ha of native forest per ownership. Recent studies have observed increasing concentration of native forest in larger ownership.

With regard to plantations the situation is quite similar. In 1994, the Government Research Institution, Instituto Forestal (INFOR) estimated that 2,600 small landowners (less than 100 ha) owned only 7.5% of the radiata pine plantations area. They account for 72% of the total number of ownerships.

The availability of denuded land suitable for forestry purposes and government incentives, has allowed the establishment of about 2 million ha of fast-growing forest trees, mostly radiata pine and eucalyptus species. Plantations with these species were started late in the 19th century, and until 1930 the average annual plantation was only 160 ha nation-wide. In 1931 a Forest Law, promoting afforestation and management practices through tax incentives, was implemented, and the afforestation annual rate increased to an average of 16,000 ha until 1973. Since 1974, when DL (Decree Law) 701 introduced an incentives system for afforestation in denuded forest lands, the rate of annual planting increased to an average of 80,000 ha until 1990, to more than 100,000 ha during the period 1991 to 1995, then decreased to 55,000 ha in 1997.

Current area covered with plantations amounts to a total of 1.88 million ha, with 1.4 million ha planted with radiata pine (INFOR 1998a) (Table 2).

Table 2. Forest plantations by species, year 1997 (INFOR 1998a)

Especie	ha	%
TOTAL	1,881,925	100.0
Pino radiata	1,420,015	75.5
Eucalipto	317,212	16.9
Atriplex	49,320	2.6
Tamarugo	20,632	1.1
Pino oregón	12,620	0.7
Alamo	4,115	0.2
Other species	58,011	3.0

Plantations are made mainly with radiata pine (currently 69% of annual plantations) and eucalyptus (21%); other species account for only 8%. Reforestation of harvested land accounts for 35% of annual plantations; eucalyptus accounts for 7% and other species 3%. The remaining area (55%) is all afforestation on denuded, eroded lands, and less than one-half of this afforested area qualifies for the incentives program. The reason is that DL 701 applies only to lands technically qualified as "forestlands," excluding agricultural lands.

During the last decade, an important issue has been plantations established on lands with native forest, thus replacing it. Instituto Forestal (INFOR), and non-government conservation organizations, have estimated that between 3% and 6% of annual plantation have been established on land with some type of native vegetation. This estimation range is mostly explained by different concepts of a "forest" definition. Thus, the existence of some substitution reflects the fact that, with or without a plantations incentives program, for a landowner it will be more attractive to harvest his degraded native forest and then to plant fast-growing trees whenever the lack of control or an ambiguous definition for "native forest" allow him to do so.

Also observed, particularly during the last seven years, is the conversion of marginal agricultural lands to plantation forests. The economics of agricultural crops and livestock is outperformed by plantation forestry in some sites, even though these plantations cannot obtain a subsidy. Therefore, either forest companies purchase the land, or the landowner manages his financial requirements to invest in forest plantations.

In general, it can be said that forest plantations in Chile have allowed the recovery of eroded land with inadequate vegetation cover. Recent estimates by Corporación Chilena de la Madera (CORMA), the Chilean forest industry association, and Instituto Forestal demonstrate that 43% of plantations have been made on severe- and very-severe-erosion soils, 44% on moderately eroded soils, and 11% on slightly eroded lands. The study could not recognize previous soil conditions in 2% of the area.

At the same time, industrial plantations have become the source of raw material for an expanding economy based on forest resources, giving some relief to the pressure on native forests. Table 3 illustrates the growth versus harvest balance of forest resources in Chile. Estimate for native forest of 7.6 million ha includes all production forest and part of the protection forest (Table 3).

Table 3. Yield and harvest from Chilean forests

		Native forest	Radiata pine	Eucalyptus	Total
Area	(thousand ha)	7,616	1,387	308	9,311
Volume	(million m^3)	941	198	39	1,178
Growth	(m^3/ha/year)	2.7	20	25	
Yield	(million m^3/year)	20.6	27.7	7.7	56.0
Industrial harvest	(million m^3/year)	3.0	18.0	1.7	22.7
Firewood	(million m^3/year)	6.1	2.4	0.8	9.3
Losses	(million m^3/year)	3.3	2.0	NA	5.3
Balance Yield - Harvest	(million m^3/year)	8.3	5.3	5.2	18.7

Source: INFOR (1998a) estimates

With regard to the resource base for forestry activities, equity has become critical to forest policy. Different landowners, many of them with restricted access to capital, or low-value standing timber because of accessibility constraints constitute a social problem calling for innovative policy approaches.

2.2 Institutions

Chilean institutions related to the forest sector are essentially based upon private ownership of resources. Government agencies control legal regimes and subsidize activities where private agents are not economically attracted.

The main institutions related to the forest sector are under the Ministry of Agriculture: Corporación Nacional Forestal (CONAF), Servicio Agricola y Ganadero (SAG) and Instituto de Desarrollo Agropecuario (INDAP). The government research institution (INFOR) is a private corporation linked to Coporación de Fomento and to the Ministry of Agriculture.

CONAF is the Chilean forest service dealing with forest law enforcement, forestry promotion, administration of state-owned wildland protected areas, and fire and pest control. Created in 1973, it has played a very important role mainly in promoting afforestation and forestry in general.

The relation of SAG with forest activities refers to sanitary monitoring of products and quarantine surveillance. Also related to forestry, but to a minor extent, hunting and fishing licenses are issued by SAG.

INDAP is a governmental agency facilitating access to capital and to technology for small rural landowners. In a country with equity problems not solved by economic growth, its role is often quite relevant to marginal areas with small ownership. These marginal areas are typically on forestlands.

To deal with all environmental problems and issues, the government has recently created the Comisión Nacional del Medio Ambiente (CONAMA). The Law on Basis for the Environment was approved last year and the regulations are currently under final review. With the new legal background, CONAMA has become crucial when authorizing large-scale industrial or forest management investments.

Forest owners and industries are associated in CORMA, the traditional organization of the forest industry.

During the last five years, as a result of a political and economic evolution of the country, the legal status of forestry has been under complete revision. New social demands, a better knowledge of forestry externalities, and the recognition of market failures for socially optimal resource allocation have been the causal ingredients of this institutional revision.

The rationale imposed on design of the new forest legislation and institutions appears to be the maximization of land rent subject to equity constraints and given uncertainty on parts of the forest production function, particularly those related to non-market outputs.

Land is considered a fixed production factor, and forestry is an economic activity with increasing average costs. Therefore, the long-run behavior of competitive forestland owners will cause them to allocate their resources so that the land rent is maximized.

Since externalities of forestry activities are not accounted for in the market, a degree of government intervention is needed. The problem has been how to design the most efficient and effective package of policy tools to allocate scarce fiscal resources and to satisfy social demands of non-market outputs, some of them public goods, related to forestry.

The approach currently taken by government recognizes that a different set of policy tools (incentives, taxes, standards, or quotas) needs to be chosen, under efficiency and effectiveness criteria, for the different combinations of soil and vegetation conditions.

As shown in Figure 1, soils need to be characterized according to fragility and stability, whereas vegetation needs to be evaluated according to its contribution to bio-diversity, landscape, production, and economic welfare.

For a complete approach to the different dimensions of forestry, the government is reviewing a set of four laws. The wide scope they cover will constitute a sort of "forestry code":

- **DL 701.** This Law regulates a set of incentives to forestation. It is an extension of Decree Law 701 of year 1974, which ended in 1994. The main modification with respect to the former is that recipients of incentives must be only "small landowners" and afforestation is subsidized only on fragile or eroded lands. The category of "small" landowners has been defined in a way that, at least, prevents large companies and exporters from receiving the subsidies.
- **The Wildland Protected Areas.** This is a law prepared in 1984, never implemented, that regulates categories and administration of National Parks, Natural Reserves, and Biological Monuments. The main objective of the revision is to include regulations and incentives for private lands that may be candidate to any category. The original text was devoted completely to public lands.
- **The Promotion and Conservation of Native Forests**. This is a complex law that has been worked out over the last four years. It is focussed on regulating the use of productive native forests growing on stable and productive lands, so that its degradation or substitution can be avoided. It will also regulate incentives for the recovery of native forests and the promotion of native species.
- **Organization of Government Agencies.** To efficiently administrate all the aspects implied by the former three laws, the new social preferences, and the increasing challenges of an active production sector, the authority has recognized the necessity of restructuring its government

agencies. A new institutional design is being analyzed and developed to establish it in 1999 (Figure 2).

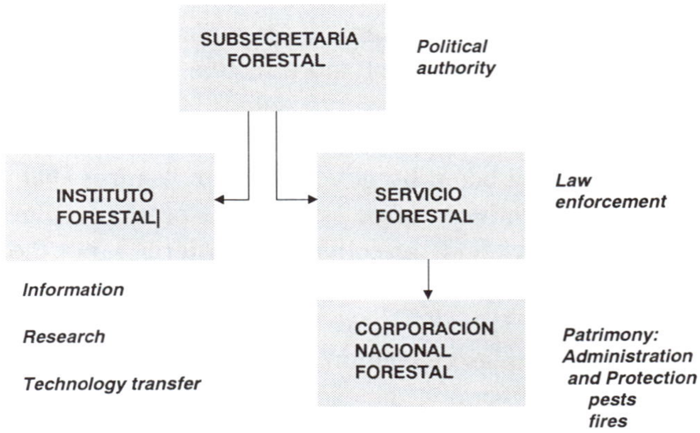

Figure 2. State agencies in Chilean forestry: a government proposal for year 1999

2.3 Production and exports

The forest sector has been one of the most dynamic sectors in Chilean economy, showing an average growth rate of 9.4% between 1984 and 1997, higher than the 6.2% average growth of the total economy. Nevertheless, its relative importance in Chilean economy is small, accounting for only 3% of total Gross Domestic Product.

In 1984, the value of total forest production was 575 million US$. Thirteen years later, in 1997, the value of total forest production has nearly increased more than four times to 2,440 million US$, and 74% of this value is exported.

In terms of volume, the industrial harvest in 1984 was 8.9 million m^3, while in 1997 it reached a volume of 23.6 million m^3 with 78% of harvest being radiata pine logs (Table 4).

Table 4. Evolution of total production (INFOR 1998a)

		1985	1997	variation (%)
Pulp	(million ton)	837	2,078	148
Chips	(milion m^3)	5,530*	6,033	9
Sawnwood	(million m^3)	2,191	4,661	113
Boards	(million ton)	163	745	357
Paper	(million ton)	172	184	7

* figure in 1991

Not only in terms of area planted, volume harvested, or product tonnage the forest industry has evolved in the country. Significant changes have occurred in the production structure. In 1980 the number of exported products was 54, increased to 385 by 1991, thus showing a horizontal expansion of forestry activities. Table 5 describes the evolution of the main forest products exports in the last decade.

Table 5. Evolution of forest exports by products (INFOR 1998a)

	1985	1997	variation (%)
		million US$	
Pulp (ton)	503	1,575	213
Chips (ton)	-	2,778	
native species	-	1,775	
eucalyptus	-	756	
Radiata pine	-	247	
Logs (m^3)	1,273	1,729	36
Pulplogs	14	252	1,700
sawlogs	1,259	1,477	17
Sawnwood (m^3)	706	1,244	76
Boards (ton)	43	282	556
Remanufacturing (ton)	42	299	512
Paper (ton)	145	231	59

Changes in production are not only in the diversification of products and increase in volumes, but also--and mainly--in the technological changes of production processes to improve efficiency and quality of end products, and to meet higher environmental standards imposed by international trade and stimulated by national government.

This structural change in production has also been accompanied with a significant diversification of market countries. In 1970 exports reached only to 17 countries, and during the last decade the number of destination countries for Chilean products has increased from 49 to 76. These figures are the result of a large marketing effort and successful trade. They show the healthy and diversified portfolio of commercial opportunities.

Table 6 shows the value exported for selected countries and years of last decade.

Table 6. Destination countries of forest exports (INFOR 1998a)

	1985	1990	1997
		million US$	
Japan	30	220	362
United States	21	50	318
Korea	22	55	206
Belgium (to EEC)	12	83	107
Argentina	18	10	90
Taiwan	5	14	92
Germany	22	76	28
Other	203	348	627
Total	335	855	1,830

As observed in Table 6, the share of total exports for the seven most important countries has remained around two-thirds in the last decade.

Chilean forest products are distributed to many countries, and none of them takes more than one-fourth of the total.

On a regional basis, the main increase on market destinations has been observed in the Asian and North America regions, as is shown in Table 7.

Table 7. Forest products exports, evolution by region (INFOR 1998a)

Region	1985	1997	variation (%)
	million US$ FOB		
Asia	117	821	601
Europe	64	356	456
South America	108	299	177
North America	32	345	978
Secondary Products			
Other	15	9	-40

A significant increase of forest product imports has been observed in recent years as the domestic economy expanded, however the total trade balance for forest products is highly positive and increasing (Table 8).

Table 8. Trade balance of forest (INFOR 1998a)

Item	1980	1990	1997
	million US$		
Exports	468	855	1.830
Imports	86	126	515
Balance	382	729	1.315

Chilean forest production and export products represent only a small fraction of the world's total production and trade; however, in some products such as pulp, its share is about 5% of total exports. Table 9 describes the share of main forest products of total production and exports. In terms of value traded internationally, Chile shares only 1% of the total.

Table 9. Participation of Chile in world production and exports, 1996 (FAO 1998)

	production (%)	Exports (%)
industrial logs	1	5.4
sawnwood	1	1
pulp	1.4	5
paper	0.3	0.8
boards	0.6	0.9

The potential for further expansion of the forest industry is still very high. Considering the annual plantation rate of fast-growing species, their growth rate, and management status of plantations, INFOR (1998a) estimates that timber supply from radiata pine plantations can still grow up to a level of 25 million m^3 by the year 2000. All industrial plantations in Chile are managed on a long-term sustainable basis; therefore timber supply from other species, such as eucalyptus, is also expected to increase.

This additional expansion of Chilean production is most likely to be traded as manufactured products rather than raw material. The cost of transport to the main country destinations, relative to the value of the products, suggests this is a long-run tendency of the forest industry. Table 10 shows the transport cost-to-product value relation for those products exported to the northern hemisphere.

Table 10. Transport cost and FOB values for main products exported 1997

Item		FOB value	Transport cost
chips (Radiata pine)	\$/GT	39	35
Pulplogs (Radiata pine)	\$/m^3	55	36
sawlogs (Radiata pine)	\$/m^3	62	32
sawnwood (dimensioned)	\$/m^3	175	36
millwork	\$/m^3	500	40-45
Particleboard	\$/m^3	239	52
MDF	\$/m^3	393	69
plywood	\$/m^3	934	76
pulp	\$/ton	438	51

Given these free on board (FOB) prices and the transport cost to the main markets, and considering the cost structure of Chilean forest industries, it becomes clear that the in long-run, forest firms would consider evaluating capital investments to further process the raw material. This is particularly straightforward for a firm that manages its resource base on a sustainable basis, and under normal economic and international trade environments.

In terms of wood supply, the next five-year period will show the first volumes of radiata pine pruned logs, even though the harvest of intensively managed plantations will still be at a low level--less than one thousand ha per year. Figure 3 shows future annual supply from radiata pine and eucalyptus plantations.

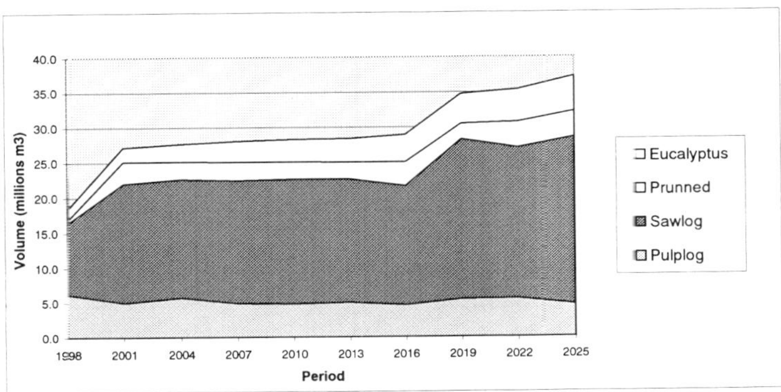

Figure 3. Timber supply next three decades radiata pine and eucalyptus (INFOR 1998b)

After the year 2000, the relative importance of radiata pine supply will decrease slightly as a result of the current scenario for long-term land investments: higher land costs in general; crisis in traditional agriculture; restricted incentives (new scheme for government incentives will reach a larger number of small landowners but a smaller total area); and increased restrictions on converting lands with degraded, non-productive native trees. In this scenario, eucalyptus plantations are expanding rapidly on agricultural lands where traditional crops are no longer economically viable. A likely projection of forest exports considering these observations is shown in Table 11.

Table 11. Projection of forest production exports

Item			1980	1992	1997	2010
Chips						
	Radiata pine	m^3	-	423	247	500
	Eucalyptus	m^3	-	943	756	2600
	Native species	m^3	-	2140	1775	300
Sawlogs		m^3	1052	1029	1729	2000
Sawnwood		m^3	1259	998	1243	3500
Millwork		m^3	8	157	226	800
Pulp		ton	415	1260	1575	2500
Value		1000$	636	1126	1830	2600

Source: INFOR (1998a) and estimations.

As the forest industry increases the production exported to other countries, the forest companies themselves become more international as a result of some, or a combination of the following actions: ownership shared with foreign companies, emission of ADR (American Depositary Receipts), investments in the stock market, and long-term joint-ventures or supply agreements.

3. Issues of International Trade of Forest Products

As forestry becomes more global because of increasing international trade, internationalized ownership of forest resources and industries, and better communications, the issues for both exporters and importers become global too. Sustainability and risks of importing pests and disease are likely the most important aspects involved. Competitiveness is also an important issue, however it is not considered here.

3.1 Sustainability

Sustainable resources management was a general declaration of intention at the United Nations Conference on Environment and Development in Rio de Janeiro in 1992. "Forest resources and forest lands should be managed to meet the social, economic, ecological, cultural and spiritual needs of present and future generations. These needs are for forest products and services, such as wood and wood products, water, food, fodder, medicine, fuel, shelter, employment, recreation, habitats for wildlife, landscape diversity, carbon sinks and reservoirs, and for other forest products. Appropriate measures should be taken to protect forests against harmful effects of pollution, including airborne pollution, fires, pests and diseases in order to maintain their full multiple values." Sustainable management has been, however, an old and traditional concept used in forestry, and it is now being adapted to new social concerns and values, and to new scientific knowledge.

The Forest Declaration is a general intention that does not include quantitative, operational goals or a definitive schedule. However, Chilean authorities and economic agents have given a high priority to following the Declaration, as soon as possible, considering the positive effects it has on the international trade of Chilean forest products, and thus on economic welfare.

For practical purposes, it has also been recognized that: (a) sustainable forest management has important spatial and temporal dimensions; (b) concerns for sustainable forestry should be balanced against other kinds of land use; and (c) forest policies should be integrated with development policies into a coherent legislation for all sectors.

The space-time dimensions of forest management imply that, even though efforts should be made for the individual forest and forest properties to fulfill all criteria for sustainability as far as possible, it must be recognized, that it is not possible to reach sustainability in every respect, everywhere, and at every moment.

For a given forest it is necessary to consider its former and future treatments, its size and site homogeneity, and the conditions of nearby stands. To fulfill sustainability, it should be possible to manage some areas for intensive production, of pulpwood for instance, while others are managed extensively for preservation purposes, for instance national parks or natural monuments.

In this sense, Chile has already established 29% of its national territory as protected wilderness areas--more than any other developed country as compared, for instance, with the 8% for New Zealand, 6% of Japan and 3% in the United States. In spite of this effort, it is recognized that, at a lower resolution-- forest type--more areas need to be preserved in order to maintain the biological diversity of specific habitats, currently not represented in the Protected Wildlands System.

With regard to the balance of concerns with other kinds of lands uses, Chilean forestry has to acknowledge a large-scale recovery, throughout plantation forests, of lands moderately to severely eroded by past agricultural practices.

Even though 7 to 10% of these cultivated forests may have been established replacing native vegetation, they have, overall, allowed for protection of native forests by supplying over 90% of the national domestic consumption and exports of forest products. Furthermore, cultivated forests of radiata pine and eucalyptus also contribute intensively to reducing the greenhouse effects by absorbing, each hectare, up to 9 metric tons annually of carbon from the atmosphere.

The recent crisis of agricultural crops and a homogeneous governmental policy treatment for both sectors, have encouraged an important conversion to forest crops on lands where agriculture is only marginally viable, thus improving soil conservation.

The necessity of integrating forest policy into a wider and coherent legislative framework, as one of the crucial recommendations pointed out by the 1987 Bruntland Report (World Commission on Environment and Development (WCED 1987)), has become an implicitly accepted policy framework for Chilean authorities.

The forest policy approach, selected by government authorities, seems to have considered the implicit uncertainties of the long-term relations between both domestic and foreign capital. Then an adaptive approach has been chosen based upon the social needs for welfare growth, the establishment of minimum safe standards, and avoidance of irreversibility's in natural capacity by considering the opportunity costs of preservation. Obviously, this approach has not fully satisfied the demands of deep ecologists and ecological economists, on one extreme, and of orthodox free-market oriented economists, on the other extreme.

One topic, still unsolved by this approach to a sustainable forest policy, is that of equity among forest landowners. In establishing safe minimum standards, the immediate economic effects are different depending on the access to capital of different landowners. Some landowners have direct access to capital in developed countries, while others have more restricted access to capital in Chilean financial markets, and some do not have access at all. Under these conditions an environmental standard applied to the same forest, in the same land, and for the same operation could be a viable option for someone while not viable for another landowner.

3.2 Sanitary risks

Chilean forestry, as well as agriculture, has a long-standing tradition of protecting itself against pathogens that can be brought into the country. Vineyards, kiwis, peaches, apples and many other vegetable or fruit crops have their natural origin in other parts of the world. These crops enjoy the good climate and soil conditions of Chile, in an environment with a low risk of pathogens. The same holds for forest plantations with eucalyptus, radiata pine, Douglas-fir and poplars.

Therefore, it has been easy for any Chilean producer or exporter to understand the concerns about pest risks in countries receiving forest products as raw materials. Given the non-discussed, short- and long-term beneficial effects of international trade, the problem then becomes how to design and to scale correctly the efforts to be undertaken to avoid pest transmissions associated with trade and communication in general. The following are important aspects to be considered for a correct design of mitigation and control efforts.

- Most, if not all, of the pathogens that could become a risk in raw materials from Chile, are natural species (Scollytidae for instance) in the Northern Hemisphere. Therefore, not always can they be viewed as introduced pests, or, as in other cases, there are more relevant flows of raw materials that can introduce the pest to an importing country. The volume of material flows between countries of the Northern Hemisphere is one or two orders of magnitude larger than the flow with Southern Hemisphere countries.
- To correctly dimension and select the efforts to be made with regard to pest risk assessments and mitigation activities, it is necessary to consider that the long-run trend of Chilean forest products exports is to evolve gradually from unprocessed raw materials to more elaborate processed products. Integrated processing then allows for easier, more effective and more efficient monitoring of pathogen presence in the product. Chips can be considered, for these purposes, to be a processed product, as chipping involves a physical transformation of the log in a single facility, thus facilitating its inspection, monitoring, treatment, and certification.
- From a scientific point of view, pest risk assessment for pathogen introduction from one country to another is a difficult task. The objective is to determine which organisms would be able to survive the trip and then adapt to the new habitat and use local flora to complete their life cycle.

These assessments are, necessarily, non-experimental methods, mainly based upon comparison of life cycles, environmental requirements (limiting temperature, for instance), and availability of potential host species in the destination country.

Objective pest risk assessments require strong linkages between researchers and scientists (entomologists, pathologists, and epidemiologists) that have a strong knowledge of the situation in both origin and destination countries. Fortunately for Chilean exporters to the United States, and for U.S. importers of Chilean products, there are well-reputed research groups in both countries that, working together, can objectively quantify the risks and select the more effective and efficient mitigation actions.

More importantly, the results of combined research teams published in scientific journals will allow clarifying many hypotheses and assumptions with regard to pathogen insects and fungus transmissions.

4. References

CONAF-CONAMA-BIRF. 1997. Catastro y Evaluación Recursos Vegetacionales Nativos de Chile: Resultados Finales Síntesis. 12p.

FAO. 1998. Anuario de Productos Forestales,1992-1996. Roma.

INFOR. 1998a. Estadísticas Forestales 1997. Boletín Estadístico Nº 61. Instituto Forestal - CORFO, Chile. 123p.

INFOR. 1998b. Disponibilidad de Madera de Pino Radiata en Chile, 1998-2027. Informe Técnico en prensa. Instituto Forestal - CORFO, Chile. 102p.

WCED. 1987. Our common future. Oxford University Press. Oxford.

The Development Strategies of Forest Resource and Forest Industries in Korea

Byoung Il Yoo

Key words: Forest resources, forest industry, forest products, demand, supply, Korea

Abstract: During the last 3 decades, Korea created 4 million ha of artificial forest; about 50% of these plantation areas remain. As of 1996, total stock volume was estimated at 296 million m^3. The average was 50 m^3 per ha--much higher than the 1960s figure, 10 m^3 per ha. The total stock volume is estimated to be 563 million m^3 in 2010. For management of these resources, Korea has formed four forest development plans from 1973 to now. Korea used 25.3 million m^3 for the domestic market in 1995. Domestic timber production was about 1 million m^3, or about 4% of the domestic market. In 2010, Korea is expected to produce 4.8 million m^3 of timber, 16% of an expected 30.7 million m^3 domestic market. In 1995, the production of sawn timber, plywood, particleboard and fiberboard was 3.4, 1.0, 0.5, and 0.5 million m^3, respectively, and in addition import amounts were 1.0, 1.3, 0.5, 0.1 million m^3. The import ratio of these products was 22.8, 56.5, 50.0, 16.7%, respectively. The production of board has increased sharply in the 1990s. The demand of sawn timber, plywood, and boards is expected to be 4.7, 3.0, and 2.2 million m^3 in 2010. Recently the sawn timber and plywood industries have been depressed, and the board industry was activated in the 1990s. These trends will continue in the near future.

1. Forest Resource in Korea

1.1 Forest resource

In Korea, as of 1995, forest land was about 6.5 million ha, representing 65% of the total land area. However, forestland per capita is very low at 0.15 ha, only 20% of the world average. The total stock volume is 309 million m^3 (47 m^3 per ha), and the annual growth rate is estimated to be 4.9%. Compared to other countries, the stock volume seems pretty poor but is much higher than the 10 m^3 of the 1960s.

By the 1960s, forestlands in Korea were severely depleted as a result of population pressure, fuelwood demand, and the Korean War. In 1961, the unstocked forestland accounted for 41% of the total forestland area. Between 1961 and 1995, stocked forestland increased from 4 million ha to 6.3 million ha, as a result of a large-scale reforestation program. It was possible to accomplish the massive planting project in a short period by the participation of forest owners, village residents, and the government and other public organizations. Such a large-scale project based on the participation of the village residents is a rare case in the world. In Korea, it is possible because of the village community concept called "Sanrimgae" which was organized for forest protection several hundred years ago.

However, after reforestation, forest owners have harvested the fuelwood forest and fast-growing trees like black locust and Japanese alnus, and changed forestlands into other purposes to increase the benefit of land use. Because of these reasons and silviculture failure, the remaining reforestation area is about 2.50 million ha (2.04 million ha in mountain zones) or about 32% of total forest. Recently the annual artificial reforestation area has been reduced because of a policy change in favor of natural forest tending, and due to the economic difficulties such as shortages of forest labor and slump of log prices. The government target for artificial reforestation is for plantations to make up 50% of forests.

Now forest lands are classified into national, public, and private forest by ownership and are also divided into reserve (for production and public welfare) and semi-reserve (for industrial) forests by utilization. National forest, which is 21% (1,382 million ha) of the total forestland, can be subdivided into permanent and harvestable forests. At the end of 1993, the average stock volume of the national forest was 68 m^3 per ha, which was greater than the national average, 43 m^3 per ha. Public forest owned by local governments composes about 8 % (0.491 million ha) of the total forest land. Public forest contains only 7% of the total stock volume. Private forest occupies 71% (4,586 million ha) of the total forest lands, but the number of private forest owners is more than 2 million. Therefore 96% of the owners own less than 10 ha of forest for a total of about 52.5% of forest area, so the ownership scale is very small; this trend is accelerating. The main purposes of private forests are for burial places and as an investment.

1.2 Trend of forest products' trade flows

In 1995, the export value of forest products was 505 million US$, and its share was 0.4 % of the total value of export. The main export items in forestry were chestnut, pine mushrooms, oak mushrooms, and plywood. The plywood export quantity increased 70% compared to 1994: the export value of plywood increased 22%. The value of all wood and wooden exports was 70.9 million US$, down 3% since 1994. In 1995, Japan received 77% of Korea's forest products exports at a value of 388 million US$.

In 1995, the total import value of forestry products was 2,779 million US$, and its share among the total imports of 135,118 million US$ was 2.1%. The main import items are logs, sawn timber, plywood, and non-timber products such as oak mushrooms. The trend of wood and wooden product imports is the same as in the past. Log imports decreased from Malaysia and the United States, and increased from New Zealand and Chile

2. Forest Industry in Korea

Timber product industries had been one of the main exporting industries in Korea, and plywood and sawn wood industries led exports from the late 1960s to the early 1980s. At the same time, they played a leading role in national economic development by offering a stable supply of construction and housing materials. From the 1980s, Korean timber industries have suffered from declining competitiveness both domestically and abroad, which resulted from restriction of log exports by tropical countries, increased prices of imported logs and of domestic labor cost, and development of substitutes for timber.

2.1 Logs

Annual domestic log production remains around 1 million m^3. Currently domestic timber is mainly low-grade and small-diameter softwood, which is generally unsuitable for sawn timber production. The majority of domestic timber is used for pit props and pulpwood. Domestic log production is around 15% of Korea's log consumption. The production capacity in Korea is very limited in the near future because of young and premature forest stands.

However, the total log consumption has increased annually owing to higher domestic log demand for general-purpose logs. In 1995, the total log consumption in Korea was about 9 284 thousand m^3, nearly 96% of them for domestic use.

Log demand in 2010 in Korea is estimated to be about 19.5 million m^3, and domestic supply to be 3.1 million m^3; domestic share of total supply is expected to increase to about 15.9% from 12.0% in 1990. Therefore in 2010, Korea will import around 16 million m^3 from overseas, mainly for the domestic demand in Korea.

2.2 Woodchips

The total consumption of woodchips in Korea was 3,058 thousand m^3 in 1995. The production of woodchips in Korea is mainly for pulp and board manufacture, and of the woodchips for pulp, most is for chemical pulp. In 1995, of the total of 1,039 thousand m^3 for pulp, 968 thousand m^3 was used for chemical pulp. Imports of woodchips increased rapidly. They were 1,181 thousand m^3 in 1995 (equal to about 656 thousand metric tons).

2.3 Sawn Timber

The production of sawn timber in Korea was 1,148 thousand m^3 in 1970, and the production increased every year. The production of sawn timber peaked at 5,949 thousand m^3 in 1988, and consumed 8,842 thousand m^3 of logs.

The share of domestic logs for sawn timber dropped from 12% in 1980 to 6% in 1991, and slightly increased to 9% in 1995. However, usually it stays around 10% of total sawn timber production. In domestic sawn timber, the share of softwood has increased. The species composition of imported logs, which dominated log sawn timber production, has changed: in 1985, tropical hardwood logs were 31.5% of total log imports, but this had dropped to 12% in 1995. Recently, imported softwood logs from temperate zone countries have increased.

In Korea, about 73% of sawn timber is consumed in construction and engineering work. The construction and engineering companies order the sawn timber for their needs and their favorite sizes. About 73% of domestic logs and 83% of hemlock imported from North America is cut to special order.

In the future, the consumption of sawn timber will change depending on the domestic construction and engineering work, and packing for export goods. A lot of substitutes will be developed according to sawn timber price increases.

2.4 Woodpulp

Korea produces 3 kinds of pulp: chemical, mechanical, and chemi-mechanical pulp. The utilization of pulp mill capacity is about 60%. In 1995, domestic supply of pulp was 23% of demand, and the rest was imported. The production of woodpulp in Korea depended on mechanical pulp before the mid 1980s, but after that, the production of chemical pulp increased and surpassed mechanical pulp. From 1975 to 1995, chemical pulp increased from 7 thousand metric tons to 319 thousand metric tons, the mechanical pulp increased from 87 thousand metric tons to 181 thousand metric tons.

Pulp is usually used for paper production. The consumption of pulp increased 24 times from 92 thousand metric tons in 1965 to 2,170 thousand metric tons in 1994, and the consumption of waste paper also increased 182 times from 46 thousand metric tons in 1965 to 4,704 thousand metric tons in 1994. In the early 1970s, the share of pulp for paper production was greater than that of waste paper, but since then waste paper has become more popular than pulp as raw material to produce the pulp. The share of pulp in raw material for paper production was about 35% in 1994.

2.5 Paper and paperboard

In 1995, the production capacity of paper was 7,285 thousand metric tons, and it had increased 39.3% compared to 5 years earlier. Printing paper production increased 59%, and newsprint 58%. Korea consumed 7,417 thousand metric tons in 1995, of which 6,877 thousand metric tons of paper was supplied domestically, and 697 thousand metric tons was imported. In 2010, paper demand is estimated at 11,939 thousand metric tons by Korea Rural Economic Institute (1994) and 12,617 thousand metric tons by Food and Agriculture Organization (1994).

In 1995, Korea imported 697 thousand metric tons paper, an increase of 24% compared to 1974. This is because of increases in imports of newsprint and other special-purpose paper. But due to increased local production capacity, imports decreased in 1996.

The export trend of paper and board is slowly increasing. In 1995, Korean exported 952 thousand metric tons, an increase of 4.4% compared to 1994. Of that amount, board exports were 555 thousand metric tons (58%), and printed paper was 308 thousand metric tons (32%). The value of paper export was 1.125 million US$, which is an increase of 42.5% compared to 1994.

2.6 Particleboard and fiberboard

Due to their recent increased use as plywood substitutes, the production of particleboard and fiberboard is steadily growing. In case of fiberboard, it is expected that supply will temporarily exceed the demand. The production of particleboard increased from 13 thousand m^3 in 1966 to 524 thousand m^3 in 1994, and fiberboard also increased from 10 thousand m^3 in 1966 to 506 thousand m^3 in 1994.

2.7 Plywood and veneer

Plywood manufacture led the timber-related industry in 1970s and 1980s, but because of the difficulty of tropical log import and the labor cost increase, its competitiveness declined in the 1990s. The production of plywood in 1966 was 445 thousand m^3, and the production amount increased steadily. It peaked in 1978 at 2,557 thousand m^3, and after that it decreased slowly. Plywood production in 1995 was only 974 thousand m^3 or 38% of the peak volume. The thickness of major plywood products in the past was 6 mm for furniture, but recently it changed to 12 mm for construction, engineering, and concrete panels for the domestic market. The production for export is very limited.

The plywood industry has depended on imported logs for raw material. In the 1970s and 1980s, the species of logs imported for the plywood industry was mainly lauan, and after that it changed to kapol, keruing, and other mixed tropical hardwood species. Recently softwood has also been imported for plywood raw material.

3. Forest Development Strategies in Last 3 Decades

The Forest Law of 1961 required the Forestry Administration to set up the national forest plan every 10 years in order to establish and manage the forest resources efficiently. Local governments should set up local forest plans on the basis of the direction of the national forest plan.

3.1 The First 10-Year Forest Development Plan (1973-1978)

A turning point in Korean forestry was witnessed in 1973 with the initiation of the First 10-Year Forest Development Plan, which targeted the reforestation of denuded forest lands as early as possible; 1 million ha was to be planted within 10 years under the following objectives: first, to implement national tree planting through the participation of all people in various reforestation projects; second, to develop new economic zones of forest lands directly connecting the goals of land conservation and income enhancement by reforestation and forest production; third, to achieve rapid reforestation of denuded forestland through the planting of fast-growing tree species; and fourth, to stabilize the shifting cultivation based on fire. In particular, the government has appointed the month of April as the "National Tree Planting Period" for the best time to plant trees. The government also encouraged various groups such as villages, families, and schools to participate in the reforestation program. In this period, remarkable progress was made in forest protection minimizing damage to forests by prohibiting access to the mountains and by the formation of rural fuelwood forests. Through the implementation of this movement, the First 10-Year Plan, which was originally planned to be completed in 1982, accomplished the reforestation of 1.08 million ha by 1978.

3.2 The Second 10-Year Forest Development Plan (1979-1987)

The basic objective of the Second 10-Year Forest Development Plan was to build large commercial forest zones for production of industrial timber. The government initiated various forest policies: the strengthening of the national reforestation plan, the expansion of forest protection activities, the enlargement of forest development funds for supporting private forest management, grouping and enlarging the national forests, and conducting forest conservation projects for improving the public benefits from forests. Financial and administrative support by the government were especially to encourage the reforestation activities and the cooperative management of small private forests.

The major accomplishments in this period were the reforestation of 966 thousand ha with the formation of 80 large, group commercial forest zones (325 thousand ha). Continuous management of natural forests and juvenile plantations was also carried out. Erosion control works were undertaken in accordance with the principle of regional restoration of denuded forest areas.

3.3 The Third-10 Year Forest Development Plan (1988-1997)

The objective of the Third Plan was to harmonize the goals of increasing the economic development of forests and improving public benefits from the forests. Although reforestation was nearly completed by the First and Second Plans, Korea depended on imported timber for more than 85% of domestic timber use. At the same time, there was also an increasing demand for the conservation of the environment and for outdoor recreation in forests.

The fundamental targets of the plan are maximizing the efficiency of forestland use, building a foundation for forestland management, creating forest income sources in rural areas, and improving the multiple public benefits of the forests. Under these objectives, the government has established the following priorities.

First, forestry development promotion zones of about 1.52 million ha have been designated in private forests on which government investments will be concentrated. Simultaneously the management system should be modernized through the mechanization of forest works and the designation of adequate management units in national and public forests.

Second, the new concept of multiple-use management was introduced, and this new idea promoted harmonious relations between timber production and the other functions of the forest.

Third, the government planned to enlarge urban forests in city areas and recreation forests in rural areas to provide people clean air, water, and rest spaces.

Additionally the government planned to increase the efficiency of forest activities by improving forest road infrastructure, amending forest laws and the forest taxation system, raising forest development funds to implement private forest cooperative management, and finally operating research projects in various fields.

4. The Fourth 10-Year Forest Development Plan (1998-2007)

The fourth 10-year Forest Development Plan is based on article 6 of the Forest Law. The Plan is: first, providing the basic frame for the forestland management system--conservation and development of forestland, second, establishing and managing of forest resources, third, controlling the demand and supply of forest products and support of the forest industry, fourth, increasing the environmental and cultural function of forests, and finally, managing wildlife and forest ecosystems.

In the 21st Century, the situation surrounding Korea is expected to change very rapidly, as are ideas and concepts for the new forest policies. The range of the plan will be expanded from the nation to the globe, the main point of the plan will be enlarged from forests, and forest owners, to land, people, and industry. Targets of the plan will be diverted from mountain forests to urban forests; the main body of the plan will be changed from central government to local government and local people; and the importance of the forest's value should be turned from economic and environmental resources to future, culture, and aesthetic resources.

1) Improvement of reasonable forestland use system

As a result of rapid economic development, the demand for other uses of forestland has accelerated. Therefore it is important to meet various demands for forestland use resulting from socio-economic progress, to reserve the forestry production area as a timber supply base, and to manage forest lands based on harmonizing the balance between conservation and development in the program for complex land development in a small country.

2) Acceleration of establishment of new forest resources, and protection and fostering of existing forest resources

Establishment of commercial forests is emphasized at Forestry Promotion Development Areas. The plan is to increase the proportion of man-made forests to total forests from 31% in 1992 to 47% in 2030 and to increase self-sufficiency for logs from 13% in 1993 to 46% in 2030. In addition, it is intended to emphasize protection activities against forest fire, disease and insect pests, and the diversion of forest for other industrial uses.

3) Enlargement of forest management infrastructure and improvement of forest management structure

The Korean government plans to construct more forest roads, increasing from 7,114 km (1.10m per ha) at the end of 1994 to 56,000km (8.6m per ha) in 2010. The mechanization of forest harvest has also been accelerated to increase productivity and to reduce management costs.

The national forest has been enlarged to secure greater timber supply and to promote public benefits from the forest. National forestlands is scheduled to expand to 2,063 thousand ha, 33% of Korea forest lands by 2040. Forest cooperatives, which intend to expand forest management units by as much as 3,000 ha through mutual cooperation among small-scale private forest owners, are encouraged. Forest development funds need to be increased to encourage forest owner's investment motivation.

4) The importance of environmental conservation

It is expected that harvesting and export of logs from countries rich in forest resources will be restricted by the global environmental conservation movement. Forest development may also be restricted by national economic limits. Therefore the enlargement of the national forest stock and construction of a sustainable forest management system is necessary.

5) Change of social structure and the human desire for an improved quality of life

It is expected that demand will increase for recreation, environment, culture, and relaxation to improve the quality of life. Therefore enlargement of the recreation and cultural functions of forests will be requested.

6) Enlargement of overseas forest development and improvement of the forestry industry

To respond to environmental conservation and log-export bans in timber export countries, the government is also trying to shift its strategy of timber imports from strictly purchase imports to "development imports" by expanding overseas reforestation projects from which timber supplies can then be imported. Also plans are scheduled to promote the restructuring of the timber industry to better adapt to economic change in domestic and overseas markets. Timber storage and processing factories for forest products are planned to increase from 29 places in 1992 to 120 places in 2001 to increase forest owners' income and to stabilize the price of forest products.

7) International competition under the World Trade Organization system

Traditional boundaries will be diminished, and the exchange of technology, capital, and goods will be unrestricted globally. The system of competition will change from nations to city and regional units, so it will be necessary to increase the competitiveness of the forest industry and to revitalize regional forestry.

8) Industry technical innovation and the development of information and communication

The construction of an ultra-high-speed network of information and communication and advanced technology will be wide spread. The importance of forests as future resources will also be increased by the development of new materials using the diversity of the forest ecosystem. So it is important to develop a forest information network system that links the national information system with the Korean-style forest management model.

9) Maturity of local self-government and the progress of localization

The desire for local area development and the guarantee of self-regulation of forest management by local governments will increase, as will the importance of cooperation between central and local governments. Therefore it is necessary to develop a forest management system that considers the characteristics of the local area and forest industry policy.

10) Formation of Northeast Asian economic zone and the enlargement of exchange between South Korea and North Korea

It is expected that a Northeast Asian economic zone will be developed as a globally important economic block. The possibility of reunification of the Korean Peninsula will be increased as will the opportunity for forestry exchange between South Korea and North Korea. Therefore the plan is to help form a Northeast Asian economic zone, to participate as a member, and to make the blueprint for the Korean Peninsula's forest management.

5. References

Korea Rural Economic Institute. 1994. Basic study for forest 100 years planning stablishment (in Korean). Seoul.

FAO Planning and Statistics Branch. 1994. Pulp and paper trends 2010. An Executive Summary. FAO Rome. ISBN 92-5-103540-7 10p.

Socio-Economic Importance of Eucalyptus Plantations in Morocco

Abdellatif Khattabi

Key words: Eucalyptus, Morocco, forest economics

Abstract: The Moroccan forest covers almost 12% of the whole territory, and contains about 9 million ha of natural woodlands and 0.5 million ha of plantations. The productivity of the natural forest is very low owing to harsh weather conditions and to a great physicochemical fragility of soils. Even though this forest is important to the environment and to the rural population whose economy is biomass dependent, it is not as important for the forest industry. In order to reduce the deficit in lumber production, in the late 1940s, the Moroccan government started big plantations of fast-growing species such as eucalyptus and pines. Nowadays, the eucalyptus species, mainly *Eucalyptus camaldulensis* and *E. eomphocephala* constitutes 40% of all plantations, and are concentrated in the northwest part of the country where a pulp mill was installed in the mid-1950s. This mill was initially the only user of wood from eucalyptus, but presently many other uses of this raw material have been developed, like particleboard and woodpacking. This paper deals with the production of timber from eucalyptus plantations in the northwest of Morocco. The analysis is concerned with resource production and its socio-economic importance for the country and for the region. The results show that eucalyptus produces 48% of Moroccan timber production, and generates more than 1 million days of jobs and 13.5 million US$ return per year.

1. Introduction

Morocco is located in the southern Mediterranean zone, in the extreme northwest of the African continent. Although its climate is very diverse, ranging from sub-humid to Saharan, more than 90% of the territory is classified as semi-arid or arid. Owing to harsh weather conditions and to a great physicochemical fragility of soils, the productivity of its natural forest for industrial roundwood is very low. However, these ecosystems are very important to protection of the environment and to subsistence of the rural populations.

Besides the option and existence values, and the direct and indirect use values accruing to intangible goods and environmental services and amenities, the forest sector provides a variety of products contributing to the socio-economic welfare of people. This contribution is estimated to be 10% of agriculture's Gross National Product (GNP), if we account for all the forest resources directly gathered by local populations (AEFCS 1996).

However, the Moroccan forest sector is facing some problems, mainly an excessive forest harvesting of trees leading to a yearly deforestation of about

30,000 ha, a domestic supply of forest products covering less than 30% of consumer demand, a deficit in the trade balance of forest products greater than 70% on average, and a poor technical level of some wood processing mills.

With the main objectives of increasing lumber production and protecting soils from erosion, in the late 1940s, the Moroccan government started to develop big plantations of fast-growing species such as eucalyptus and pines. The objective of this paper is to evaluate the socio-economic importance of eucalyptus plantations in Morocco, and particularly in the northwest part of the country.

2. Forest Resources and Forest Products

The forested land covering 9.5 million ha, representing about 12% of the whole country's area, is mostly State owned and is managed by the Ministry in Charge of Waters and Forests. Of this total forest area, 61.2% is naturally wooded land--three-quarters of which are broad-leaved species--5.2% is plantations composed of 54.3% hardwoods and 45.7% softwoods, and 33.5% is grassland dominated by alfa grass (*Stipa tenacissima*).

Table 1. Main species of the natural forest

SPECIES	%
Hardwoods	
- Holm oak (*Quercus rotundifolia*)	23.4
- Cork oak (*Quercus suber*)	6
- Acacia (*Acacia spp*)	19.4
- Argan Tree (*Argania spinosa*)	14.3
Softwoods	
- Thuya (*Tetraclinis articulata*)	10.3
- Juniper (*Juniperus spp*)	4
- Cedar (*Cedrus atlantica*)	2.3
- Pines (*Pinus spp*)	1.4
- Cypress (*Cupressus spp*)	0.1
Other species (Shrub)	18.8

The productivity of the natural forest for industrial roundwood is very low. Except for cedar and pines, all the remaining species are used mainly as fuelwood. However, this forest plays an important role for environmental protection in helping to stabilize natural systems, and in offering opportunities for recreation. The Morrocan land is mostly dry and precipitation is irregular. The role of forests in reducing soil erosion and in regulating runoff and groundwater is very important in the maintenance of the productivity of soils, and in the protection of downstream areas from sedimentation and flooding.

Forest-related activities employ around 114,000 persons, including 5,800 government civil servants, 42,300 persons working directly either in forest nurseries or in commercial harvest and transformation of forest products, and the remaining are active family members that collect fuelwood for domestic use or rearing livestock. Also, the forest sector provides a variety of products contributing to the socio-economic welfare of people. This contribution is estimated to be 10% of agriculture's GNP, if we account for all the forest resources directly gathered by local populations (AEFCS 1996).

The average production of timber sold by the Forest Administration, for the period 1990-1997, is around one million m^3 per year. More than 50% of this volume is fuelwood, and the remaining is industrial roundwood. The non-wood forest products harvested were 15,000 metric tons of cork, 10,000 metric tons of alfa grass, 1.5 billion Fodder Units (1 F.U.= 1kg of barley), 900 metric tons of locustbeans, 245 metric tons of lichen, 770 metric tons of mushrooms, 100 metric tons of heath's stumps, and 7 metric tons of rosemary.

The forest products industry counts about 240 plants for timber processing and wood products manufacturing, eight cork mills, one pulp factory, and nine paper and cardboard plants (MCI 1995). The timber processing and wood products manufacturing plants include 25 sawmills, 5 plywood and veneer mills, 132 industrial carpentry, 59 furniture-maker plants, and 19 packing manufacturers.

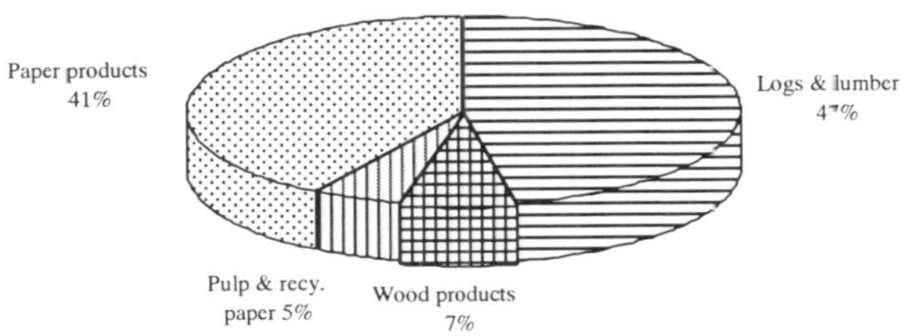

Figure 1. Composition of imports value - 1996 -

Except for cork, which is abundant enough to cover industry's demand, all other wood manufacturing plants are dependent on imports. The local supply of industrial roundwood, including pulpwood, represents only 30% of the domestic demand. The average value of imports of timber and woodproducts for the period of 1990-1996 amounted to 317 million US$, and was ranked in second place, after petroleum products, in total Moroccan imports. The value share of the different components of imports for 1996 (ODC 1997) are shown in Figure 1.

3. Plantations

The operations of plantations were started on a big scale in Morocco in 1949 by the Forest Service, as a result of the creation of the Forest National Fund. Prior to this date, only 5,000 ha of mostly private plantations existed, and consisted of reforestation of sandy lands using eucalyptus species (Ahizoune and Omerani 1997). In 1970, the National Reforestation Plan was launched, the principal objectives are to increase timber production, to protect watersheds against erosion, and to promote recreation areas. Its goal was to reforest 662,000 ha by the year 2000, with an average planting of 22,000 ha per year. By 1996, the total reforested area reached 495,326 ha, 75% of which is State forest, 7% is private forest, and 18% is collective forest.

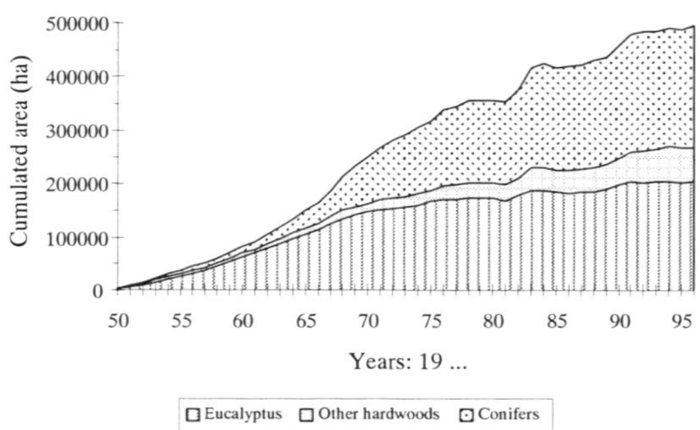

Figure 2. Evolution of plantations

The historic evolution of these plantations is pictured in Figure 2, and their distribution by groups of species and by economic region is shown in Table 2 (Mekki 1998).

The majority (66%) of plantations were installed in the five northern economic regions of Morocco, where climatic conditions are favorable. Seventy-two percent of conifers and eucalyptus, and 39% of other non-conifer species are present in these regions.

Eucalyptus species represent 40% of the total reforested area, the major species is *Eucalyptus camaldulensis* which covers 47% of the eucalyptus area. Second place is *E. gomphocephala* with a share of 32%, and the rest is other species such as *E. sideroxylon*, *E. grandis*, *E. globulus*, *E. occidentalis*, etc. Figure 3 gives the distribution of the total area covered by various eucalyptus species in 1996.

Table 2. Distribution of reforested area by groups of species and by economic regions

	CONIFERS		NONCONIFERS Eucalyptus		Others		TOTAL	
ECONOMIC REGIONS	(ha)	%	(ha)	%	(ha)	%	(ha)	%
Gharb-Cherarda-B.Hsine	11,380	5	75,388	38	4,766	7	91,534	18
Oriental	47,428	21	13,018	6	12,054	19	72,500	15
Tanger-Tetouan	42,852	19	11,505	6	2,600	4	56,957	11
Rabat-Salé-Zemour-Zaer	14,074	6	39,250	20	2,654	4	55,978	11
Taza-Houceima-Taounate	47,547	21	3,566	2	3,349	5	54,462	11
Doukala-Abda	3,095	1	19,440	10	16,756	26	39,291	8
Meknes-Tafilalet	26,009	11	2,057	1	1,108	2	29,174	6
Marakech-Tensift-Haouz	9,623	4	6,804	3	12,300	19	28,727	6
Chaouia-Ourdigha	6,523	3	18,624	9	3,404	5	28,551	6
Souss-Massa-Deraa	4,060	2	4,893	2	3,170	5	12,123	2
Tadla-Azilal	8,781	4	2,300	1	857	1	11,938	2
Fes-Boulemane	8,696	4	955	0	252	0	9,903	2
Grand Casablanca	776	0	2,143	1	127	0	3,046	1
Guelmime-Semara	0	0	511	0	518	1	1,029	0
Layoune-Boujdor-Sakia	0	0	0	0	95	0	95	0
OuedDahab-lagouira	0	0	1	0	17	0	18	0
TOTAL	230,844	100	200,455	100	64,027	100	495,326	100
%		47		40		13		100

Most eucalyptus plantations (64%), are located in the northwest part of the country, with *E. camaldulensis* as the most dominant species (72%). These plantations were primarily developed to supply raw material to a pulpmill, which was installed in the region in the early 1950s.

The National Plan of Reforestation came to a standstill in 1994 and underwent revision because of certain difficulties in implementing its actions owing to financial, social, and climatic constraints. At present, a new Directive Plan of Reforestation is being developed to satisfy various needs, within a framework of participation and partnership.

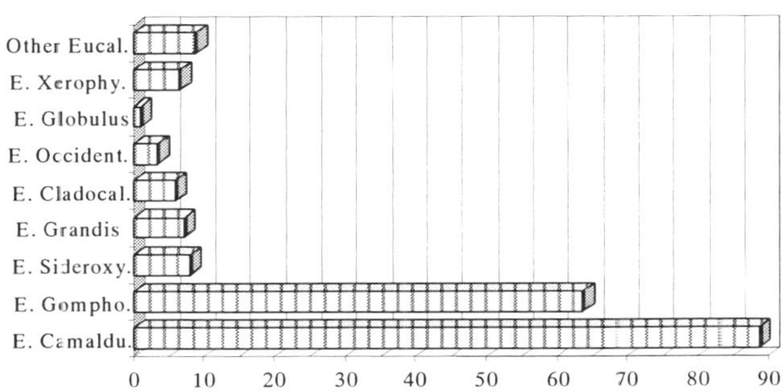

Figure 3. Eucalyptus plantations by species

4. Production and Utilization of Eucalyptus

Eucalyptus species were introduced to Morocco by private farmers in the beginning of this century. They were primarily used as ornamental trees or for planting low-productivity agricultural lands in the northwest of the country (Menager 1952). In the early 1950s, the Forest Service decided to expand these plantations in order to supply raw material to a pulpmill installed in the region. The objective at that time was to reforest 100,000 ha by 1966. In 1996, these plantations were estimated to be 200,455 ha.

According to ORT (1992), the forest's total possible production is around 2.5 million m^3 of wood per year, 65% of which is fuelwood. From 1990-1997, the average official real production was only 1 million m^3, and was composed of 17% sawlogs, 32% other roundtimber (pulpwood, poles, posts, studs, piles), and 51% fuelwood. However, besides this official production, more than nine million m^3 of wood were extracted from the forest directly by local populations, and used as fuelwood. This non-controlled direct exploitation lead to reductions in the forest area by more than 30,000 ha per year.

Even though eucalyptus plantations represent only 3.2% of the total timber land, they contributed 48% of total wood production from 1990-1997 (Figure 4). This share is composed of 1% sawlogs, 30% other roundwoods, and 17% fuelwood. If we account for the wood directly gathered by local populations, and of stumps coming from the removal of eucalyptus trees at the end of the last rotation, the eucalyptus fuelwood production is around 1.5 million m^3, representing more than 15% of the total domestic fuelwood consumption.

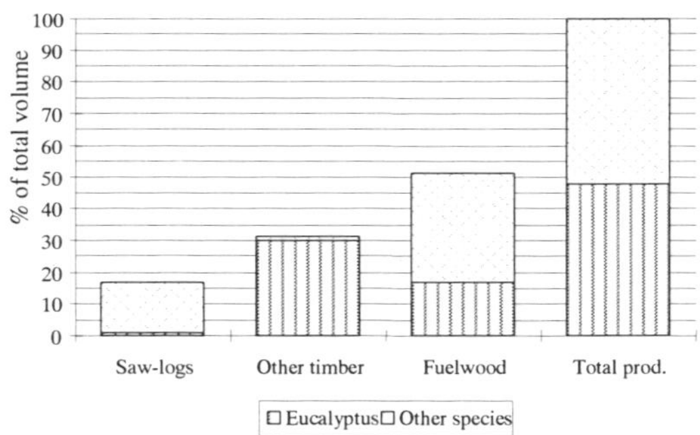

Figure 4. Contribution of eucalyptus in timber production

The principal use of eucalyptus timber in Morocco is to produce bleached chemical pulp in a mill installed in the region of the Gharb, in the northwest of the country. This mill started production in 1957 at 7,500 metric tons per year, and has been expanded many times to reach 100,000 metric tons per year in 1976, requiring almost 400,000 m^3 of bark-free boles.

To insure a continuous supply of timber, the mill signed an agreement with the Forest Service in 1973, stipulating that all loggers of the region of the Gharb are required to provide a quota of 55% of the volume of timber harvested to the mill, and this pulpwood should be bark-free, have a length of 1 or 2 meters, and a diameter between 5 and 33 cm. The selling price of logs is fixed annually by the mill, according to the price of pulp in the international market.

In 1980, the Forest Service started to have difficulties in meeting the wood supply agreement, because of many constraints related mainly to low productivity of eucalyptus plantations. The mill was then foreced to purchase wood from the local free market, and in 1987 it started importing wood to complement the local production (Jakko Poyry 1993). The agreement expired in 1993, and was renewed in another form, which consists of subsidizing the mill to grow its own forests on rented collective lands.

Wood not used by the pulpmill in the Gharb region, and most eucalyptus harvested in other regions of Morocco is used in many other sectors. It is used in agriculture as posts for fencing, as piles in the construction of greenhouses, or as lumber in the manufacturing of packing for domestic use.

Eucalyptus wood is also used for the production of particle board, as power or telephone poles, as studs in phosphate or coal mines, as lumber in the construction of boats, as piles in rural homes or in scaffoldings, or as fuelwood for direct combustion or for charcoal making. An experiment in slicing and peeling eucalyptus boles, and in using eucalyptus lumber in furniture building has been done (Fechtal 1985), but without real potential application.

5. Socio-Economc Importance

Eucalyptus species were introduced in Morocco primarily to reforest lands with low agricultural productivity (sandy soils and wetlands) in the northwest part of the country. Then, more than 130 species were planted experimentally in arboretums at various ecological stations. Some of these species have given good results in terms of adaptability and productivity, and they were used in large plantations either to produce pulpwood, or to protect soils from wind or water erosion.

Some argue that the environmental effect is controversial. These plantations consume a lot of water, lead to impoverishment of soils, do not let other species grow underneath, and are sensitive to pathologic attacks.

However, their socio-economic importance is significant for the country, mainly for the northwest part of it, where more than 60% of the plantations are concentrated. The most important sociological effect of eucalyptus plantations was observed in the Gharb region, where the population was originally nomadic and shepherds. The extension of eucalyptus plantations, mainly to collective lands and to scattered quercus oak forest, reduced the pasture's potential (Chakiri 1992). Also, eucalyptus lumber production and transformation created new professions, e.g., woodcutters, charcoal-makers, transporters, forest and pulpmill laborers, timber dealers, etc. As a result, the local population settled and relatively large communities were developed in the region.

An average of 7,500 ha per year was affected by eucalyptus plantations, if we consider that from 1949 to 1993 a total area of 767,000 ha was planted, and 41% of this area was covered by eucalyptus species (Ahizoune and Omerani 1997). The number of working days in the nursery and in the forest for planting this area is estimated to be 382,500, calculated on the basis of an average of 51 working days per ha (Chakiri 1992).

The average volume of eucalyptus wood harvested during the period 1990 to 1997 is around 482,442 m^3, composed of 2% sawlogs, 36% fuelwood and 62% of other roundwood (pulpwood, poles, piles,...). The process of felling trees, debarking boles, cutting them up to required dimensions, and to load logs into a truck and unload them, requires 0.8 working day for 1 m^3 of industrial roundwood, and 0.6 working day for 1 m^3 of fuelwood (Chakiri 1992). The total working days created by harvesting eucalyptus stands is estimated to be 351,216 working days per year.

The pulpmill employs an average of approximately 400 permanent employees, and 100 temporary laborers, year-round. The total working days directly created by the manufacturing of pulp is estimated to be 175,000, but two-thirds of them can only be attributed to domestically produced timber and one-third to imported wood. Because more than 90% of the pulp manufactured is exported, and because paper making is highly technology intensive, we can neglect without making significant bias the number of jobs created by the employment impact of 10% of the pulp that is locally consumed is not significant.

Half of the quantity of fuelwood produced is processed into charcoal. In traditional charcoal making, as is the case in Morocco, the transformation of 10 m^3 of wood necessitates 12 working days (Khattabi 1992). The quantity of wood that was transformed into charcoal for the period 1990 to 1997, is an average of 86,840 m^3 per year. The number of working days generated by this activity were approximately 104,208 for that period.

The trade of eucalyptus roundwood, other than pulpwood, e.g., saw-logs, piles, studs, etc., engenders many jobs like wholesalers, retailers, laborers in sawmills, in particleboard industry, in the manufacturing of boats or packing, in the construction of houses; or in green-houses, etc. The total number of persons working in these enterprises is difficult to estimate, but can be counted in thousands of working days.

A rough estimate of the working days generated by the activities related to eucalyptus plantations is over one million per year. If we consider an average wage rate of 50 dh (5 US$) per day, these activities engender an average annual income greater than 5 million US$ for the local populations. Most of this income is injected into the northwest region, where more than 60% of the eucalyptus plantations and the pulpmill are located.

The tax-free returns to the forest owner, which is the State, from selling eucalyptus timber, are estimated to be an average of 8.5 million US$ per year for 1990 to 1997. The stumpage price of timber for that period was 180 dh per m^3 (almost 2 US$ per m^3).

Pulp is the most important industrial product manufactured from eucalyptus timber, and is among the most significant provider of hard currency for the country. The average quantity exported from 1990 to 1996 is around 775,000 metric tons, for an approximate value of 37 million US$. In that period of time, imported long-fiber pulp, and recycled paper amounted to 46,000 metric tons, for a value of 18.5 million US$. Figure 5 (ODC 1997), illustrates the rank that eucalyptus pulp occupies in forest products exportations, evaluated at 75 million US$ for 1996.

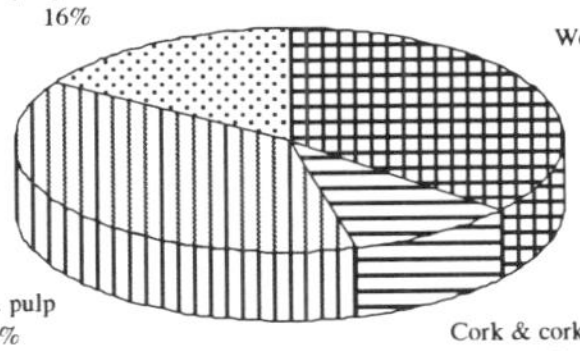

Figure 5. Composition of exports value -1996 -

Many other products derived from eucalyptus plantations, e.g., honey, essential oils, tannins, etc, are very important to the local economy.

6. Conclusion

Eucalyptus plantations constitute 40% of the reforested area. Even though, they represent only 3.2% of the total timberland, they produce almost half of the country's wood production, having tax-free stumpage value of 8.5 million US$. They also contribute to the creation of jobs by more than one million working days per year, estimated to be 5 million US$ yearly income. The principal industrial product derived from eucalyptus is bleached chemical pulp, which is mostly exported and contributes a large share of total forest products exportations.

7. References

AEFCS. 1996. Atelier no. 1: Eléments pour une stratégie nationale forestière. Colloque National sur la Forêt (Ifrane, 21-23 mars 1996). Admin. des Eaux et Forêts et de la conservation des Sols, Rabat. 28p.

Ahizoune, A. et Omerani, A. 1997. Bilan de 25 années d'application du Plan National de Reboisement. Journée d'information, 25 Avril 1997. Admin. des Eaux et Forêts et de la Conservation des Sols, Rabat. 25p.

Chakiri, B. 1992. Le role des eucalyptus dans le développement socio-urbain de Sidi Yahia du Gharb. Thèse de 3ième cycle INAU. 257p.

Fechtal, M. 1985. contribution à l'étude technologique des bois des eucalyptus de l'arboretum de Oued Cherrat. Note interne. Station de Recherches Forestières, Rabat. 31p.

Khattabi, A. 1992. Comparaison de la carbonisation du bois en meule traditionnelle et en four. Ann. Rech. For. Maroc. 26:76-83.

Jakko Poyry, S.A., 1993. Etude des prix du bois d'eucalyptus. Rapport national. Administration des Eaux et Forêts, Rabat. 76p.

MCI. 1995. Situation des Industries de Transformation. Ministère du Commerce, de l'Industrie et de l'Artisanat. Rabat. 30p.

Mekki, M. 1998. Personal communication. Service des reboisements, Ministère Chargé des Eaux et Forêts, Rabat.

Menager, L. 1952. Les eucalyptus dans le Gharb. Revue Inter. Bot. Appl. Agri. Trop. 7.

ODC. 1997. Annuaire Statistique du Commerce Extérieur. Office de Changes, Rabat. Vol. I: Importation.

ODC.1997. Annuaire Statistique du Commerce Extérieur. Office de Changes, Rabat. Vol. II: Exportations.

ORT. 1992. La filière bois; étude sur la production et la commercialisation du bois et ses dérivés. Projet Gharb Maamora, Direction des Eaux et Forêts, et de la Conservation des Sols, Rabat. 80p.

Cameroon's Logging Industry

- Structure and contribution to the country's economic development -

Richard Eba'a Atyi

Abstract: Cameroon is now considered to be one of the most important producers of tropical hardwoods in Africa, however its logging industry has remained unknown until recent years. A study was conducted under the sponsorship of CIFOR and TROPENBOS to describe the current structure of Cameroon's logging industry and its recent development. It is now known that, industrial logging in Cameroon is performed by 440 enterprises owned by both Cameroon nationals and foreigners. Cameroon nationals own the largest number of enterprises (70% of the total number of accredited enterprises). However, their involvement in logging has occurred only recently and can be related to the beginning of the economic crisis in 1986 to 1987. Before the crisis started, Cameroon nationals owned 29% of all accredited logging enterprises. Since then, the proportion of logging enterprises owned by nationals has been increasing to reach 90% in 1996. Furthermore, more than 80% of timber products exports are under control of foreign enterprises. Log production fluctuates between 2.8 and 3 million m^3 per year with a harvesting intensity of $7m^3$ per ha. Industrial logging contributes about 7% to Cameroon's Gross Domestic Product (GDP), and timber products exports contribute more than 20% to the balance of payments. The economic context of the last 10 years characterized by an economic crisis followed by structural adjustment policies has had a great effect on Cameroon's logging industry. Log production and exports have increased substantially, especially after currency devaluation. This has brought up concerns about the sustainability of current management strategies of Cameroon forests. Although harvesting intensity does not seem to be a threat to the sustainability of Cameroon forests, the current extend of yearly logging rights should be a matter of concern.

1. Introduction

Cameroon, a country located in the Congo Basin of Central Africa, has a total forest cover estimated at about 22.5 million ha of closed forests. Of these, productive forests on drained land are estimated to cover 17.5 million ha, while very degraded forests and swampy forests cover 4.5 and 0.5 million ha, respectively (Côté 1993). With these forest resources, Cameroon plays a major role in tropical timber production in Africa.

Until the mid-1980s, logging played a rather secondary role in the national economy, which was mostly based on the exports of crude oil and agricultural cash crops. But since 1986, logging has become increasingly important for the economy of Cameroon as its contribution to the balance of payments and the Gross National Product has increased more and more. In the meantime, Cameroon started to be considered as a leader in tropical timber production because of both a severe degradation of forest resources in other important producing countries of Africa and the improvement of public infrastructure needed for timber transportation.

Still, the timber sector of Cameroon has remained poorly known both inside and outside the country. A need to put forth reliable data on this sector has led to an analysis, results of which are presented in this paper. The paper includes a description of the structure of the logging industry in Cameroon, an evaluation of the contribution of the logging industry to the national economy, and new developments and trends in the logging industry of Cameroon.

2. Structure of Cameroon's Logging Industry

2.1 Administrative and legal framework of logging in Cameroon

Administratively, the Cameroon government has a Ministry of Environment and Forestry, which is responsible for designing and implementing forest policies. The Ministry of Environment and Forestry supervises logging activities in accordance with the existing regulations concerning logging. To carry on its duty, the Ministry of Environment and Forestry has provincial and divisional offices that supervise field forestry activities in their respective jurisdictions.

Two main documents define the legal framework within which logging occurs in Cameroon. These documents are the Forestry Law of January 1994 (Government of Cameroon 1994) and its Decree of Application signed in 1995 (Government of Cameroon 1995). The Law defines different types of forests and their respective ownership regimes. In general, it categorizes the national forest estate into a permanent forest estate and a non-permanent forest estate. All permanent forests are under state (or sometimes local council) ownership and cannot be converted to land uses other than forests. The non-permanent forest estate belongs to the national community and can be converted to other forms of land use. However, tracts of lands within the non-permanent forest estate can be set aside to produce forest goods according to locally defined objectives; these are called community forests.

In accordance with these two documents, logging activities in the field are regulated by two preliminary administrative steps, namely: an official accreditation in the logging profession and the acquisition of exploitation rights. Cameroon nationals as well as foreign entrepreneurs registered in Cameroon can apply for accreditation to the logging profession. An official accreditation is obtained from the Ministry of Environment and Forestry. Once the accreditation is obtained, the interested enterprise can apply for logging rights. Two types of logging rights can be granted in Cameroon:

- Sales of standing volumes (in French *ventes de coupe*), which can be small tracts of lands (not more than 2,500 ha) or a given volume of standing timber. Sales of standing volumes within both the permanent and the non-permanent forest estate can be granted only to Cameroon nationals. This gives Cameroon nationals easier access to timber resources compared to foreign investors.

- Concessions are granted to both Cameroon nationals and foreigners for 15 years. Concessions are bigger tracts of land covering up to 200,000 ha within the permanent forest estate.

There is a zoning plan in Cameroon (Coté 1993, The Government of Cameroon 1996), which demarcates the permanent forestry estate on the map, and distinguishes production forest in which logging is allowed from other types of permanent forests. The production forests of Cameroon cover 6 million ha and are subdivided into 90 forest management units. A concession can include one or more forest management units.

2.2. Ownership and distribution of logging enterprises in Cameroon

There are about 440 accredited logging enterprises in Cameroon. Of these 73% are owned by Cameroon nationals, 22% by foreigners, and 5% are joint ventures. But not all these accredited enterprises are currently active; only half of them have logging rights and harvest timber in the field. Table 1 gives an overview of the ownership of accredited and active logging enterprises. Based on the number of enterprises, Cameroon's logging industry is dominated by Cameroon nationals who own more than 70% of all accredited enterprises. However, the proportion of national enterprises in the logging business is higher for accredited than for active businesses, while the reverse is true for foreign enterprises. This suggests that fewer national enterprises have had prolonged success in the logging industry.

Table 1. Distribution of logging enterprises by ownership and level of activity

	Accredited		Active		Inactive	
	Number	%	Number	%	Number	%
Domestic	322	73	154	70	168	76
Foreign	96	22	58	26	38	18
Joint venture	22	5	8	4	14	6
Total	440	100	220	100	220	100

A number of reasons explain the relatively high number of accredited but inactive logging enterprises. For enterprises owned by nationals, there is poor access to capital, which makes it difficult for them to acquire heavy equipment necessary for timber harvesting in tropical forests where the public road infrastructure is less developed. However, a good number of these inactive enterprises are just entering the industry. These new enterprises, although accredited, have never been granted logging rights and are thus considered inactive. Even the enterprises considered active here are sometimes not harvesting timber in the field, but are just involved in log trading or processing although they are accredited for timber harvesting.

The territory of Cameroon is divided into ten administrative provinces. Logging occurs in five of these provinces, which are located in the most southern part of the country. However, as indicated in Table 2, the headquarters of these enterprises are located in six provinces.

Table 2. Log production and distribution of headquarters of active logging enterprises in Cameroon in 1996

Province	Number of enterprises headquarters	Log production (m^3)	Share of national production (%)
Centre	96	447,404	16
East	5	1,600,163	57
Littoral	86	225,388	8
South	15	451,841	16
South-West	12	81,134	3
West	6	0	0
Total	220	2,805,930	100

As one may note, the biggest share of logs is harvested in the East province, which produces 57% of logs, but only about 3% of logging enterprises have their headquarters there. The largest number of enterprises have their headquarters in the Littoral province where the most important seaport facility is found or in the Centre province where the country political capital is located. Most enterprises prefer to have their headquarters where the best communication infrastructure can be found rather than close to their logging site. The choice of the location of headquarters is an illustration of the importance of exportation in the logging industry of Cameroon. Industrial logging is primarily oriented towards timber product exports and for that reason, management services of logging enterprises tend to be close to export and administrative facilities.

2.3. Industrial log production and processing

As already shown by Table 2, log production in Cameroon varies between 2.8 and 3 million m^3 per year. Table 3 shows timber production by ownership of logging enterprises. Although few in number, enterprises of foreign investors are responsible for the most important share of log production in Cameroon. On average, an enterprise owned by a foreign investor harvests about five times more timber than one owned by a Cameroon national.

Statistics in Table 3 show that in contrast of the number of enterprises shown in Table 1, effective timber harvesting in Cameroon remains mostly in the hands of foreign (mostly European) companies with enough capital. However, proportions shown in Table 3 may change rapidly because the industry is in a transitional phase and the interest of Cameroon nationals in logging is still growing.

Table 3. Log production per average enterprise and ownership regime

Ownership	Log production (m^3)	Average log Production (m^3)	Proportion (%)
Domestic	1,018,553	13,058	36.3
Foreign	1,748,094	61,575	62.3
Joint venture	39,283	5,611	1.4
Total	2,805,930		100.0

2.4.Timber processing capacity and the market of timber products

In Cameroon, industrial timber processing is mainly primary processing, which includes lumbering as well as veneer and plywood manufacturing. Secondary processing has remained mostly at small local scale. In addition, there is currently no pulp or industrial board manufacturing.

The types and number of timber processing firms found in Cameroon are shown in Table 4. There are currently 60 processing units in the country. The total installed processing capacity is about 2,000,000 m^3 of roundwood, but only about 1,550,000 m^3 are effectively processed, which represents around 55% of the total log production. It should be noted that, Cameroon's official forestry policy has defined a target for processing at least 70% of the total log production within the country. Thus the current volume processed is both below the nominal capacity of established firms and the forest policy targets. Reasons for this may be that some of the existing factories are obsolete and the forestry administration has not put in place adequate incentives to encourage the installation of effective processing units.

Table 4. Primary wood processing units in Cameroon

Ownership	Type of Processing firm	Number
Domestic	sawmills	20
	rotary-cut veneer	1
Foreign	sawmills	26
	rotary-cut veneer	4
	plywood	4
Joint venture	sawmills	3
	rotary cut	1
	sliced veneer	1
Total		60

Timber products, both logs and processed products, are exported from four seaports: Douala in the Littoral province, Kribi, and Campo in the South province and Limbe in the Southwest province. However, Douala remains the main seaport from which about 90% of all timber products are exported. The current level of timber products exported from Cameroon can be estimated at about 1,980,000 m^3 of roundwood equivalent representing more than 70% of the total log production. This underscores the importance of export markets for the logging industry in Cameroon. In fact, industrial logging is most oriented toward the export market, which is why most logging companies have their headquarters in the Littoral province (Table 3) where the largest seaport facility is located (Douala).

As shown by Figure 1, log exports in Cameroon is mostly done by foreign companies. In fact, a closer look in the statistics available in the Cameroon seaport authority shows that the six biggest enterprises export more than 50% of all timber products shipped from Cameroon. These enterprises are well informed about external markets, and they export products harvested by themselves, as well as timber bought from small enterprises within Cameroon.

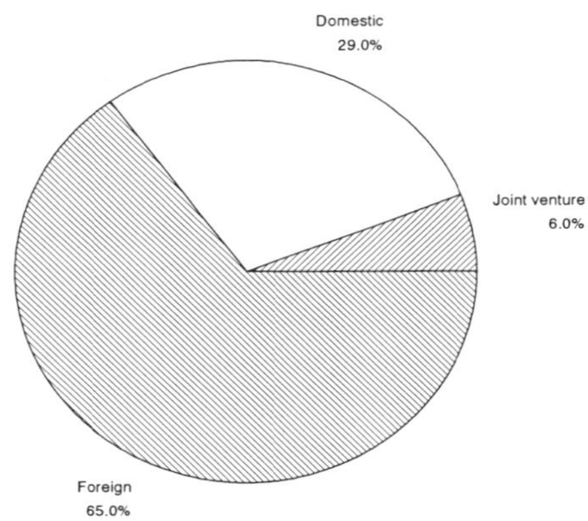

Figure 1. Log exports by ownership of enterprises

The domestic market consumes processed products--mainly lumber, plywood, and matches--all of which represent less than 30% of total industrial log production. These products are mostly of lower quality and cannot be considered for export. An important share of the national market is out of the control of the forestry administration and is supplied by local and illegal sawn wood.

3. Contribution of the Logging Industry to Cameroon's Economy

3.1. Contribution to the balance of payments

In 1996, the total value of exports from Cameroon was 811.02 billion Cameroon (CFA) francs (1.65 billion US$) (Ministry of Industrial and Commercial Development 1996). The contribution of timber products to total exports amounted to 160.53 billion CFA francs, representing 19.8% of the total value of exports from Cameroon. In fact, timber products rank second in export value behind crude oil. The most important share of this contribution comes from log exports (Table 5).

Table 5. Contribution of different timber products to export revenues

Type of product	Contribution to export revenues	
	Value (millions CFA)	%
Logs	89,308	55
Lumber	60,369	38
Veneer and plywood	10,850	7
Total	160,527	100

3.2. Employment in the timber industry

The timber industry of Cameroon employs about 34,000 people. However, logging is dominated by the lowest skilled personnel. Laborers account for 92% of the total number of employees. Figure 2 shows the distributions of employees by skills in the Cameroon logging industry. During the study, it appeared that all laborers were Cameroon nationals, less than 2% of the supervisors were foreigners, and more than 38% of executives were foreigners. In fact, the bigger foreign-owned enterprises employed more than 55% foreign executives.

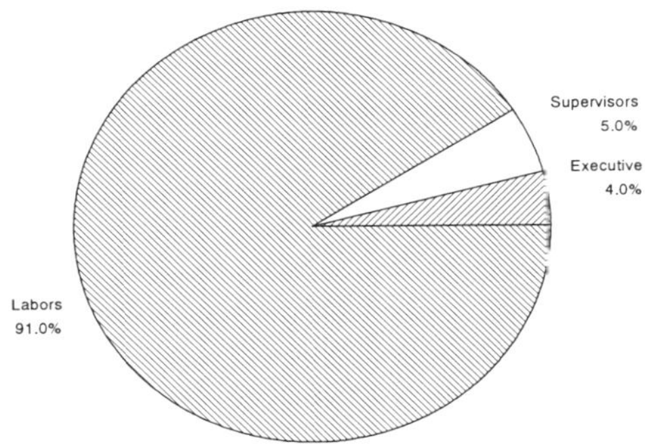

Figure 2. Distribution of employees by category in the logging industry of Cameroon

3.3. Contribution to the gross domestic product and government revenues

Logging contributes about 7% of the GPD of Cameroon. It should be noted that these figures include only timber harvesting and not primary processing because the accounting system places processing under the secondary sector.

As a contribution to government revenues, the logging industry provides income through felling taxes (*ad valorem*), area-based taxes, miscellaneous (including penalties and special permits), and export taxes. The first three types are collected by the forestry department while the last one is collected by the customs department. For 1997, the contribution of logging to the government budget amounted to about 30 billion CFA for a total budget of 660 billion CFA. This represents a relative contribution of 4.5% to government revenues. For a total log production of 2.8 million m^3, the mean contribution is about 10,714 CFA of taxes per m^3 (approximately 20 US$ per m^3). This contribution appears low compared to the contributions to the balance of payments and to the GDP, suggesting that returns from the logging industry benefit more the private sector.

4. Recent Development in the Cameroon Logging Industry

During the last ten years, the forestry sector of Cameroon and especially the logging industry has undergone important changes. These changes result from two main sources, one is the economic crisis that the country has experienced since the mid-1980s and the other is the quest for sustainable forest management. The changes have been concerned with both the institutional context in which logging is conducted and the structure of the logging industry itself.

4.1. Institutional changes

In 1988, the tropical forestry action plan of Cameroon was published (FAO-UNDP 1988). The document reported a detailed diagnosis of Cameroon's forestry sector and concluded that the forests in Cameroon were not managed sustainably. Since 1998, and with pressure and assistance of the international community, Cameroon has made a great deal of reforms of its forestry sector. The most important institutional reforms concern the creation of a Ministry of Forestry and Environment, the adoption of new forestry legislation, and the drawing of a zoning plan.

The creation of an entire ministry responsible for forestry in 1992, has highlighted the forestry sector in general and especially logging in the national life of Cameroon. The ministry has extensions throughout the country, and the means to monitor and control logging have increased substantially both in terms of personnel and logistic equipment. However, despite all the increase they are still not enough.

The new forestry legislation consists of a forestry law adopted in 1994 and its decree of application signed in 1995. The new forestry legislation has made some substantial changes in the institutional framework within which logging is conducted. Among the important reforms is the duration of logging grants. Before the adoption of the law, logging was conducted in licenses granted to companies for five renewable years. With the current legislation, the forestry administration defines logging concessions (or forest management units) that are granted for 15 years. Lengthening the duration of grants was done to give an incentive for logging companies to adopt more sustainable management practices for their concessions. In addition, logging companies are now requested to draw up management plans for the concessions granted to them and obtain approval of the plan from the forestry administration before logging activities in the field can proceed. Such a regulation has brought new needs for companies in terms of qualified personnel and equipment. Because the process of implementing all these new regulations is complex and slow, the tangible effect on logging practices has been little until now. However, this effect is becoming visible progressively. For example, it was only in 1997 that concessions really started to be granted according to new regulations, and the concessionaires have three years to submit their management plan. If the new regulations are applied rigorously, they may result in a decline in log production from Cameroon, because the total area opened every year for logging will be less than in recent years. The law also targets a ban on log exports starting from five years after its adoption (so from January 1999). If applied, such a regulation may lead to less log production too, because the current installed processing capacity is below the current log production.

The zoning plan is a national land-use plan, which does not cover the whole territory of Cameroon yet, but covers most of the forested area. The plan defines areas to be included in the permanent forest estate and sets primary objectives of their management. It distinguishes production forests from protection forests. According to the zoning plan, timber production is an objective on a total forested area of 6 million ha. The indications contained in the plan are proposals, and the procedures for finalizing the boundaries of different permanent forests include negotiations with the local populations. Given the experience gathered for the early stages of the implementation of the plan, it can be expected that the size of the production forest will be less than the proposals contained in the plan. This, in turn, will lead to even less timber production.

4.2. Recent development of the industry characteristics

The changes within the logging industry itself were noticeable beginning in 1986 when Cameroon was first hit by an important economic crisis. The economic crisis resulted mainly from the simultaneous drop in crude oil and agricultural export products prices (Alibert 1992). The country has since then been battling the fall of its economy with the adoption of structural adjustment programs. One spectacular measure of structural adjustment was the currency devaluation that occurred in 1994. Some of the most important developments are concerned with the number of logging enterprises, timber production, timber processing capacity, and timber product exports.

Although the number of logging enterprises has continuously increased from 1986 to 1996, the increase was particularly spectacular after the devaluation in 1994 (Figure 3). From 1994 to 1995, the total number of enterprises accredited for logging increased by more than 80%. The currency devaluation has attracted the business community to logging, certainly because the profit margin (devaluation surplus) of the logging industry has improved as well as the competitiveness of Cameroon timber products in international markets.

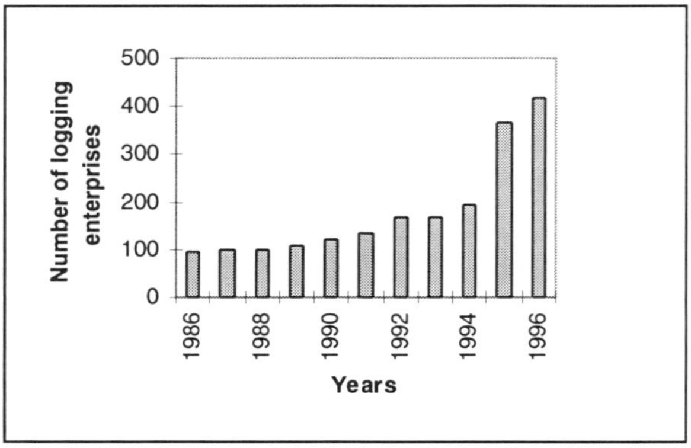

Figure 3. Evolution of the number of logging enterprises from 1986 to 1996

Timber production was quite stable from 1986 to 1993, but after the devaluation of 1994 it increased substantially by about 34% in one fiscal year (from July 1993 to June 1994) and remained about the same thereafter. The increases came mainly from the harvesting of new species, which were not harvested in the past because of their low market prices as compared to logging costs (Eba'a 1998). The currency devaluation has widened the difference between prices and costs, and it is now possible to make a profit on those species. Traditionally, timber exploitation in Cameroon has centered around ayous (*Triplochiton scleroxylon*), sapelli (*Entandrophragma cylindricum*), azobe (*Lophira alata*), iroko (*Milicia excelsa*) and sipo (*Entandrophragma utile*). New species such as frake (*Terminalia superba*) and tali (*Erythrophleum ivorensis*) now play an important role in Cameroon timber harvesting. All of the increased production has been oriented toward exports.

5. Outlook

From the preceding synthesis of recent developments in the forestry sector of Cameroon, some anticipation can be made about the future of Cameroon's logging industry. Because of both the foreseen ban of log exports and a more careful planning of logging activities in production forests, timber production may decrease in Cameroon in the near future. The decrease of timber production is also likely to occur because of decreased demand for timber products as a result of the economic crisis in Asia. It should be noted that Asia has become an important destination of Cameroon's timber products exports for the last five years. Traditionally Cameroon timber was mostly exported to Europe. The importance of the logging industry for Cameroon's economy will also decline not only because of the anticipated decline in timber production, but also agricultural products are regaining importance in exports. However, timber production is expected to be carried out in a more sustainable way with careful planning.

6. References

Alibert, J. 1992. Le Cameroun à l'heure de l'ajustement. Marchés tropicaux. Décembre 11: 3287-99.

Coté, S. 1993. Plan de zonage du Cameroun méridional. Prepared for the Ministry of Environment and Forestry. Poulin Theriault Inc. Quebec Canada.

Eba'a, A. R. 1998. Cameroon's logging industry: structure, economic importance and effects of devaluation. CIFOR occasional Paper, No. 14. CIFOR, Bogor, Indonesia.

FAO-UNDP. 1988. Tropical Forestry Action Plan, Cameroon. FAO, Rome.

Government of Cameroon. 1994. Loi No 94/01 du 20 janvier 1994 portant régime des forêts, de la faune et de la pêche. Yaoundé, République du Cameroun.

Government of Cameroon. 1995. Décret No 95/531/PM du 23 août 1995 fixant modalités d'application du régime des forêts. Yaoundé, République du Cameroun.

Government of Cameroon. 1996. Décret No 95/678/PM instituant un cadre indicatif d'utilisation des terres en zone forestière méridionale. Yaoundé, République du Cameroun.

Ministry of Industrial and Commercial Development. 1996. Le commerce extérieur en chiffres. MINDIC. Yaoundé, Republic of Cameroon.

The Economic and Environmental Impacts of Biofuel Taxes on Heating Plants in Sweden

Runar Brännlund, Bengt Kriström and Artur Sisask

Abstract: Sweden's energy policy is currently based on a large-scale introduction of biofuels. Following a 1980 nuclear power referendum, the current plan is to phase out nuclear power, replacing nuclear energy with renewable energy sources. This policy is supported by various tax breaks for biofuels. There is an ongoing discussion about a restructuring of the energy tax system, which will have far-reaching impact on the markets for biofuels. This paper evaluates the impact of changes in current energy taxation by analyzing a panel of approximately 150 district heating plants in Sweden. We estimate plant-specific production functions and derive the economic repercussions of the tax. We also estimate the resulting changes of emissions of sulfur, NO_X and particulates and assess the externality costs.

1. Introduction

Sweden's energy policy is currently based on a large-scale introduction of biofuels. The most important issue in future energy policy in Sweden is the fate of nuclear power. The current interpretation of the 1980 referendum is that nuclear power should be abolished by 2010, a goal that may conflict with unilateral climate policy goals. The future of nuclear power in Sweden is therefore unclear. A Commission on Future Energy Policy has been reviewing the matter and reported in 1995. In essence it recommended *against* a complete shutdown in 2010, generating considerable debate in policy circles. Bergman (1991) calculated the costs of closing nuclear power within a binding CO_2 constraint. His calculations, based on a small CGE (Computable General Equilibrium) model, suggested significant costs of closing nuclear power while stabilizing (or reducing) CO_2 emissions.

A possible way to phase out nuclear energy, which will not affect the CO_2 target, is a large-scale introduction of biofuels. Beginning in 1997, the government has devised a program that stimulates a significant increase of the use of biofuels. The program includes subsidies to households and various tax-breaks for biofuels in the current energy taxation system. The plan involves keeping electricity prices constant, or at least not rising, for the competitive industry. The plan comes amidst an overhaul of the whole energy taxation system that may entail a relative cost increase of using biofuels in heating plants. In the new energy taxation system, all energy sources may be subject to energy taxes, disallowing particular exemptions for biofuels. Hence, if the overall goal to increase the use of bioenergy is to be successful, a comprehensive package directed towards heating plants and their energy mix will be needed. This paper sheds empirical light on this issue. It sets out to detail the impact on district heating plants of a comprehensive change in energy taxation. Using a detailed panel-data set that describes heating plants during the period 1989-1996, we estimate the technology for each plant and simulate the choice of fuel-mix for several policy packages. Because the combustion of fuels results in a vector of air pollutants, it is important to analyze these kinds of reforms from a broader perspective. We do this here by using detailed data on plant technology with regard to environmental performance. By combining our estimates of the physical environmental impact with studies of the external costs, we are able to shed light on important social costs and benefits associated with the proposed policy change.

The paper is structured as follows. In order to understand the motivation for this paper more clearly, it will be necessary to introduce some background about the complex Swedish energy taxation system. Hence, section 2 details the current energy taxation system and the proposals to restructure it. Section 3 provides a theoretical model of a cost-minimizing power plant. Section 4 describes the econometric specification and presents the estimation results, while section 5 includes the results of a number of simulations. Section 6 displays the projected emission changes as well as the estimated externality costs. Section 7 offers some concluding remarks.

2. Energy Taxation in Sweden[1]

Sweden has used taxes on energy since 1929, when a tax on gasoline was introduced. Electricity has been taxed since 1951, followed by a broadening of the energy taxes in 1957. The motivation underlying these taxes was purely financial. In the 1970s, propelled by the global energy crisis, energy taxes were increasingly motivated by a desire to discourage consumption of fossil fuels. Thus, increased taxes on oil products were coupled by a significant expansion of electricity supply in order to promote a different profile of energy consumption.

Environmental concerns entered the discussion in the 1980s, manifested by the introduction of a tax differentiation of leaded gasoline in 1986. The Environmental Tax Commission then recommended a rich array of environmental taxes in their final proposal (see SOU 1990:59). This investigation led the government to propose taxes on emissions of CO_2 and sulfur, *inter alia*, in 1991. While this was not the first official body in Sweden to discuss environmental taxes, this mission was unique in that it was coupled with a major overhaul of the Swedish tax system in the beginning of the 1990s. The general tax reform included a reduction of income taxes, to be financed partially by an increased use of energy and

[1] This section is based on Harrison and Kriström (1998) and SOU 1997:11

environmental taxes (including the introduction of VAT on energy consumption).

For the purpose of harmonizing Swedish energy taxes with those prevalent within the most important competing countries, another reform of energy taxation passed on January 1, 1993. This reform was closely tied with the international competitiveness concerns that have been a recurring issue in the design of Swedish energy policy. It meant that manufacturing industry no longer paid energy tax on the use of fuels and electricity in their production processes. In addition, there was a reduction in the CO_2 tax for the manufacturing industry, as detailed below.

2.1. Industry exemptions

In an international context Swedish energy taxes are high. Because export-oriented industries are competing on markets with significant price elasticities, it is not surprising that several tax exemptions are being used. Beginning in 1974, through the law on (partial) exemptions of the general energy tax, energy-intensive manufacturing industries and the horticulture industry have escaped some part of energy taxes. This, of course, is not unique in Europe. Similar exemptions have also been used in Denmark and Norway for manufacturing.

In the tax system prior to 1993 approximately 100 energy-intensive firms were granted reduced tax rates on fuels and electricity. In 1992 the reduction for energy-intensive industry was worth 1.3 billion SEK. The new energy and carbon tax system introduced in 1993 resulted in significantly reduced tax rates for industry. The total amount of energy and carbon tax collections dropped from 3.8 billion SEK in 1992 to just 0.5 billion SEK in 1994.

Before the 1993 change of the energy tax system, tax exemptions were essentially granted on a case-by-case basis. Thus energy-intensive industries could apply for a reduction of the energy tax on electricity and fuels. With a zero energy tax on electricity and fossil fuels, such applications are now redundant. There are still possibilities for deductions for fuel use, some of them of considerable importance for individual firms (see SOU [1994:85; p.106]). These deductions are only possible for firms producing cement, lignite and glass. They only apply to the carbon tax on coal and natural gas, and not on the use of oil products. In 1995 less than 10 energy-intensive firms could benefit from this rule, and the value of the reduced tax was less than 50 million SEK.

By far the most important of the environmental taxes introduced as the result of the Environmental Tax Commission is the carbon dioxide tax. Introduced in January 1991, the tax of 0.25 SEK per kilogram of emitted CO_2 was followed by intense controversy. Eventually, a reform of energy taxes in 1993 led to significant reductions for manufacturing industries, as explained above. The government argued that it was important to reduce Swedish energy taxes to European levels for internationally competitive industries. Carbon taxes in Sweden in 1997, the base year of the model's representation of the tax system, are generally about 0.37 SEK per kilogram of emitted CO_2 for non-exempted sectors and 0.09 SEK per kilogram for manufacturing sectors.

2.2. Comments on the current energy taxation system

In order to understand the background for the simulation below, it will be useful to comment on some salient features of the current energy taxation system. A first observation is that fuels used for energy purposes are taxed differently depending on where they are used. The general principle is that an energy tax, a carbon dioxide tax, and a sulfur tax, is charged on fossil fuels used for heating and as vehicle fuels. In addition, an energy tax is charged on the consumption of electricity.

A second observation that fuels used in the production of electricity and heat are taxed according to different principles. All fuels used in electricity production, including plants that produce both electricity and heat, are exempted from the energy tax and the CO_2 tax. On the other hand, fuels used for heating are taxed differently, depending on whether heat is produced by the industry, by district heating plants, or in combined power and heating plants. If heating production takes place within a district heating plant, fuels are charged the energy tax and the full CO_2 tax. Fuels used in combined power and heating plants are charged 50% of the energy tax and the full CO_2 tax, for that part of the fuels which is considered to be used in the generation of heat. Heating production in the manufacturing sector is exempted from the energy tax, and the CO_2 tax amounts to 25% of the general level.

A third observation of interest here is that peat and biofuels are exempted from both the energy tax and the CO_2 tax. On the other hand, peat is subject to a sulfur tax.

A fourth observation that is vital for understanding the current system is that fuels are taxed differently depending on how they are used. In general, motor fuels are charged a higher energy tax than fuels used for heating purposes. For example, motor fuels used by the industry are charged with full energy and CO_2 tax. One exception is the mining industry, where diesel used in trucks is taxed by the same amount as if it were used for heating purposes.

A fifth observation is that fuels are taxed differently within the same area of use. The energy tax is not proportional to energy content. For instance, the energy tax on coal is only 60% of the energy tax for oil. As a consequence, the total tax on coal and oil is similar, in spite of the higher carbon content in coal per unit of energy. There is also a large variation between different fuels for the energy tax on motor fuels. The petrol energy tax is twice that of the diesel energy tax[2]. Furthermore, the energy tax on diesel and gasoline is differentiated according to an environmental classification system. Diesel or petrol with unwanted environmental properties is charged with a higher tax. The fact that the CO_2 tax differs between different fuels, expressed in SEK per unit of energy, follows from the definition of the CO_2 tax. This means that fuels with a relatively high carbon content per unit of energy are charged a higher CO_2 tax per unit of energy. For example, coal has the highest CO_2 tax per unit of energy, and this reflects that coal has a very high content of carbon in relation to the energy content.

Our *sixth* and final observation is that fuels used in electricity production are exempted from the energy tax and the CO_2 tax. These fuels are only subject to the sulfur tax. Consumption and production of electricity is taxed. Production taxes on electricity produced in nuclear power plants and in hydro power plants, may be interpreted as specific energy taxes on the crucial production inputs uranium and the drop.

2.3. A reformed system of energy taxation

The work on a reformed energy taxation system is in progress, although it is known that it will most likely be based on a proposal made by the Green Tax Commission in 1997 (SOU 1997:11). In any case, our simulations are based on a switch to a system based on the following principles:

Table 1. Starting points for a new system of energy taxation

	Energy-tax per kWh	CO_2 tax per kg CO_2	Sulfur tax per kg/S	"Traffic and environmentally related tax"	Total tax
Energy products	E	K	S	T	E+K+S+T

The point of departure is that all primary use of energy products – fossil fuels, biofuels and uranium, should be taxed according to the sum of the following four components:

- The carbon dioxide tax, which remains proportional to carbon content.
- The energy tax, which is restructured such that it is proportional to energy content.
- The sulfur tax, which remains proportional to sulfur content.
- A traffic and environmental tax, which is allowed to vary between fuels to accommodate concerns about particular environmental impacts and other external costs.

[2] The vehicle licence duty, on the other hand, is considerably higher for diesel cars.

The energy-tax is isolated to become a pure fiscal tax in the new system, while the carbon dioxide, sulfur and traffic and environmental tax are viewed as environmentally related taxes with a fiscal component. The traffic and environmentally related tax component can also be used to correct externalities originating from motor-traffic.

The new tax model implies that all energy products are taxed equally, independently of whether the products have been used for generation of heating or electricity. However, motor fuels are subject to a higher tax compared to fuels used for heating purposes. There are no special rules for manufacturing industry. This provides a simple and transparent system that eliminates all delimitation problems of the current system of taxation.

3. Theory

We use a straightforward production theoretic approach where production of heat may be written as a function of the input of a number of fuels, labour, and capital, and the state of the technology. While there are combined heating plants, producing both heat and electricity, we study only the single-output firm here. In the empirical analysis we abstract from any other inputs than energy inputs, excluding e.g., labor and capital. We might interpret the inputs of labor and capital as quasi-fixed factors, being constant over the time frame implicitly involved in this analysis. With additional data, we will be able to restrict these assumptions in subsequent analyses. Nevertheless, our focus in this paper is on the short-run.

The restricted production function is written as;

[1] $Q_{it} = F_{it}(x_{it}^{bio}, x_{it}^{oil}, x_{it}^{el}, x_{it}^{rest}, z_{it}),\ i = 1, \ldots, N,\ t=1, \ldots, T$

where Q is production of heat, , x^{bio}, x^{oil}, x^{el} and x^{rest} denote inputs of biofuels, oil, electricity and other fuels, respectively, z denotes a vector of plant-characteristics, and $F(\bullet)$ is a production function. Furthermore there are $i = 1, \ldots, N$ plants and $t = 1, \ldots, T$ periods of time.

Given an estimate of the elasticity of substitution between fuels and assuming cost minimizing behavior, it is possible to derive the price-elasticities. Hence for fuels j and k with elasticity of substitution σ_{jk}, we have that the price elasticity is given by (see Berndt 1991 for details);

[2] $\varepsilon_{jk} = \sigma_{jk} \cdot S_k,$

where S_k is the cost-share of fuel k.[3] The number σ_{jk} in equation [2] is the Allen elasticity of substitution (see Chambers 1989).

[3] Note that we have dropped the time subscript, t, and the firm subscript, i. This is purely notational convienience. In general the substitution elasticities, the cost shares, and hence the price elasticities differ between firms and time periods.

Two empirically useful functions have been used to estimate the production function, the Cobb-Douglas and the Translog. The latter can be written as;

$$[3]\ \ln Q_{it} = \alpha_0 + \sum_j \alpha_j \ln x_{it}^j + \sum_j \sum_k \alpha_{jk} \ln x_{it}^j \ln x_{it}^k + g(\beta_{it}, z_{it})$$

j = *bio, oil, el, rest*

The function $g(\bullet)$ signifies that production of heat depends on individual plant characteristics. We return to this fact later. Similarly, we see that the Cobb-Douglas is a special case of equation [3], since it can be written as;

$$[4]\quad \ln Q_{it} = \alpha_0 + \sum_j \alpha_j \ln x_{it}^j + g(\beta_{it}, z_{it}),\ j = \textit{bio, oil, el, rest}$$

Given estimates of the parameters in equation [3] or [4], σ_{jk} and ε_{jk} can be estimated, which in turn gives us a simple estimate of the change in demand for any fuel as;

$$[5]\quad \Delta x_j = \sum_k \varepsilon_{jk} \cdot \Delta p_k,\quad j = \textit{bio, oil, el, rest},\quad k = \textit{bio, oil, el, rest}$$

where Δ denotes percentage change.

The change in demand calculated according to equation [5] should be viewed as a first order approximation of the "true" change. The reason is that ε_{jk} in general is a function of input quantities and can thus not be treated as a constant, at least not for large changes in p. It should also be noted that we have omitted the time and firm subscripts. However, in general the change in demand will vary between firms and time periods.

The price-change for each fuel is computed here as;

$$[6]\quad \Delta p_i = \frac{(\bar{p}_i + t_i^1) - (\bar{p}_i + t_i^0)}{\bar{p}_i + t_i^0}$$

where $\bar{p}_i$ is the price before tax, and t_i^0 and t_i^1 denote the sum of taxes for fuel i before and after the change, respectively. We include the carbon-dioxide tax, the energy-tax and the sulfur-tax in our computations. Given estimates of the parameters, we can simulate the impact of a large set of potential tax-reforms. We shall return to this issue below, but begin by describing the current energy taxation system.

4. Empirical Specification and Results

4.1. Specification

Given the choice of functional form, all inputs are necessary for production to be positive. Given the fact that there is a large variation between plants in terms of fuel-mix, we need to make some sort of aggregation. Because our primary purpose is to study the substitution between biofuels and oil and electricity, we have aggregated fuels into four groups, as suggested by the derivations in section 2. We further restrict our study to those plants that use all fuels simultaneously.

Equation [3] and equation [4] imply that production of heat is conditional on individual plant characteristics, such as capital stock, technology and so on. Our data contain no information on investments or other characteristics of technology. We therefore assume that those "omitted variables" are strongly correlated with the size (production) of the plant and time. Hence, we allow for difference in technology between the "small" and the "large" plant. In addition, plant technology may differ across time periods. These assumptions provide us with an estimatable production function;

$$[7] \quad \ln Q_{it} = \alpha_0 + \sum_j \alpha_j \ln x_{it}^j + \sum_k \gamma_k \cdot d_{ik} + \sum_l \delta_l \cdot T_{il}$$

where *i (firms) = 1, …, N, j (inputs) = bio, oil, el, rest, k(class size) = 1, 2, 3, 4, 5, l(time period) = 1, 2, 3, 4, 5, 6, 7, 8*

$$d_{i1}\begin{Bmatrix} =1, \text{if } 0 < Q_{it} < 10 \\ =0, \text{else} \end{Bmatrix}, \quad d_{i2}\begin{Bmatrix} =1, \text{if } 10 < Q_{it} < 50 \\ =0, \text{else} \end{Bmatrix}, \quad d_{i3}\begin{Bmatrix} =1, \text{if } 50 < Q_{it} < 250 \\ =0, \text{else} \end{Bmatrix}$$

$$d_{i4}\begin{Bmatrix} =1, \text{if } 250 < Q_{it} < 1250 \\ =0, \text{else} \end{Bmatrix}, \quad d_{i5}\begin{Bmatrix} =1, \text{if } 1250 < Q_{it} \\ =0, \text{else} \end{Bmatrix}$$

Similarly, $T_{il} = 1$ in period 1, otherwise it is zero.

The specification in equation [7] suggests that the marginal product in plant i depends on size-class and the time period. Assume, for example, that $\delta_1 < \delta_2 < \delta_3 < \delta_4 < \delta_5$. This means that productivity is increasing over time. In a similar manner, we can investigate differences between firms by size-class.

Furthermore, the specification in equation [7] is of a Cobb-Douglas type which means that the price elasticities are constant and equal for all firms. However, since we do not explicitly derive the demand equations, which corresponds to the production function in equation [7], equation [2] is used directly. When applying equation [2] observed values of inputs and outputs are used, which implies that our elasticities will vary between groups[4].

The data we use are a panel data set covering almost every heating plant in Sweden for the time period 1989-1996. The data set includes production levels of heat, the amount of fuels used in the production, and the location of the plant. Unfortunately the data set does not include investments or other

[4] The elasticity formula we implicitly are using can in principle be expressed as $(dx/dp)(p/(x+v))$, where x is the fitted value, while $x+v$ is the observed value of an input. It means that the elasticities presented here will vary between different firm's since the error term, v, differs between firm's. In addition, the elasticities for each input will not sum to zero.

firm characteristic information. Descriptive statistics is given in Table A1 and A2 in the appendix.

Table A3 in the appendix has detailed estimation results. In summary, there is a strong correlation between fuel input and production, as suspected. Most of the parameters are significantly different from zero, although the time-specific effects, δ, are not significantly different from zero. Thus, there seems to be no significant productivity difference over time, according to the model. Size-specific effects are all significantly different, suggesting that larger plants are more efficient than small plants.

The results in Table A3 in the appendix have been used to calculate the price elasticities displayed in Table 2. The elasticities are evaluated at the 1996 levels of production and input levels.

As we can see, all own-price elasticities are negative, as expected from the theoretical considerations. There is an interesting pattern across size-class, regarding the price-elasticity of biofuels; the (absolute value of) elasticity increases. It is also apparent that biofuels are substitutes for other fuels (in fact all off-diagonals are positive for this particular technology and set of estimates). It is of particular interest to note that an price-increase of "other fuels", implies a relatively large effect on the demand for biofuels, oil and electricity (this effect is not symmetric). The explanation is that the cost-share is large for this group of fuels.

Table 2. Price elasticities of demand for different fuels in Swedish heating plants

Q < 50 GWh	p_{bio}	p_{oil}	p_{el}	p_{rest}	50 GWh < Q < 250 GWh	p_{bio}	p_{oil}	p_{el}	p_{rest}
Biofuels.	-0.32	0.10	0.06	0.72		-0.32	0.11	0.04	0.72
Oil	0.12	-0.35	0.06	0.72		0.13	-0.32	0.04	0.72
Electricity	0.12	0.10	-0.32	0.72		0.13	0.11	-0.36	0.72
Other	0.12	0.10	0.06	-1.38		0.13	0.11	0.04	-1.25
250 GWh < Q < 1250 GWh	p_{bio}	p_{oil}	p_{el}	p_{rest}	**Q > 1250 GWh**	p_{bio}	p_{oil}	p_{el}	p_{rest}
Biofuels.	-0.45	0.11	0.02	0.76		-1.45	0.11	0.02	0.84
Oil	0.11	-0.57	0.02	0.76		0.03	-1.01	0.02	0.84
Electricity	0.11	0.11	-0.12	0.76		0.03	0.11	-1.80	0.84
Other	0.11	0.11	0.02	-0.78		0.03	0.11	0.02	-0.12

5. Simulations

This section presents a number of simulation results, that we also further use to calculate emission impacts. We caution that we use only a subset of the existing plants. We still believe that the chosen sample is fairly representative. The interested reader is refereed to Table A2, where a descriptive analysis of the data is used to buttress this claim. We need to comment on another matter, before proceeding to the simulations and this is how to treat the "other fuels", when subjecting fuels to new tax levels.

Indeed, "other fuels" contain a disperse set of fuels (coal, peat, waste and others). In principle, we need to solve an index problem, that is, creating an efficient "price index" for the group.

We have chosen to calculate the induced price change on "other fuels" according to:

$$[8] \quad \Delta p_{rest} = \Delta p_{coal} \cdot \frac{x_{coal}}{x_{rest}} + \Delta p_{peat} \cdot \frac{x_{peat}}{x_{rest}} + \Delta p_{n-gas} \cdot \frac{x_{n-gas}}{x_{rest}}$$

Thus, any change in the taxes that hits coal, peat and natural gas will have an impact on the fuel-mix.

We consider 3 different policy packages. In each package we vary the carbon-dioxide tax. Across packages we vary the subsidy given to biofuel consumption. The first package entails lowering the carbon-dioxide tax to the industrial level in the heating plants, without giving subsidies to biofuels. The second package keeps the lower carbon tax rate, while allowing a subsidy rate that has been popular in the Swedish discussions. The third package we simulate adds to the first by introducing a subsidy that will keep current consumption of biofuels constant. Thus, given the lowered carbon tax, we estimate the subsidy (per giga watt hour (GWh) of input) that will guarantee constant consumption of biofuels in heating plants. In tabular form, our simulations take the following form;

Table 3. Simulations

Scenario	Carbon Tax	Biofuel subsidy
SIM1,SIM4	-70% or –50% of current	0
SIM2,SIM5	-70% or –50% current	0.05 SEK/kWh
SIM3,SIM6	-70% or –50 % of current	Biofuel consumption is constant (about 0.01 SEK/kWh)

Baseline taxes are reproduced below in Table 4 (as of 1/1 1997).

Table 4. Baseline fuel taxes SEK/kWh

Fuel	Carbon tax	Energy tax	Sulphur tax	Sum
Biofuels	0	0	0	0
Oil	0.10	0.066	0.01	0.176
Electricity	0	0	0	0
Coal	0.122	0.037	0.02	0.179
Peat	0	0	0.01	0.01
Natural gas	0.073	0.02	0	0.093

Table 5 summarizes the results of the simulations

It is seen that a carbon tax corresponding to 30% of the general level (0.37 SEK/kg) is equivalent to (in a partial equilibrium sense) a lowering of the price of oil and coal of 22 and 38% respectively.

According to the results of SIM1, consumption of oil is increasing, while the consumption other fuels is reduced (at the chosen level of aggregation).

SIM1 entails a reduction of biofuels by 2.7%, while oil consumption increases with 7.2%. Government revenue is decreased by 0.5 billion SEK. SIM2 complements, as explained above, SIM1, by introducing a subsidy to biofuels equal to 0.05 SEK/kWh. The subsidy means that the price of using biofuels (everything else being given) decreases by 49%. Hence, according to the simulations, there is a significant increase of both biofuel and oil consumption. This scheme costs about 1.1 billion SEK, in terms of lower tax revenues and the cost of the subsidy itself. SIM3 uses a subsidy that maintains consumption for biofuels at the initial level, We calculate this subsidy to be about 0.008 SEK/kWh (a price reduction of about 8%). SIM4-6 gives structurally similar results, although a smaller substitution towards oil.

Table 5. Simulation results

Price-changes, %						
	Δp_{bio}	Δp_{oil}	Δp_{el}	Δp_{coal}	Δp_{peat}	Δp_{natgas}
SIM1	0	-22	0	-38	0	-0.3
SIM2	-49	-22	0	-38	0	-0.3
SIM3	-8	-22	0	-38	0	-0.3
SIM4	0	-16	0	-27	0	-0.2
SIM5	-49	-16	0	-27	0	-0.2
SIM6	-6	-16	0	-27	0	-0.2
Quantity changes, % (GWh)						
		Δx_{bio}	Δx_{oil}	Δx_{el}	Δx_{other}	
SIM1	-2.7	(-208)	7.2 (524)	-2.8 (-42)	-1.7 (-498)	
SIM2	12.7	(1468)	3.7 (275)	-6.3 (-95)	-4.5 (-1356)	
SIM3	0.0	(0)	6.6 (482)	-3.4 (-50)	-2.2 (-644)	
SIM4	-1.9	(-220)	5.1 (374)	-2.0 (-30)	-1.2 (-356)	
SIM5	13.5	(1556)	1.7 (126)	-5.5 (-83)	-4.0 (-1213)	
SIM6	0.0	(0)	4.7 (344)	-2.4 (-36)	-1.5 (-8)	
ΔTax revenue(million SEK.)						
		Biofuels	Oil	Electricity	Sum	
SIM1		0	-454	-3.8	-458	
SIM2		-650	-481	-8.6	-1139	
SIM3		-98	-458	-4.6	-561	
SIM4		0	-316	-2.7	-319	
SIM5		-655	-348	-7.5	-1010	
SIM6		-69	-321	-3.3	-393	

6. Externality Costs

There is today a significant and expanding literature on valuation of external costs of energy generation. Health economists have carried out a large number of studies on the welfare impacts of energy use. In particular, there exist a large number of studies on mortality and morbidity connected to energy generation. Research at the Stockholm School of Economics show that the Value of a Statistical Life (VSL) is about 4.5 million US$ for the average Swede (see Johansson et al. 1997). In Oettinger el al. (1990) a similar estimate is provided, implying a VSL of about 4 million US$. Pearce et al. (1992) use a VSL of roughly 3.5 mill US$ using the average of values reported in the Ives et al. (1991) review of 46 studies. Additional examples of the external costs of biofuels are listed in Table 6.

Table 6. External costs of biofuels, some examples

Energy type	Comment	Value	Source
Bioenergy	7.6 deaths/Twh	1.6 US Cents/kWh	[1]
Bioenergy (New York area)	0.8-6.5 deaths/Twh	0.3 US Cents/kWh	[2]
Waste incineration		1-2.5 US Cents/kWh	[3]
Biomass in CHP		0.0625 (NO_x+SO_2) US Cents/kWh	[4]
NO_X Händelöverket (Swedish heating plant)	ExternE	2.625 US$ per kg NO_X (Health impacts)	[6]
PM10 Händelöverket	ExternE	5.88 US$ per kg (Health)	[6]
Sulphur Händelöverket	ExternE	2.81 US$ per kg SO_2 (Health)	[6]
Sulphur Energysystems (USA)	Survey	0.25-4.5 US$ per kg	[7]
Ozone (USA)	Los Angeles region	2.6 US$ per kg (Health)	[5]

[1] "Skador av Bioenergi", Analysgruppen vid Kärnkraftsäkerhet och Utbildning (KSU), Nyköping (http://www.apec.se), Sweden

[2] "New York State Environmental Externalities Cost Study: Report & Computer Model" Edited by Empire State Electric Energy Research Corporation (ESEERCO) samt Rowe, R. (1995) cited in Radetzki (1997)

[3] Miranda, L.M., and B. Hale. 1997. Waste not, want not: The Private and social costs of waste-to-energy production. Energy Policy. 25(6): 587-600

[4] Edin, K.A. 1997. The future of biofuels: A comment on the paper by Herman Vollberg. Energy Policy. 25(6): 623-627

[5] SOU 1996:177, p. 156.

[6] EXTERNE, Externalities of Energy: Swedish Implementation of the ExternE Methodology, SEI, 1998.

[7] Koomey, J., and Krause, F. 1997. Introduction to environmental externality costs. CRC Handbook on Energy Efficiency, CRC Press.

Table 6 shows a large variation between different externality cost estimates. This is presumably due to the fact that different approaches have been used, and that the figures do not refer to the same country. Thus, income and other differences between countries may explain part of the disparities.

In order to construct the mapping from energy combustion to emissions, we have scrutinized a substantial number of plants, via their environmental reports. We have also assembled a number of survey reports, summarizing the input-emission mapping for Swedish heating plants. We assume that there is a constant relationship between the fuel input and the emissions in the relevant data-range. We also assume that the regulatory constraint is not binding. Indeed, if the price vector facing the plant changed substantially, some plants may choose an infeasible point in the output-emission space, at least from a regulatory point of view. It is, at least in principle, possible to complement our econometric model with plant-specific regulations (which do vary between plants). However, due to the complexity of the regulations and other constraints, we have not been able to implement all constraints facing an individual plant, at this stage of our research.

- *Emission Changes*

Combustion of fossile fuels, such as biofuels, leads to a large number of different emissions. We will limit our attention here to three important emissions to air; sulphur, NO_X and particulates, leaving other parameters for future research. The advantage of restricting our analysis to these three parameters is that much is known about their impact on health and the externality costs. Indeed, a substantial number of studies have been carried out on the social costs of energy systems, see e.g., equation [7] for a review.

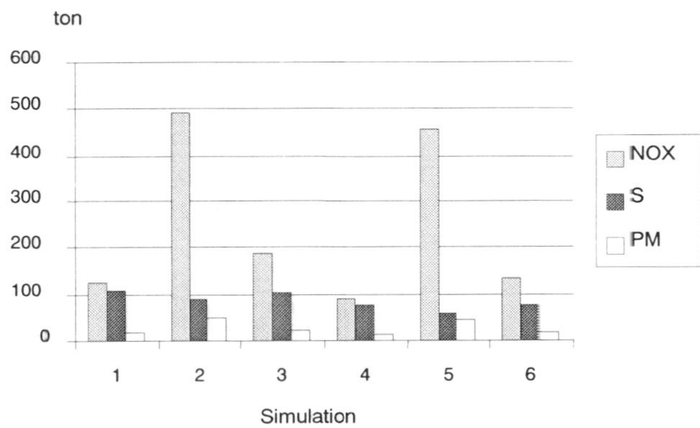

Figure 1. Calculated emission changes under scenario SIM1-SIM6

We now present a series of tables that display all necessary details for replicating our calculations on emission changes and the externality costs.

Table 7. Total emissions from Swedish heating plants, 1996 (ton)

Sulphur	2000
NO_X (NO_2)	6000

Source: tab 3, sid 13, Na 18 SM 9701, 1996, ton

Table 8. Average emission from Swedish heating plants (ton/GWh) for biofuels and oil

	NO_X	S	Pm
Biofuels	0.24097	0.0193	0.0227
Oil	0.39469	0.1622	0.024

Source: Yearly environmental reports from a number of heating plants and surveys of emission technologies

Given the data in Table 7 to 9, and the estimates of the change in the fuel mix for each scenario, the change in emissions can be calculated. The result is displayed in Figure 1.

Table 9. Externality costs used in calculations (US$/ton)

Parameter	Used cost	Basis
NO_X	4500	Current Environmental Charge
S	3750	Current Environmental Charge
PM	32500	Literature Survey

As is to be expected, emissions of NO_X, sulphur and particulates increase in all simulations. The heating plants substitute away from other inputs, towards oil and biofuels. SIM2 leads to a fairly significant emissions increase in case of NO_X in relation to current emissions from this sector, but this number is still low in comparisons to total emissions.

The costs of the emission changes are summarized in Figure 2. According to our simulations, the largest social costs of the policy packages we have simulated are due to an increase of NO_X emissions. Because the overall emissions from heating plants comprise only a small part of the total emissions, we cannot expect large externality costs in this case. When heating plants are already paying environmental taxes for sulphur and NO_X-emissions, we cannot claim that the reform is socially costly, despite the emission increases. In some sense, the costs are already internalized. Even so, the impact of this kind of reform of the Swedish energy taxation system will be heterogeneous across the country. Roughly, it is the large heating plants in the larger cities that are likely to increase their emissions significantly. This raises the issue whether or not the environmental charges should be uniform across the country, as it is today; if the damage is local, there is a case in favour of using a differentiated tax. This argument loses some of its force, because each plant is subject to a local regulation. Even so, our simulations suggest that those regulations may have to be re-assessed, if the government proceeds with a policy stimulating biofuels.

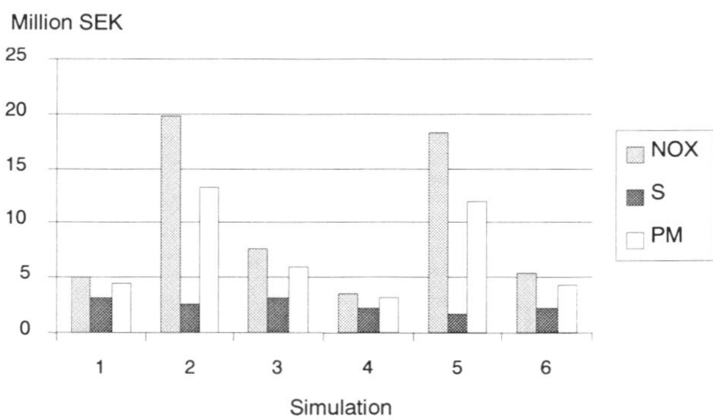

Figure 2. Estimated externality costs

7. Conclusions

The purpose of this study is to analyze possible impacts on the Swedish heating sector and externality costs due to a reformation of the Swedish energy tax system. The basic scenarios analyzed are reductions of the CO_2 tax and subsidies to promote biofuels. To do this we link three different modules; a production module which gives us the change in fuel mix due to a change in taxes and subsidies; an emission module which includes emission parameters for each fuel type; an externality cost module which provides us with a monetary value of the health effects caused by various emissions.

Our empirical results shed some light on the impacts of various changes of the tax system. The policy packages we have simulated, imply changes in the fuel mix, which in turn impose a change in externality costs. In fact, since all policy packages include a reduction of the CO_2 tax, externality costs increase to some extent, as expected. A policy package where a reduced CO_2 tax is combined with a subsidy to biofuels induces a shift towards more use of both biofuel and oil, which increases the externality costs. We also note that the distribution of the externality costs is likely to be uneven in the sense that they will be concentrated to large cities. Because NO_X and VOC emissions are related and of particular interest in large cities, our results seem to suggest that regulations need to be scrutinized, if the current policy of stimulating biofuels is continued. Our results also raise the issue if the tax system needs to be complemented with additional environmental taxes, covering, say, emissions of particulates. If we assume away the potential suboptimality, i.e., taxing only one of many sectors that contribute to "untaxed externalities", it would seem reasonable to buttress the feasibility of introducing new environmental taxes in the system. However, because the geographical variation of damages is likely to be substantial, an overall assessment of current regulatory schemes seems preferable. The current system of using both taxes and regulations needs to be re-structured, in particular in the case of global environmental problems. It is inefficient to use both a carbon dioxide tax and a regulatory cap on carbon dioxide emissions. It makes sense, however, to tax some locally important emissions, at the same time keeping the regulatory safeguard.

8. References

Bergman, L. 1991. "Tillväxt och miljö" (Environment and Growth), Supplement to LU-90, Ministry of Finance, Stockholm.

Berndt, E.R. 1991. The practice of econometrics, classic and contemporary. Addison Wesley.

Chambers, R.G. 1989. Applied production analysis: A dual approach. Cambridge University Press.

Harrison, G. W., and Kriström, B. 1997. Carbon taxes in Sweden. *in* SOU 1997:11, Skatter, Miljö och Sysselsättning, Ministry of Finance", Stockholm.

Johannesson, M., Johansson, P-O., and Löfgren, K.-G. 1997. On the value of changes in life expectancy: Blips versus parametric changes. Journal of Risk and Uncertainty. 15(3): 22_-39.

SOU 1997:11 "Skatter, Miljö och Sysselsättning", Final Report from the Green Tax Commission. Summary in English available as "Taxes, Environment and Employment", Ministry of Finance, Stockholm, Sweden.

Appendix

Table A1. Descriptive statistics for the Swedish heating sector. Whole population. All numbers are in GWh

	1989	1992	1994	1996	NUTEK 96
Q (summa)	32125.0	38485.0	40717.0	46084.0	
Q (medel)	223.1	253.2	271.4	301.2	
Q (std avv.)	470.4	544.1	589.2	692.5	
X_{bio} (summa)	3087.0	5132.0	8440.0	11937.0	12400
X_{bio} (medel)	21.4	33.8	56.3	78.0	
X_{bio} (std avv.)	49.5	73.3	107.8	141.3	
X_{olja} (summa)	4254.0	4115.0	5547.0	7282.0	8700
X_{olja} (medel)	29.5	27.1	37.0	47.6	
X_{olja} (std avv.)	60.1	50.0	120.3	128.1	
X_{el} (summa)	5077.0	5014.0	2250.0	1492.0	1600
X_{el} (medel)	35.3	33.0	15.0	9.8	
X_{el} (std avv.)	105.2	87.8	32.2	40.0	
$X_{övrigt}$ (summa)	26286.0	29503.0	28408.0	29938.0	
$X_{övrigt}$ (medel)	182.5	194.1	189.4	195.7	
$X_{övrigt}$ (std avv.)	445.6	497.7	486.1	522.2	
NOBS	145	152	151	153	

Table A2. Descriptive statistics for the Swedish heating sector. Subsample used in estimation. All numbers are in GWh

	1989	1992	1994	1996
Q (summa)	14067.0	18427.0	23343.0	27531.0
Q (medel)	334.9	376.1	395.6	509.8
Q (std avv.)	635.4	694.8	804.9	1047.8
X_{bio} (summa)	2348.0	4217.0	6050.0	7766.0
X_{bio} (medel)	55.9	86.1	102.5	143.8
X_{bio} (std avv.)	78.3	109.5	128.3	186.0
X_{olja} (summa)	1892.0	2163.0	3627.0	4055.0
X_{olja} (medel)	45.0	44.1	61.5	75.1
X_{olja} (std avv.)	93.1	67.8	182.0	192.2
X_{el} (summa)	2842.0	2766.0	1420.0	999.0
X_{el} (medel)	67.7	56.4	24.1	18.5
X_{el} (std avv.)	173.2	116.1	45.5	64.8
$X_{övrigt}$ (summa)	9975.0	12355.0	15075.0	17850.0
$X_{övrigt}$ (medel)	237.5	252.1	255.5	330.6
$X_{övrigt}$ (std avv.)	522.6	613.1	646.7	782.5
NOBS	42	49	59	54

Table A3. Estimation results

Parameter	Estimate	Std.error	t-statistic	
0	1.47202	0.102565	14.35	SSR = 27.86
1	0.14159	0.011188	12.65	R2 = 0.96
2	0.139954	0.013316	10.51	
3	0.068246	0.011344	6.01	
4	0.235369	0.012242	19.22	
1	0.749004	0.104247	7.18	
2	1.4945	0.110303	13.54	
3	1.99866	0.128365	15.57	
4	2.91515	0.162193	17.97	
1	0.034882	0.057774	0.60	
2	-0.02123	0.055797	-0.38	
3	0.060778	0.056188	1.08	
4	0.04978	0.054776	0.90	
5	0.045467	0.054734	0.83	
6	0.05844	0.054708	1.06	
7	0.080044	0.058419	1.37	

End User or End User Oriented Timber Supply in A Mechanized Northern Coniferous Wood Chain

Jan Sondell

Key words: Timber demand, end user, harvester, merchandising, bucking computer, simulation, apportionment bucking

Abstract: In Sweden the logging of Norwegian spruce and Scotch pine is almost fully mechanized by cut-to-length harvesters with on-board bucking or merchandizing computers. The computers are able to buck the trees according to a demand log tally and that makes it possible to produce log dimensions ordered by the sawmills. Therefore, development is going on to cut the logs according to the end-user demands. The actual wood chain is planned to work as follows: To find tree dimensions that fit different demands, it is necessary to collect inventory data in advance from actual stands to be logged. Based on the field data, a test stand is computer generated to be used for simulation of the harvesters' production. The sawmill collects the customers demands on sawn timber and expresses them as demand tallies of saw logs and sends that information to the supervisor of the harvesters. He carries out simulations and sets up an allocation plan for the harvesters that is best suited to the stands. The wood is then logged and within two weeks delivered by trucks to the actual sawmills. When setting up the saw log demand, the mill takes into account all inventories along the wood chain: the stock of sawn timber and saw logs at the mill, saw logs at the truck roads, and at the logging sites of the harvesters. SkogForsk has a leading role in the development presented earlier, by setting up operational demands for the harvesters and to verify the actual functions, and at present by developing and testing computer programs for inventory data collecting, simulation of harvester production, setting up log demands at the sawmill, planning of the harvesters' logging routes based on the simulation results, etc. SkogForsk also coordinates a joint European Union (EU) Research Project named 'PROMOTE' (FAIR-CT 96-1837).

1. Introduction

The primary logging system in Sweden is almost fully mechanised using cut to length harvesters and forwarders to haul logs to roadside. The clear-cut stands, mainly Norwegian spruce and Scotch pine, are medium sized (DBH 25-35 cm) and the terrain is fairly flat. The ground surface is often covered with small and sometimes big stones. The logging system is able to get access to most of the forest land. On soft ground logging takes place in the winter when the ground is frozen. From the landings at the roadside, the logs are brought by trucks and sometimes railways to different pulp and saw mills. The key unit of the logging operation is the harvester with its on-board bucking or merchandizing computer. The computer is able to buck and crosscut the trees according to a demand log tally, and that makes it easy to produce log dimensions ordered by the sawmills.

In the following, I will describe an ongoing development that aims to link the wood chain from stump to end user by integrating the actual bucking decision. Information Technology (IT) and user-friendly computer programs are used to define and verify industry demand and the stocks of timber in the forest, at roadside, and in the mill yard. Most of the actual development work is carried out within the frame of the European Union project PROMOTE (Production and Marketing of Timber in Europe, *http://www.skogforsk.se/eng/default.htm*), and I will refer to some of the Swedish reports.

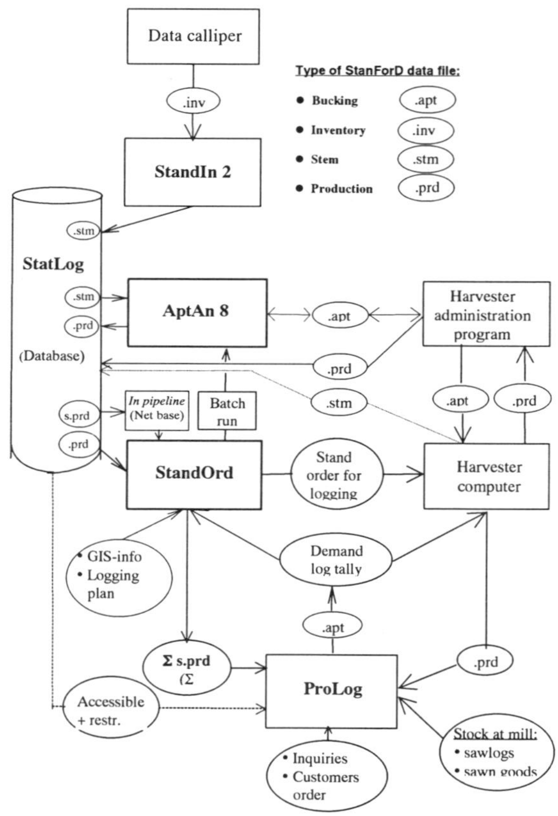

Figure 1. System of programs and IT connected to an ordering system of saw logs

2. System Description

The actual ordering system (Möller et al. 1998, Ogemark et al. 1998) can be illustrated with the flow chart in Figure 1. The stands to be logged are surveyed in advance using a data calliper with a special program designed for collecting the relevant inventory data variables. The stand data are stored in StanForD standard format (SkogForsk 1997) and transmitted to a PC.

In the PC a number of stem data files are generated in the newly developed program StandIn 2 (*Stand Inventory routine*, Ogemark and Sondell 1998) based on the actual data measured in the stand. The files are then stored in a database (StatLog, *Status on Logs*). When a number of stands have been surveyed a batch of simulations can be conducted by using the program AptAn 8 (*Bucking analysis*, Sondell and Ogemark 1998) and the results are stored stand by stand as simulated production files (log tallies, values, etc.) in the database. An administration program for the harvesters is used for handling information to and from the bucking computer.

The program for selecting the stand order is called StandOrd *(Stand Order for logging)*. The sawmill demand is generated by a program called ProLog *(Prognosis on Logs)* as a demand log tally (apportionment bucking) and transmitted to StandOrd. The harvester crews will be informed on the order in which the stands have to be cut. This order is determined with support from StandOrd. The demand log tally is also transmitted to the harvester.

The daily production of the harvesters is transmitted to the forest office and to the actual mill(s) to be used as decision support when updating the demand tally. Planning as described will be conducted at least monthly but could be made weekly or even daily.

3. Structure of StandOrd

The information flow and calculations in StandOrd are illustrated in Figure 2. StandOrd operates in close connection to the database, StatLog. For every logging object surveyed, a number of stem files corresponding to an area of one ha is transferred to StatLog. In StandOrd information on logging conditions, harvester production, etc., is added to the timber data and then a printout of accessible stands is provided *(Stand register)*. When a production plan is required, the planner starts by selecting all or a restricted number of accessible stands for simulation. He also selects a specific apt-file containing timber prices and the desired log tally for the actual period.

Based on the data selected he starts the simulation of a batch of logging units. StandOrd handles the simulations in AptAn 8 and presents the results on a spreadsheet. It is also possible to get a printout of the summarized production *(Simulated production)*. A production plan for the harvesters is settled based on manual selection *(Production plan)* (In the near future it will be possible to optimize this selection by taking all significant factors into consideration and producing the production plan automatically.). In the program there are possibilities both to reselect stands and recalculate the production plan. The computer program StandOrd was ready this summer and has been preliminary tested. The description of the program ProLog (below) is based on a specification but it will be programmed this autumn.

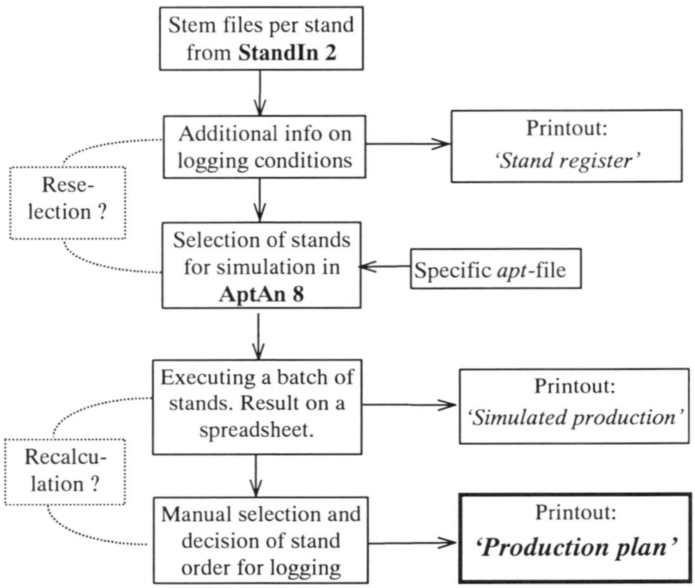

Figure 2. Information flow in StandOrd

4. Structure and Demand of ProLog

The program ProLog will be designed to be used by persons responsible for (1) the log supply of a sawmill and (2) for the marketing of sawn timber. The output of the program will be a demand log tally in StanForD format (apt-file) which can be used directly as a specification from the sawmill of the demand log distribution to the harvesters with on-board bucking computers.

The program will be able to handle three levels of market demand, according to the following specification:

Orders (1) are defined by the variables volume, number, length, width, thickness, quality, surface finish (planed-not planed), and moisture content.

The time factor must be considered. Inquiries (2) are in principle handled as orders, but as they are more uncertain, ProLog must handle the inquires separately from orders and be able to rank them. In a market situation with few buyers, the program must be able to also handle other market signals (3). The three categories would be kept apart from each other.

In the demand log tally it would be possible to allocate certain diameter and quality classes to specific buyers of saw logs from the forest. It would also be possible to cut part of the log tally (for example the biggest diameter classes) without any specified dimension demand.

The market demands will be handled both as volumes and as number of pieces. When an order for sawn goods is specified, a program function is used to convert it to saw logs. It will also be possible to handle volume and pieces after conversion.

In the conversion stage, it will be possible to handle at least 25 timber-diameter classes in mm per quality and tree species and 15 length classes. For each timber diameter and length class it will be possible to define different conversion options (e.g., 2 boards or 3 boards in a saw log, etc.) and to add a priority to each alternative.

When converting from sawn timber to saw logs it must be possible to give a correction figure, which will move, as logs, a certain number of a certain dimension to the next higher diameter and-or length class. It will be possible to adjust these correction figures from "follow-up data."

The sawmill supply is normally met by cutting whole stands (stand by stand) down to a set minimum diameter for saw logs. Therefore it is important to establish a relevant apportionment demand for every diameter class. With respect to dimensions and quality of the forests in the sawmill's supply area, limits are set within which the demand is allowed to fluctuate. Reasonable limits are probably best deduced by studying the variations in the delivery for the last year(s).

The database StatLog is connected to ProLog and will contain production files from the harvesters. They will be taken into account together with the other stocks in the wood chain from the stump to the delivery of sawn timber from the mill.

The flow chart of the demand for sawn timber with respect to the accessible wood supply is shown in Figure 3. The differences may be expressed as saw logs or of sawn timber, and this information is of great importance when buying more timber or marketing the sawn goods.

5. Field Trials at Södra Skogsägarna in 1998-1999

The described end-user-oriented supply system will be tested at Södra Skogsägarna, *the Forest Owners Association of southern Sweden*, in the winter of 1998 to 1999 within the PROMOTE project. The actual mill, mainly exporting sawn timber to the European Union (EU) but also to the Far East, will (via the forest department at Södra) be connected to 3-5 harvesters by wireless transmission. The demand log tally may be sent out from the office to the bucking computers as often as necessary.

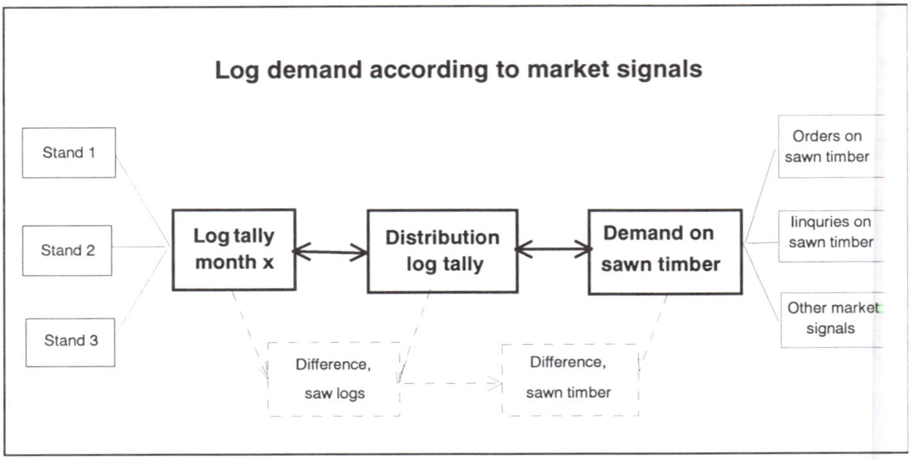

Figure 3. Flow chart of the demand for sawn timber with respect to accessible wood supply and market signals. The differences may be expressed as saw logs or as sawn timber

The information technology within the Södra environment is presented in Figure 4. ProLog, for example, will be connected to the program AlexPlan for planning of the timber processing.

The following data will be presented:

- Stock of logs-raw material
- Prognosis on logs-timber under way
- Allocated logs
- Demand log tally

A simple spread sheet with the number of logs per timber class will be displayed when opening up ProLog. The volume per timber class will be presented split in a number of columns as follows:

- Stock of timber at the mill
- Cut logs in the forest
- Delivery logs from small forest owners
- Logs according to prognosis based on standing timber
- Total

The user decides which columns will be displayed and the program only shows a sum for the other logs. For example, a column for the stock of raw material, another for logs in the forest, and a third for the remaining volume.

The user can also decide if he wants a graphical display or a detailed presentation of figures, including lengths per diameter class.

The period for a prognosis will be adjustable accordingly:

- Start date
- Period length (number of days, weeks, or months). It will be able to pre-set 'Period length' to a default value for each mill.
- Number of periods (e.g., 12 periods of a month gives a year).

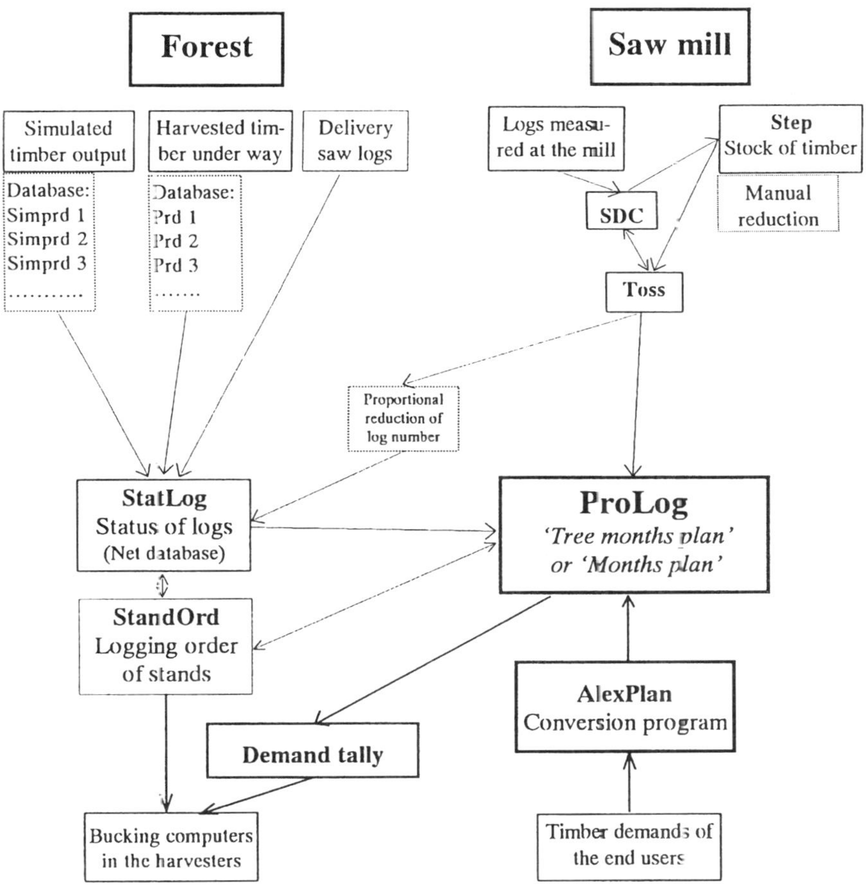

Explanation to abbreviations used

SDC: The Forestry Data Centre in Sweden.

Step: Program that handles timber data from the sawmills measuring frame

Toss: Program that administers the measurement orders and log payment

AlexPlan: A program from for planning of sawing activities.

Figure 4. ProLog and its IT-connections at Södra Skogsägarna

For each prognosis generated, the share of logs available in each stock will be presented, e.g., 25% at the mill, 25% cut in the forest, 15% delivery logs, 15% as standing timber, and 20% as general prognosis based only on figures from the last year.

These percentages can be regarded as reliability indications for the sawmill and are thus of great help in production planning. If the share of logs and timber cut based on inventory data is low, the prognosis on logs will be more uncertain. The totals will always be given.

Data in the table 'Allocated logs' will be separated into four parts showing the order status compared to the available raw material:

- Logs in the yard, stock at the log yard and delivery from the forest
- Planned production (planned 1=confirmed orders, planned 2=uncertain orders)
- Net sum of timber. Remaining accessible timber to book for processing or the amount overbooked.
- Production for the actual period including future shortage according to predictions. In this part, shortage in raw material for each dimension is specified for periods further ahead, and the net sum indicates saving of shortage to the next period.

The result will be presented both as tables and graphically. Different volumes will be presented in different colours, related to the accuracy: timber class is overbooked or is according to schedule. Timber allocated for a later period will be separately indicated. There will be a table in which the lengths are presented in detail. In this table the situation for each timber class is presented as shortage or surplus.

Orders will be registered according to sawing priority for the actual period. A secure order gets priority 1 and an uncertain order priority 2. Uncertain order does not influence operations in the forest.

Two types of demand tallies will be generated. Demand tally 1,with 2 to 3 months time, is normally used.

The tally is based on the length demand in registered orders and a specified basic demand valid for the part of the volume where no orders or length demands exist. The basic need can, for example, be specified by the demands from the last period. Actual lengths for an order will be correlated with normal length adjustments at the mill. The expected delivered timber is subtracted to give a demand log tally for the harvesters.

This demand tally is normally created 6 to 7 weeks in advance and is valid for the next 2 to 3 months. The continuous stock change is calculated by ProLog based on saw log volume at the mill and volume cut in the forest. Normally a sawmill has enough timber stored at the mill for 3 weeks production and 2 to 3 weeks stored in the forest. The administration of an apt-file (with the demand tally) will normally take a week, giving a total of 6 to 7 weeks.

The program will also be able to generate a tally of missing dimensions, based on booked volumes. This tally will be used to cope with a situation of shortage within 3 to 4 weeks (first period). The logging supervisor will possibly be able to find logging sites to compensate for the missing dimensions, for example big pine stands, long slender spruce trees >25 cm, etc. In the diameter classes where the stock is in balance, the basic flow remains the same.

6. Conclusions

The presented system for end-user supply in a mechanized wood chain will hopefully be fully working in the beginning of 1999 in the actual sawmill. The system will, if it is handled and updated properly, have full information of the timber supply in the chain from the stump to the delivery to each end user. If this information is used efficiently, it will be much easier than today to optimize logging and processing activities to meet desires of the customers.

The described system is the first one in Scandinavia trying to collect and tie all available information together in the wood chain from stump to end user. Within a couple of years this approach will probably be the main behavior in this field.

7. References

Möller, J.J., Ogemark, T., and Sondell, J. 1998. Practical working routine for a sawmill to create a demand list on logs, 'ProLog'. Milestone Report from the PROMOTE project, FAIR CT 96 1837. SkogForsk.

Ogemark, T., and Sondell, J. 1998. StandIn 2, Användarhandledning. SkogForsk.

Ogemark, T., Sondell J., and Thyr, U. 1998. Development of large stand selection routine, 'StandOrd'. Milestone Report from the PROMOTE project, FAIR CT 96 1837. SkogForsk.

SkogForsk. 1997. Standard for forest data and communications - StanForD. Last English version dated 1997-04-21. SkogForsk.

Sondell, J., and Ogemark, T. 1998. AptAn 8, Användarhandledning. SkogForsk.

Employment Impact Multipliers and the Economic Role of Timber in the Small Forest Communities of Southeast Alaska

Guy Robertson

Key words: Economic impacts, employment multipliers, input-output models, small-area economics, timber sector, Southeast Alaska

Abstract: Recent harvest declines in the Western United States have focused attention on the question of economic impacts at the community level. This paper uses community specific, time series employment data to estimate linear employment impact multipliers resulting from exogenously induced changes in timber employment and other basic employment in the small, semi-isolated communities of Southeast Alaska. Estimates were derived for each of the fourteen communities. When viewed as a whole, the study results provide no evidence for the existence of positive relations between changes in basic employment and other employment at the community level; changes in timber employment were seen to have no impact on other employment in Southeast Alaskan communities. While contrary to common perceptions regarding the workings of small scale economies, these results are not inconsistent with standard economic theory. However, they do highlight the need to consider the broader economic context within which economic impacts occur and the need to assess the validity of simplifying assumptions commonly used in models designed to estimate economic impacts at smaller spatial scales.

1. Introduction

Recent harvest reductions in the United States Pacific Northwest have served to focus forest policy debates on the issue of adverse economic impacts. The fact that much of the declining harvest volumes are the result of policy decisions regarding public lands in the region, rather than strictly economic factors relating to timber supply and demand, has further intensified the debate. Since timber-related economic activity is largely concentrated in rural areas in the region, much of the conflict has played itself out along geographical lines, with urban populations who see little economic gain from timber production pitted against rural populations who see logging and timber processing as an important part of their cultural heritage and economic livelihood. Moreover, as a result of both harvest declines and long-term growth in other sectors, timber activity comprises an ever smaller proportion of aggregate economic activity in even rural portions of the region. At the same time, however, certain communities maintain relatively high proportions of timber employment in their total employment mix. An important consequence is that economic impacts which are insignificant at the regional level, may nonetheless be quite large at smaller spatial scales. As a result, the United States Forest Service and other public planning agencies have increasingly been called upon to address the local impacts of their policy decisions, particularly at the community scale.

Community level impacts have been a perennial concern for forest policy for many decades. Initially, the discussion centered on the relationship between various definitions of community stability and harvest scheduling policies designed to achieve an even flow of timber from public lands (Waggener 1997). Here, much of the concern was centered upon the direct impacts of timber flows on logging and sawmilling activity. More recently, the focus has shifted to measuring and understanding adverse impacts on local economies from harvest reductions, and new concepts such as economic resiliency and diversity have been introduced (USDA Forest Service 1996). The central concern here has been the overall impact of reductions in timber activity on the local economy at large. This focus is further increased by the need to include an economic impact assessment in the environmental impact statements which must now accompany major forest planning decisions. In determining this impact, an overarching model which organizes and codifies our understandings of local economic interactions is essential. Various models, both quantitative and qualitative, have been used to understand and predict local economic impacts. Perhaps the most prevalent of these is the export base model and its more formal derivative, the input-output model. In their simplest formulations, these models assert a linear relationship between export oriented economic activity ("basic") and local support activity ("non-basic"). The central idea here is that the basic sectors are the sole source of new income to the local economy and thus drive activity in the non-basic sectors. While many of the assumptions in these models are certainly open to question, the existence of a strong relationship between basic activity, such as timber harvesting and processing, and non-basic activity is widely upheld both in public and professional debates.

Owing in great part to a scarcity of data at smaller spatial scales, the actual magnitude and dynamics of the relationship between basic and non-basic activity has not been subject to much empirical investigation. Rather, a common approach has been to adjust broader scale models for perceived local conditions, and it is rare that the resulting estimates are evaluated against actual economic performance over time. This is unfortunate since the strength of the relation between basic and non-basic sectors is a key component in the accurate estimation of economic impacts in particular, and the overall role of forest sector activity in the sustenance and development of forest communities in general. In recent years, however, economic data at smaller spatial scales is becoming increasingly available, and empirical tests of this relation based on time series estimation techniques are becoming increasingly feasible. This report provides an example of such an investigation, using the small, isolated forest communities of Southeast Alaska as its laboratory. The overall approach is to use community level, time-series data to test for the existence and magnitude of a linear relationship between basic employment (of which timber employment comprises a substantial proportion) and non-basic employment at small spatial scales. The results indicate that no significant linear relation exists. While perhaps unexpected, these results are not all that surprising when viewed in light of the few similar studies which have been conducted in the past and the various qualifications which accompany the model. Nonetheless, the absence of a significant relation should help us to focus on the important elements driving local development and change, as well as point out the shortcomings of common modeling techniques.

2. Secondary Impacts and Their Estimation

In concentrating upon communities as a unit of analysis (or any other spatial unit for that matter), secondary impacts are a key concern. The focus of the impact analysis shifts from a consideration of directly affected industries and sectors to the ways in which these direct impacts further affect the local economy at large. Indeed, in the absence of secondary impacts, it makes little sense to speak of impacts in a spatial context, and individual industries or even firms become the appropriate unit of analysis. The fact that so much attention is directed to the question of economic impacts at local and regional levels attests to the importance of spatially concentrated secondary impacts in the eyes of concerned publics and professionals.

Several techniques exist for estimating these impacts. One such technique is the application of neoclassical economic concepts in regional modeling exercises. Here, the numerous behavioral equations characterizing various local markets are parameterized using econometric techniques, and the new equilibria following an exogenous shock are used to gauge economic impacts. Recent elaboration's of this approach have resulted in extremely complex econometric models applied at national and regional levels, and commonly codified in the form of a Computable General Equilibrium model. The lack of adequate data at smaller spatial scales, however, has largely precluded the broad scale application of this sort of model to the estimation of local impacts. Input-output models, on the other hand, have enjoyed greater success in this area. This is, in part, the result of readily accessible techniques for adjusting national data-sets to local conditions, and the proliferation of computer programs which simplify and automate what was once an extremely tedious process. Hybrid models combining elements from both econometric and input-output approaches also exist (e.g., Israilevich et al. 1997), as do certain other approaches, but none have experienced the same level of practical application as the pure input-output approach.

While a detailed explanation of Input-Output techniques is not needed here, a brief description of the model's key characteristics and assumptions is useful. Essentially, input-output models take a snapshot of the interactions governing a given economy at a particular point in time and use these interactions to map an exogenously generated change in one sector into expected changes in other sectors. In an impact analysis setting, this process can be summarized in the form of a linear impact multiplier for employment, income, or other variables of interest. For example, an employment multiplier of 1.7 attached to the timber sector would indicate that a total of 1.7 jobs would be generated in the local economy for every one job exogenously generated in the timber sector (one job directly generated in the timber sector and 0.7 secondary jobs generated throughout the local economy through indirect and induced effects). Two key assumptions are needed to derive these multiplier estimates. The first is the assumption of a static production function exhibiting constant returns to scale and fixed proportions of factor inputs. The second is the assumption of perfect elasticity in the supply of factor inputs, including labor. These assumptions, particularly the latter, are directly at odds with neoclassical theory. Input-output practitioners respond that, while inconsistent with mainstream economic theory, their models are the best available approach given the practical limitations of data and resources available for most impact assessments. While this may be true, it is important that analysts bear in mind the relation between modeling assumptions and perceived economic realities. Is, for example, the assumption of perfect elasticity in factor inputs viable given the size of exogenous impacts and the spatial scale considered in the analysis? If not, how might the violation of this assumption affect multiplier estimates? More generally, how effective is the use of a static input-output approach and its linear impact multipliers in predicting dynamic change? Alternative estimates of impact multipliers based on empirical analysis of actual impact processes as they occur over time are necessary in order to begin to address these sorts of questions.

3. Study Setting and Data

This report uses time-series data from the small, semi-isolated forest communities of Southeast Alaska to provide an empirical estimation of linear impact multipliers arising from exogenous changes in basic employment. From the town of Haines in the north to Ketchikan some 3[illegible]0 miles to the south, the region is comprised of numerous islands with miles of convoluted shoreline. The topography is extremely rugged and the population sparse. Currently, Southeast Alaska's population stands at about

75,000 people, 29,000 of which live in the town of Juneau. As well as being the seat for the Alaska state capital, Juneau constitutes the region's principal transportation hub and trade center. Other regional centers include Ketchikan with approximately 14,000 people living in or around the community, and Sitka with approximately 7,000 people living in the general area. Both of these communities were formerly the site of pulp mills, but both mills have closed in the last five years. The remainder of the population is divided between various small towns and settlements some of which hold over 1,000 residents, but many are considerably smaller. Many of the communities rely on air or boat traffic as their sole linkage to the other towns, and, with the exception of Haines, non of the major towns in the region have road linkages to the outside world.

Along with government activity centered in Juneau, the region's major industries include timber, commercial fishing, seafood processing, and, increasingly, tourism. Figure 1, which shows indexes for timber harvest, timber employment (including sawmills and pulpmills), and other employment, indicates several developments in the regional economy which are important within the context of this analysis. Owing primarily to growth in government activity and tourism, and increases in retirement benefits and other forms of non-wage income, total employment in the region demonstrates relatively steady gains throughout the study period. Timber harvests and employment, on the other hand, have witnessed broad fluctuations, including sharp increases in the latter half of the 1980s and equally sharp declines in the 1990s. Moreover, when viewed at the community level, these fluctuations are even more pronounced, involving large shifts in logging employment and the opening and closure of lumber and pulp manufacturing facilities. Much of the variation in basic employment examined in this study arises from these fluctuations. Of course, fluctuations in other basic categories, notably seafood processing,[1] are also important, but these fluctuations have not approached the absolute and relative magnitude of those in the timber sector nor do they follow the same general trends. It is important to note that strong increases in timber employment, as well as declines, are included in the study sample. Consequently, the estimation of local impact multipliers will include examples of both positive and negative impacts.

The data used in the analysis is supplied by the Alaska Department of Labor. It consists of quarterly employment levels in different industries (at

[1] Commercial fishing constitutes a major industry in the region, but, unfortunately, employment data for this sector are not available. Employment in the industry appears to be relatively stable, however, and the omission of this sector should not result in a large systematic bias of estimation results.

the 4 digit Standard Industrial Classification (SIC) specification) for specific communities from 1981-1996. In its combination of spatial and industry detail, these data present a unique opportunity for analyzing economic impacts at small spatial scales. Moreover, several characteristics of the study region lend themselves to this sort of analysis. Among these is the fact that almost all manufacturing production is for export, simplifying the task of delineating basic and non-basic activities in the manufacturing sector. Another is that variation in basic employment is often quite large relative to total employment in a given community, thus increasing the variability in estimating impacts. Yet another is the small size and overall simplicity of the local economies included in the study sample. It is often assumed that less developed, rural economies of this nature will more readily exhibit the sort of relations examined in this report (Mulligan and Gibson 1984). If a positive relation between basic and non-basic employment is a common feature of all rural, forest communities, then Southeast Alaska would seem to be an excellent place to look for such a relation. Moreover, the richness of available data should further aid in this search.

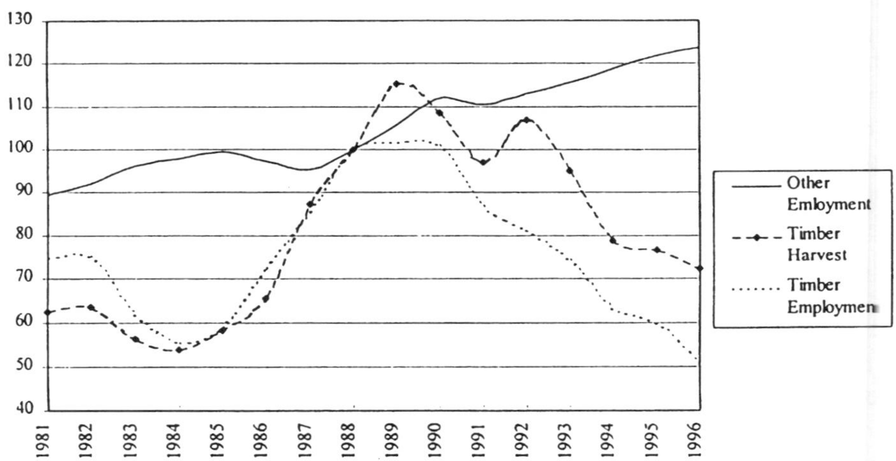

Figure 1. Indexes of timber harvest, timber employment, and other employment (1988 = 100) (Source: USDA Forest Service, Alaska Dept. of Labor). For reference, actual values in 1988 were: timber harvest: 951MMBF; timber employment: 3,499; and other employment: 27,436.

4. Estimating Impact Multipliers in Southeast Alaska

In the past, various studies have attempted to measure impact multipliers using time series data. Most early attempts to empirically estimate regional (and local) impact multipliers have relied heavily upon economic base theory. In an approach broadly analogous to a two sector input-output model, the export oriented activities which comprise the basic sector are taken to be an exogenous driver which is causally prior and linearly related to changes in the non-basic sector. This suggests a linear regression model using non-basic activity as the dependent variable and basic activity as the independent variable. Impacts in the non-basic sector, however, will not be instantaneous, and the inclusion of a lagged independent variable has been a common modification to the simple regression model. This is the general approach taken in this study. Previous studies along these lines include Sasaki (1963), Weiss and Gooding (1968), Moody and Puffer (1970), Moriarty (1976), and Henry and Nyankori (1981). Taken as a whole, results from these and similar studies are rather inconclusive, with several studies finding evidence for significant multipliers (e.g., Moriarty, and Henry and Nyankori), and others indicating no significant multiplier. Perhaps owing to the lack of consistent and convincing results, recent years have seen little activity in this sort of research, and no latter studies using time series analysis were uncovered in the course of the current analysis. One problem common to all of these studies is a lack of a sufficiently large data set in terms of both time series and cross sectional dimensions. In general, estimates are obtained for only a single town or community, and often these are based on annual data providing, extremely few degrees of freedom (especially when lagged variables are included in the regression equation). By drawing upon the numerous data points provided by quarterly data and providing estimates for numerous communities, this study attempts to provide a more robust and detailed estimation of linear impact multipliers.

Before presenting the statistical model and results, however, it is useful to visually examine the data in graphical form. Figure 2 provides an example using the town of Haines, Alaska. Basic employment was derived by summing employment in federal and state government with that in the manufacturing sectors (except in those cases, such as stone quarries, where manufacturing is obviously directed to local consumption). Non-basic employment is simply the total remaining employment. In both the figure and subsequent statistical analysis, the data was smoothed using a four quarter moving average in order to eliminate the relatively high degree of seasonality present in much of Southeast Alaska's economy. In spite of the extremely high variation in basic employment evident in Figure 2 (primarily the result of shifting logging employment and a sawmill opening and then closing), non-basic employment in Haines shows relatively little fluctuation, and it is impossible to assert the existence of a strong relationship between these two series. Similar graphs for certain other communities in the sample do exhibit some evidence of a positive relation, but others display little or no correlation and a few even exhibit a strong negative relation. When viewed as a whole, a simple visual examination of the data provides no evidence of a systematic relation between the two data series one way or the other.

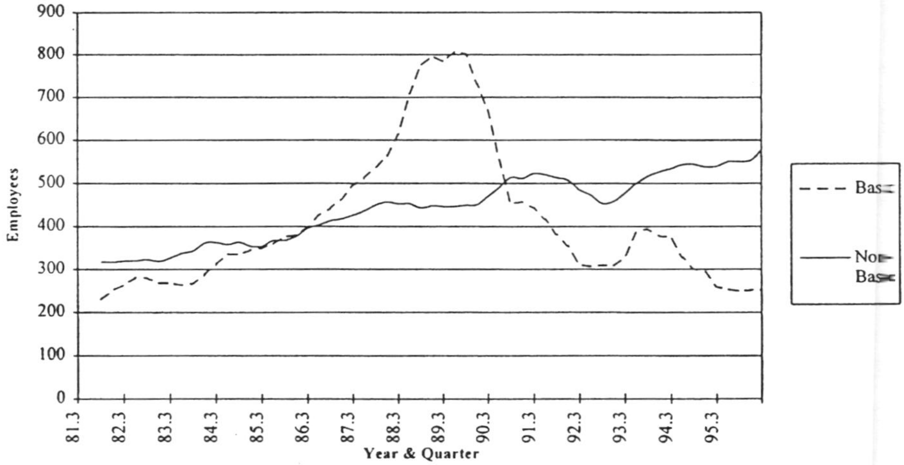

Figure 2. Basic and non-basic employment in Haines Alaska (Source: Alaska Dept. of Labor). Data smoothed with 4-quater moving average to eliminate seasonality.

The apparent lack of a linear impact multiplier can be further tested using statistical techniques, techniques which are more amenable to summarization and less liable to subjective interpretation than is a simple observation. The challenge here is to develop a statistical model which is both flexible enough to be uniformly applied to all communities in the study sample and sensitive enough to register impacts should they exist. The final model used in this analysis is relatively simple, but alternative models were also tested with quite similar results.[2] The model begins by assuming a linear relationship between non-basic employment and current and lagged values of basic employment. A simple trend variable is then added to account for growth in tourism-related activity and the influence of increases in non-wage income, factors which have experienced relatively steady growth throughout the study period. The data is then first differenced, resulting in the following

[2] Alternative model specifications included different lag structures, estimations on levels as well as differenced data, and estimations using only timber employment as an independent variable. In terms of the ranking of community specific estimates, minimum and maximum values for these estimates, and overall sample means across all community estimates, the models demonstrated a remarkable degree of stability. Much of this analysis along with a more detailed technical description of estimation techniques and issues will be published in a forthcoming US Forest Service technical report. Interested readers are encouraged to contact the author for details.

model which was estimated separately for each of the 14 communities used in the study sample:

$$[1] \quad \Delta NB_t = \alpha + \sum_{i=0}^{12} \beta_i \Delta X_{t-i} + e_t$$

where ΔNB_t is the change in non-basic employment in period t, and ΔX_{t-i} is the change in basic employment in period t-i. Note that the trend variable is now incorporated in the intercept term. Several additional modifications were made to equation [1] before conducting the final estimates. First, the β coefficients on ΔX_{t-i} were constrained to fit a fourth degree polynomial with its endpoint at $i = 12$ constrained to 0 (see Greene 1993 for a description of polynomial distributed lag techniques). Second, autocorrelation proved to be a problem even after first differencing the data--a likely result of the use of a 4 quarter moving average--and an autocorrelation correction term (AR1) was used. The resulting model states that current changes in non-basic employment are a function of a constant and changes in basic employment over the past 12 quarters. The sum of the βs provide a statistical estimate of a linear impact multiplier, except in this case 1 must be added to the estimate to obtain multipliers comparable to those produced by input-output and similar approaches. Estimation results ordered from highest to lowest are shown in Table 1.

Table 1. Regression results

		Intercept (α)		Multiplier (sum of β s)			
	Ave. Employment	Value	t-value	Value	St. Dev.	t-value	R^2
Yakutat	252	1.05	1.06	0.36	0.21	1.72	0.42
Ketchikan	6837	28.33	1.98	0.29	0.44	0.67	0.33
Angoon	284	-0.45	-0.86	0.25	0.13	1.96	0.15
Wrangell City	880	0.01	0	0.23	0.15	1.54	0.29
Sitka	3676	16.42	1.45	0.22	0.54	0.4	0.41
Hydaburg	100	0.04	0.06	0.08	0.08	0.97	0.03
Haines	849	6.12*	2.11	0.08	0.14	0.55	0.25
Hoonah	510	0.98	0.57	-0.03	0.28	-0.11	0.14
Thorne Bay	372	1.76	0.89	-0.1	0.24	-0.43	0.53
Petersburg	1491	-0.26	-0.11	-0.2	0.37	-0.53	0.3
Kake	250	1.29	0.93	-0.4	0.34	-1.18	0.46
Metlakatla	540	0.71	0.33	-0.41	0.49	-0.84	0.37
Juneau	13735	55.66*	2.09	-0.46	0.97	-0.48	0.58
Hollis	852	7.81*	4.82	-0.48*	0.17	-2.84	0.42
Mean	2188	--	--	-0.04	0.11	--	--

* indicates significance at the 95% level. Mean multiplier estimate is a simple average.

Mean standard deviation calculated using: $\sqrt{\left(\sum St.Dev^2\right)/n^2}$

When viewed on a community by community level, the results are not all that encouraging. The low r-squared (R^2) values are not surprising, given the use of first differenced data. However, the standard deviations on the multiplier estimates are quite high in spite of the 41 degrees of freedom available for each regression. Indeed, the inability to achieve a tight fit on these estimates has been a major shortcoming of many of the previous studies outlined above. When viewed as a whole, however, the estimates in Table 1 tell a more compelling story than indicated by the examination of any one community alone. Multiplier estimates are positive in seven of the communities and negative in seven. Maximum and minimum estimates are approximately equidistant from zero. The mean value for the entire sample is -0.04 with a standard deviation of 0.11. In general, these results mirror the initial conclusions obtained from the visual observation of community data, with certain communities exhibiting positive relations, others negative relations, and still others demonstrating no discernible relation at all. In the case of communities like Yakutat and Kake, which occupy opposite ends of the spectrum, the relations indicated by the multiplier estimates are clearly visible in the graphs of their employment data.

There are two possible interpretations of these results. The first is that the differences in multiplier estimates are a true indication of the differences in impact processes occurring in these different communities. Given the high standard deviations in these individual community estimates, only one community (Hollis) demonstrated a multiplier estimate significantly different from zero, and that was for an estimate at the negative extreme of the spectrum. An alternative, and more robust, interpretation would be to view the results on the whole as indicating an essentially uniform distribution around a zero mean. Here, the conclusion is that the results do not support the assumption that changes in basic employment have a positive impact on non-basic employment.

5. Discussion and Extension of Results to Other Settings

The results presented above provide no indication of the existence of a positive employment impact multiplier in the communities of Southeast Alaska during the last two decades. While this may be counter to common conceptions of the way the economies in small, semi-isolated communities operate, there are several reasons why we should not necessarily be surprised by these results. They include:

(1) Forest communities such as those examined in this study import a large proportion of the goods and services they consume. Consequently, rather than being spatially concentrated in the given community, a majority of economic impacts from changes in basic activity may occur elsewhere. This conclusion is consistent with the input-output approach and is reflected in the results of at least one input-output study (Hamilton and Jenson 1983).

(2) Employment measures may not adequately reflect changes in income. This is especially true in places like Southeast Alaska, where many residents obtain their livelihoods from a mix of activities, both part time and full time. Impacts may be manifest in changes in working hours, wages, or both, without being evident in the employment data. Once again, this is not inconsistent with the input-output approach (at least when adjustments are made primarily in the amount of hours worked rather than wages), but an income model rather than an employment model is called for. Unfortunately, income data is usually less available than employment data. Additionally, it should be noted that employment numbers are usually the primary focus of policy debates.

(3) In assuming a linear impact multiplier distributed over a finite (3 year) time period, this study ignores the possibility of long-term dynamic impacts. Through their influence on population, local government revenues, and various spatially concentrated economies of scale commonly subsumed under the general concept of "agglomeration economies", changes in basic employment may significantly alter the long-term growth trajectories of rural communities. These impacts would not be manifest in the sort of estimations presented above.

(4) The supply of productive factors, particularly labor, may be significantly constrained in semi-isolated regions and communities. This is especially true where changes in basic activity are large relative to the local population. Where this is the case, the basic industries will have to either import their labor or bid away labor from other local activities. To the extent that the latter is true, the impact of a change in basic activity on other sectors will be lessened and may even be negative. Of course, many rural communities, those in Southeast Alaska included, exhibit relatively high levels of unemployment. However, whether these endemic unemployment rates are an adequate measure of the true supply of available labor (with required skills) is still an issue.

Reasons (1) and (2) are specific to the application of input-output models and similar multiplier estimation methodologies. Reasons (3) and (4), on the other hand, entail fundamental violations of key input-output assumptions--the assumption of linearity in the case of (3), and the assumption of infinite supply elasticities of productive factors in the case of (4). More generally, these last two reasons highlight the need to view the question of impact multipliers within the broader context of overall local and regional economic conditions and dynamics. As indicated in Figure 1, the economic impacts analyzed in this report have occurred within the context of overall regional growth. Consequently, the opportunities afforded by growth may have served to mitigate negative impacts, and tight labor supply may have lessened positive impacts. In a region or era characterized by economic stagnation, the results may have been quite different.

This need to consider the broader context in which impacts occur is especially important if one seeks to extend the results of this study to other regions or countries. For the sake of argument, two examples are briefly considered here. The first is Japan, for a number of decades Japan has experienced high rates of migration from rural to urban areas, especially among working age individuals. This has resulted in increasingly severe shortages in labor and a rapidly aging workforce in many of the country's forest communities (Robertson and Waggener 1992). Under these conditions, an impact multiplier based on the assumption of unconstrained local labor supplies would significantly overestimate the impact of, for example, the opening of a new sawmill. The Russian Far East, the second example considered, presents a very different situation. The economic contractions of the 1990s have resulted in significant levels of unemployment and underemployment in the forest communities of the region (Sheingauz et al. 1996). Development projects promoting forest products production in these localities will likely find ample labor, but it is doubtful that the capital needed to generate secondary impacts will be forthcoming. Additionally, even in the case of residentiary services demanding little initial capital, it is possible that the entrepreneurial experience necessary to identify and take advantage of new opportunities is absent. Once again, an increase in basic activity may have little positive secondary impacts on the local economy. A sharp decrease in basic activity, on the other hand, may be devastating. Of course it is possible to identify conditions which would enhance rather than dampen the multiplier effect of the forest industries and other basic activities. Such conditions would include ample local supplies of labor and capital, perhaps in combination with economic expansion at the regional or national levels. In any case, the magnitude of local multipliers in particular and the overall role of forest products industries in general cannot be assessed without a thorough consideration of the broader economic context in which they exist.

6. Conclusion

In spite of an appropriate study setting and rich data-set, this analysis provides no evidence for the existence of a positive linear employment impact multiplier in the communities of Southeast Alaska. It is important to be aware of the boundaries of this result. Other types of economic impacts may be present in Southeast Alaska, but they are not apparent when analyzing employment data in a linear regression context. This, in turn, calls into question the common assumption that a positive relation between timber employment and other employment necessarily exists at smaller spatial scale. There are various reasons why this relationship may be relatively small or insignificant, each of which constitutes a potential avenue for future research. In the mean time, it is important that researchers and economic practitioners engaged in impact estimations be aware of the ways in which their models and predictions may be in error and the ways in which common measures of economic performance may fail to capture important aspects of impact processes. On a more general level, formal and informal economic models which emphasize timber activity as a driving force in local economies deserve careful scrutiny and need to be assessed within the context of broader economic characteristics and trends.

5. References

Greene, W.H. 1993. Econometric analysis (2nd. edition). Macmillion Publishing Co., New York.

Hamilton, J.R., and Jensen, R.C. 1983. Summary measures of interconnectedness for input-output models. Environment and Planning. 15:55-65.

Henry, M.S., and Nyankori, J.C.O. 1981. The existence of short-run economic base multipliers: Some new empirical evidence. Land Economics. 57(3): 448-458.

Israilevich, P.R., Hewings, G.J.D., Sonis, M., and Schindler, G R. 1997. Forecasting structural change with a regional econometric input-output model. Journal of Regional Science. 37(4): 565-590.

Moody, H.T., and Puffer, F.W. 1970. The empirical verification of the urban base multiplier: Traditional and adjustment process models. Land Economics. 46:91-98.

Moriarty, B.M. 1976. The distributed lag between metropolitan-area employment and population growth. Journal of Regional Science. 16(2):195-212.

Mulligan, G.F., and Gibson, L.J. 1984. Regression estimates of economic base multipliers for small communities. Economic Geography. 60:225-237.

Robertson, G., and Waggener, T.R. 1992. Japan's forest production for increasing self-sufficiency: A reassessment of near-term capacity and economic potential. WP 40, CINTRAFOR, College of Forest Resources, University of Washington, Seattle.

Sasaki, K. 1963. Military expenditures and the employment multiplier in Hawaii. Review of Economics and Statistics. 45:298-304.

Sheingauz, A.S., Karakin, V.P., and Tyukalov, V.A. 1996. Forest sector of the Russian Far East: A status report. Khabarovsk-Valdivostock: Russian Academy of Sciences, Far East Division, Economic Research Institute.

USDA Forest Service, and USDI Bureau of Land Management. 1996. Integrated scientific assessment for ecosystem management in the interior Columbia Basin and portions of the Klamath and Great Basins. USDA Forest Service, PNW GTR-382, September, 1996, Portland, Oregon.

Waggener, T.R. 1977. Community stability as a forest management objective. J. of Forestry. 75(11): 710-714.

Weiss, S.J., and Gooding, E.C. 1968. Estimation of differential employment multipliers in a small regional economy. Land Economics. 44:235-44.

Emerging Human Values of Forest Resources

- A Japan-United States comparison -

Theodore E. Howard, Ken Sugimura and Richard A. Cooksey

Key words: Green resources, non-timber resources, social demand, social values, Analytical Hierarchy Approach (AHP)

Abstract: In Japan and the United States, people are placing greater emphasis on green (non-timber) resources. In independent studies designed to elicit social preferences for forest functions in Japan and in the United States, survey respondents were asked to rank their preferences for two sets of forest functions and to choose between two functions in a series of paired choices. Japanese respondents indicated that land and water conservation, wildlife conservation, scenic beauty, in that order, were more important than timber production. Their preferences for specific land and water conservation measures varied by local disaster history. In paired comparisons, they weighted land and water conservation, wildlife, scenic beauty, and timber functions in decreasing order of preference. United States respondents indicated that air and water quality was the most important issue and that timber and forest products supply was second in importance. However, in paired comparisons, timber was the least important function. These respondents also indicated a willingness-to-pay approximately 32 US$ per year to protect the full range of forest functions by means of conservation easements. Respondents in each country indicated clear preferences for non-timber forest values. The extent to which forest managers respond to these preferences will impact the flows of forest goods and services. Cultural differences will affect the means by which public agencies in Japan and the United States assess and incorporate public sentiment into forest management. Private forestry will be affected through the differing objectives of landowners and regulatory climates. These, too, have important cultural aspects. Research cooperation and information exchanges across the Pacific may provide new insights for solving resource management issues.

1. Introduction

For the last twenty-five years, public and private forest management organizations around the world have struggled to understand and respond to significant shifts in the relative values their societies assign to forest functions. As in other industrialized nations, people in Japan and the United States are placing greater emphasis on functions other than timber production.

In this paper, we have adopted the Japanese term, *green resources*, to categorize those forest functions other than timber production, rather than use the North American equivalent, non-timber resources. The Japanese term encompasses the color green's association with ecology and environmental concerns. More importantly, the Japanese approach defines forest functions in a positive manner in contrast to the American approach that defines those functions by what they are not.

This paper begins by setting the context by means of a brief historical perspective, a rationale for broadly based social assessments, and the outlines of the analytical method. The next major section describes case studies in Japan and the United States. The final section discusses the similarities and differences between the case studies' results and suggests some avenues for application and further research.

2. Context

2.1 Historical perspectives

That social values change is not a new phenomenon in either Japan or the United States. The changes each nation's society has experienced are likely related to population growth, relative resource scarcity, as well as the bio-physical nature of the resources themselves. In addition, cultural, social, political, and economic trends and shocks have contributed to the shaping of social valuation of forests.

2.1.1 Japan

Japan is a crowded place, and has been so for nearly all of its 20 centuries of civilization. People's survival required manipulating nature to create rice fields for food, communal forests for fuelwood and green manure, and institutional forests for the production of timbers for castles, temples, and other structures. The land has been exploited for a broad range of goods and services for centuries and there are few forested areas that represent undisturbed (by humans) conditions. Long periods of peace in the Tokugawa era brought expanding populations, a shift toward urban living, and the creation, on a grand scale, of temples, shrines, and castles (Totman 1983). Because the expanding cities and towns were made of wood, they burned frequently and required reconstruction. The monumental works of the religious and ruling elite consumed huge quantities of the best timber. Enormous volumes of wood planks were used to set the edges of rice terraces.

Early on, Japan's rulers and villagers recognized the pressures being exerted on the forest. As fertile flatlands were stripped of trees to support a growing population, the increasing demands for timber, fuelwood, and green manure were shifted to the forested mountains (Tokoro 1980). Denuded slopes failed frequently, destroying villages and farms. By the mid-17th century, cultivation of tree plantations had initially begun as land protection measures and then expanded to deal with timber scarcity (Japan FAO Association 1994). Totman (1983) suggests that, by the early 19th century most old growth forests were gone and that Japan was "well into an era of plantation forestry" (page 276), pre-dating similar North American efforts by more than 150 years.

In the 20th century, especially after the exploitative period of the Pacific War, rising per capita wealth, increasing environmental concerns, and declining profitability of public forest timber management in Japan ushered in a new appreciation among the public and among resource managers for non-timber forest functions.

By tracking the treatment of nature in the classical literature dating from AD 700 and newspapers dating from 1880, Inoue and Ishibashi (1994) suggest that Japanese societies actually pass through cycles in terms of relative emphasis in forest use. Depending on the socio-economic and political circumstances, forest production would dominate in some periods, while spatial or spiritual uses would dominate in other periods. Spatial use in this context refers to use of the forest in place, typically as soil and water conservation forest. Spiritual use refers to aesthetics, outdoor recreation, and religious use. If spiritual and spatial uses are translated to their modern equivalents, then green resources have as much historical precedent as fiber exploitation.

2.1.2 United States

In the United States, forest resources were crucial to the Native Americans. We are only recently coming to appreciate the high degree of vegetation management practiced by many of the tribes prior to European colonization and subsequent colonial displacement and destruction of the native culture (MacCleary 1994, Williams 1989). Today's Americans have a highly romanticized view of Native Americans and their relationship with the environment. But Native Americans, like the feudal Japanese, lived as much at the mercy of nature as in harmony with it.

European colonists viewed the vast forests of the eastern United States as a source of timber and food resources and as an obstacle to their efforts to remake the New World into the image of the Old World. Forests provided construction timber and fuelwood for homes and industry as well as the land to be cleared for agriculture. As the population grew in the mid-19th century, vast acreage's in the eastern half of the country were converted to farms.

Forest exploitation accelerated when the United States industrialized in the late 19th century. Timber remained the primary construction material for the industrial age. Timber provided fuel, cross-ties, and structural support for the railroad system that spanned the continent. Devastating fires followed harvesting activities throughout the nation and helped spark the conservation movement.

The conservation philosophy emerged as part of the Progressive Era as the 20th century began. While its initial focus was the public forests, conservation ideals and scientific forestry gained support in the private sector as World War II approached. The nation's forests provided enormous quantities of wood to meet the demands associated with rapid economic growth in the postwar years.

As in Japan, the environment became a popular cause in the 1960s as Americans began to realize the full costs associated with their standard of living. In the 1970s and 1980s, environmental issues were important foci of public policy and private initiatives. As the century draws to a close, Americans are demanding that forests be managed for much more than timber. Like their Japanese counterparts, Americans want more green resources and they want their forests to be managed sustainably for the full range of forest functions.

American literature, too, suggests that the path to the present has not been as linear as the above historical summary may suggest. Like the Japanese, Americans have placed different emphasis on forest functions over the course of their history. In his book, "Americans and Their Forests," Williams (1989) notes that many American authors of the early 19th century were romantics in their writings about the forest. In the industrial age, environmental and economic concerns were important. Finally, forest preservation has been an important literary theme beginning with H.D. Thoreau and his contemporaries. The forestry profession has modified silvicutural systems to meet those changing demands (Helms 1998) and debated its social roles and the ethics of land management. These discussions are symptomatic of societal changes.

2.2 Rationale for social assessments

These days, when United States foresters talk about public participation, they are usually referring to direct public involvement in decision-making processes. The Japanese concept of public participation is different. The public's chief roles in Japan are that of volunteer data gatherers and workers. For example, in the Third National Survey of the Natural Environment, over 100,000 people collected data for the Environment Agency related to species distribution and abundance (Environment Agency 1989). City neighborhoods are organized into groups to plant and care for trees. Youth groups and company employees take charge of tending publicly-owned plantations and young suburban secondary forests (Shigematsu 1991). Conservation volunteerism is an important social activity in Japan.

The Japanese Forestry Agency is only now experimenting with public participation in forest planning. The USDA Forest Service has been working at this for 25 years or more, so that there is an increasingly broad base of citizen participation in formulating and evaluating alternative plans. This may well have reached its highest form in New England, where a town meeting approach to public decision making expects substantive public involvement. The American system is not flawless. There would likely be problems with adopting United States methods in Japan, with its history of top-down decision making. Local people and interest groups are generally unaccustomed to this type of participation. Moreover, many Japanese resource managers seem to feel that local elected officials provide the best expression of local concerns. Others are generally not willing to open the process to the public.

Japanese resource agencies are in the best Confucian tradition of a decision making elite. This heritage predates the Meiji Restoration and continues to this day. A similar system was in effect in the US during the American Progressive Era and beyond. American resource managers, armed with the scientific method and professional credentials, made the management and policy decisions affecting natural resources. Today's broader based public participation for public forest planning is a relatively new idea.

At least for the U.S. National Forests, there is a more or less well-established process for engaging interested organizations and individuals in the development and selection of a management plan for an individual administrative unit. The typical process involves outreach activities to identify issues and stakeholders, interaction between forest planners and interested parties, modeling of alternative strategies and tradeoffs, public review and agency response, and, finally, a decision by the unit administrator. Individuals and organizations that disagree with the decision or find fault in the process may subsequently initiate lawsuits to change the decision. The process, in United States cultural terms, is as important as the outcome.

In the private sector, the increasing interest in third party certification of sustainable forest management will compel private decision-makers to be sensitive to social demands. For example, the Forest Stewardship Council's Principles and Criteria require candidate landowners to incorporate social impact considerations in management planning and to consult with people and groups affected by their management (Forest Stewardship Council 1998). In fact, the entire concept of third party certification is a response to a social demand by interest groups for sustainable, holistic forestry. The Principles and Criteria represent the means by which resource managers can meet those demands. In the United States, the American Forest and Paper Association's Sustainable Forestry Initiative commits its members to environmentally sensitive management and represents another response to social demands for a broader range of forest functions.

For decision making at all levels, if social sentiments are important, then we need to assess those sentiments and to suggest ways to employ the information in the forest planning and policy process. Reliance on local officials and interest groups provides only limited understanding of social demands for forest functions. Inputs from those officials and groups are important, but there are no guarantees such inputs adequately represent the sentiments of society. Some US National Forests have recognized the need for broader input and have initiated studies of survey research applications for their planning process.

That society's valuations of and demands for forest functions change over time is clear. What becomes important for resource managers and policy makers is how to assess those values, how to anticipate future shifts, and how to use the information to modify management and policy to meet those demands.

2.3 Methodological framework

The basic techniques for eliciting preferences and values seem well developed. What seems to be lacking in the forestry sector are ways to connect what we understand about preferences and values with the management choices designed to produce those values. In this section, we outline two approaches to eliciting these preferences.

2.3.1 Ranking preferences

Through survey techniques, managers and decision-makers can determine people's preferences and concerns for a wide range of forest functions. The typical questionnaire asks respondents to numerically rank their preferred choices or concerns from a pre-determined set. In a recently published example, Takeda et al. (1997) surveyed local residents regarding their concerns about the functions associated with a recreation area. Often, the survey instrument includes additional questions about demographic characteristics and personal beliefs to provide some insight into the variation in responses. While these additional questions aid our understanding of which factors influence the rankings, we are mostly concerned with the rankings themselves. The results provide information about the ordinal rankings of social demands for the sampled population as well as information about the distribution of the rankings. Simply asking people to rank choices, however, will not provide information about the intensity of these preferences.

2.3.2 Analytical Hierarchy Approach

Because social demands are numerous and often conflicting, we believe that the AHP developed by Saaty and Alexander (1989) can be modified to provide management planning guidance to decision makers. In a classical application of AHP, survey respondents are asked to compare the relative importance of pairs of criteria that are relevant for selecting the preferred strategy. Their responses can be transformed into a vector of weights indicating the relative importance of each criterion. The alternative strategies are then evaluated in terms of their relative ability to meet each criterion; the resulting matrix is combined with the vector of criteria weights to produce a strategy evaluation vector. This final vector shows the relative score of each management regime in terms of its ability to satisfy all of the decision criteria.

The case studies discussed below focus only on establishing the relative importance of the criteria, not in the evaluation of specific management regimes. This represents the initial stage of the AHP analysis. Although standard survey methods can provide insights into the ordinal ranking of criteria, as described in section 2.3.1, the method outlined here delves more deeply into intensity of preferences and yields cardinal measures for further analytic work.

3. Case Studies

Our studies share similarities and differences in methodology. Sugimura's (1993) purpose was to explore methodology. Cooksey (1994) designed his survey to address the broad questions of the value of conservation easements; determining the relative weights of forest functions was a secondary objective. Sugimura's sample population was drawn from two urban areas in Japan with very high population densities and one less densely populated island. The citizens of these areas were relatively disconnected from the realities of forest resources management. By way of contrast, Cooksey's sample population was located in a very rural county, in extreme northeastern United States, adjacent to the border of Quebec, Canada. The county is heavily forested (95%) and dependent on timber and forest-based recreation as its economic base. Of course, there are the obvious cultural differences between the United States and Japan.

3.1 Perspectives from Japan

Sugimura (1993) asked respondents to rank the forest functions in each of two groups. Four major forest functions comprised Group A; Group B functions were specific land and water conservation functions (Table 1).

Within Group A, the four major functions were ranked similarly in the three study areas. Timber production was the least favored function. Within Group B, Amami respondents answered somewhat differently, reflecting the local history of typhoon related floods.

In the first stage AHP paired comparisons scores (Table 2), respondents in the three regions were relatively consistent in the weights assigned to each of the four Group A forest functions. Land and water conservation had the highest weights and timber production had the lowest weights.

3.2 Perspectives from the United States

As part of a research project on the value of conservation easements for protecting forest functions, we conducted a mail survey of 1,600 Coos County, New Hampshire resident households during the spring of 1994 (Cooksey 1994).

Table 1. Mean score of respondents' preferences for forest functions in three survey regions in Japan (0 indicates least preferred)

Function	Region		
	Tokyo	Kobe	Amami.
Group A			
Timber	0.00	0.00	0.00
Scenic Beauty	0.45	0.40	0.74
Wildlife	1.29	0.76	1.55
Land & Water	1.45	1.59	1.98
Group B			
Water Quality	0.00	0.00	0.00
Slope Failure Mitigation	0.56	0.56	0.38
Water Supply	0.83	0.88	0.56
Flood Protection.	0.83	0.89	0.75

Table 2. First stage AHP criteria normalized ranking scores for four forest functions in three survey regions in Japan

	Function			
	Timber	Scenic Beauty	Wildlife Conservation	Land & Water Conservation
Region				
Tokyo	0.108	0.188	0.338	0.366
Kobe	0.099	0.186	0.264	0.457
Amami	0.085	0.199	0.325	0.391

Among other questions, we asked respondents to assign one of four priority rankings to two groups of four forest benefits. The ranks ranged from 1 (most important) to 4 (least important). Group 1 benefits were a) timber and forest products supply, b) conservation of scenic beauty, c) conservation of air and water quality, and d) access to outdoor recreation. Group 2 benefits were a) wildlife and habitat conservation, b) cultural and spiritual benefits, c) educational and scientific benefits, and d) protection of threatened and endangered species.

In a second battery of questions, we asked respondents to choose between two benefit categories in a series of paired comparisons. The format of the questions followed methodology suggested by Sugimura (19[illegible]3). Respondents were asked to compare two paired benefits and indicate one of four levels of importance of the preferred benefit: 1) equally important, 2) a little more important, 3) more important, and 4) a lot more important. The idea of the paired comparison was to force respondents to choose between two benefits and then to indicate how important that choice was.

We also presented a photo, Contingent Valuation Method (CVM) scenario, and dichotomous choice (yes-no) question with a stated market value. Two situations were described: 1) continued access to market and non-market benefits with an easement and 2) loss of access to those benefits as a result of residential-industrial development without an easement. The answer to the dichotomous choice question was intended to constitute a respondent's willingness to pay (WTP) bid for the bundle of benefits-amenities described.

Of the 1,600 surveys mailed, 296 were returned (21.4%) and 287 were useable. Many respondents were confused by the paired comparison question format. The number of useable surveys available for that analysis was reduced from 287 to 155.

Within Group A, respondents favored conservation of air and water quality (Table 3). Nearly 50% of respondents ranked this function as their most important concern. While the means and standard deviations for timber supply and scenic beauty were very close, there is a difference in the distribution of responses. Recreation access ranked lowest in this group probably because there is a long tradition of public access to private forests in the region.

Within Group B, respondents favored wildlife and habitat conservation and habitat for threatened and endangered species. These results reflect a general concern in the region for wildlife issues. They are consistent, too, with another study on the willingness to support public purchase for land to protect wildlife habitat in New England (Echelberger et al. 1991). (A major wildlife preserve was recently established in the region through the cooperative efforts of the forest products industry, environmental groups, and units of the local, State, and Federal governments). Habitat conservation may have also been favored because it may be viewed as less abstract than education, scientific, cultural, and spiritual values.

Table 3. Respondents' priorities for forest functions in Coos County, NH, USA

			Level of Importance-% Rank Response			
Function	Mean	Std. Dev.	4	3	2	1
Group A						
Air and Water Quality	3.557	0.709	12.4	14.88	24.79	47.93
Timber Supply	2.411	1.166	30.86	21.81	22.63	24.69
Scenic Beauty	2.391	1.024	23.05	32.1	27.57	17.28
Access for Recreation	2.209	1.056	32.1	30.04	22.63	15.23
Group B						
Wildlife & Habitat	3.557	0.709	1.65	7.82	25.1	65.84
Habitat for Rare Species	2.558	1.022	18.18	19.42	24.79	47.93
Educ.& Science	2.296	0.883	17.7	45.68	25.93	10.7
Cultural & Spiritual Exp.	1.601	0.872	59.67	26.75	7.41	6.17
n= 287						

The results of the paired comparisons were converted to preference ratio scores that could be employed in the first stage of an AHP analysis. Due to a design flaw in the questionnaire, not all pairs were compared. For this part of the analysis, we assumed that respondents were neutral in their preferences between air and water quality and timber and between scenic beauty and recreation access. Because our purpose is to explore the potential of the method, the errors are not fatal.

Again, the Air and Water Quality criterion or function was viewed as the most important. However, unlike the group ranking results (Table 3), Timber and Forest Products Supply was ranked lowest among the criteria (Table 4).

Table 4. First stage AHP criteria normalized ranking scores for four forest functions in Coos County, NH, USA

Criterion	Weighted Score
Air and Water Quality	0.333
Scenic Beauty	0.241
Recreation Access	0.260
Timber and Forest Products Supply	0.176
n=155	

The mean WTP value estimated by logistic regression and numerical integration was 31.23 $US per county resident. Respondent income had a positive influence on WTP while years of residency, children living at home, and bid price had negative influences. Statistically insignificant variables were prior knowledge of conservation easements, participation in a conservation easement program as well as respondents' age, gender, and education.

4. Discussion

Given the above case studies, we can compare Japan and United States respondents to identify similarities and differences as well as explore some potential applications, especially of modified AHP results.

4.1 Comparing the results

Given the small samples and geographic limitations of our surveys, we cannot generalize about how all Japanese and Americans value various forest functions. However, despite major cultural differences, it was clear that our survey respondents were very supportive of green resources functions.

In the *Group A* rankings (Tables 1 and 3), some aspect of environmental protection (land and water conservation or air and water quality) was the foremost concern in both case studies. The American respondents were more sympathetic toward timber production, probably because their livelihoods were more directly connected to the forest products industry than those of the Japanese respondents.

Although the Group B rankings addressed very different forest functions, it seems that the more concrete functions were favored (Tables 1 and 3). This suggests that survey researchers should be careful to design their instruments to avoid bias against abstract categories.

In our modified AHP results (Tables 2 and 4), survey respondents were consistent in favoring green resource functions and naming timber production as the least important criterion. Because of design flaws in the United States questionnaire, we cannot explain why United States respondents were inconsistent in their valuation of timber production (Tables 3 and 4).

4.2 Applications for management

The methods employed in this study have application at the strategic, tactical, and operational levels of forest resources management. For example, early in the land management planning process, standard rankings (Tables 1 and 3) decision makers and analysts can assess social preferences to help guide the identification of issues and the development of alternative strategies. The rankings could also be used to establish the ordering of preferences for pre-emptive goal programming approaches to resource management analysis.

The weighted criteria approach (Tables 2 and 4) can be linked to land management. By multiplying a land quality index value (measuring how well a parcel of land satisfies the forest function) by the criterion weight, the land parcel can be assigned the function with the highest score. Forest level concerns can be addressed by applying the normalized weights to cumulative distribution curves of land qualities (Sugimura 1993). Finally, the weighted criteria can be used to weight objective function deviation variables in a standard goal programming formulation, assuming that the other conditions governing the use of goal programming can be met.

4.3 Concluding remarks

Respondents in each country indicated clear preferences for non-timber forest values. The extent to which forest managers respond to these preferences will impact the flows of forest goods and services. Modifying forest management approaches to incorporate a broader spectrum of values requires that managers understand what those values are and how management alternatives impact those values. Survey research can provide important insights about these social values, but more research is needed on how to incorporate the resulting information into the planning process.

Incorporating these emerging social values into our decision making implicitly recognizes that our societies have some stake in management policies governing public and private forests. Thus, whether directly or indirectly, the public is participating in resource decision making. Cultural differences will affect the means by which forest managers in Japan and the United States assess and incorporate public sentiment into forest management.

Although Japan and the United States are currently the leading industrial nations of the world, they have attained that status via very different histories and cultural contexts. Because our studies were conducted independently and for different purposes, we cannot make inferences about the effects of culture on the demands for forest functions. A single survey administered to similarly chosen samples in each country would provide insights into these effects.

5. References

Cooksey, R.A. Jr. 1994. Conservation easements as a public good: The willingness to pay for forest land benefits protected with conservation easements in the northern forest lands, New Hampshire. M.S. Thesis. University of New Hampshire, Durham. 140p.

Echelberger, H.E., Luloff, A.E., and Schmidt, F.E. 1991. Northern forest lands: Resident attitudes and resource use. USDA Forest Service. Northeastern Experiment Station. Research Paper. RP-NE-653.

Environment Agency. 1989. Aspects of nature: National survey on the natural environment, japan. Kasumigaseki, Tokyo, Japan. 18p.

Forest Stewardship Council. 1998. http://www.fscusne.org/draft/index.htm

Helms, J. A. 1998. Silviculture: Past, present, and future. Journal of Forestry 98(7): 1.

Inoue, M., and Ishibashi, S. 1994. The historical change of forest use in Japan. P. 90-95 *in* Proceedings of the IUFRO International Workshop on Sustainable Forest Management. October 17-21, 1994. Furano, Hokkaido, Japan.

Japan FAO Association. 1994. Forests and forestry in Japan. Japan FAO Association, Tokyo. 65p.

MacCleary, D.W. 1994. American forests: A history of resiliency and recovery. Forest History Society, Durham, North Carolina. 58p.

Saaty, T.L., and Alexander, J.M. 1989. Conflict resolution: The analytic hierarchy approach. Praeger. New York. 252p.

Shigematsu, T. 1991. Conservation and management of suburban forests by the citizens. Shinzansha, Tokyo, 74p. (in Japanese).

Sugimura, K. 1993. Evaluation of forest functions by local people: Development of a research method. P. 555-566 *in* Proceedings, Advancement in Forest Inventory and Forest Management Sciences, IUFRO Conference, Seoul, Korea, Sept 20-25, 1993.

Takeda, M., Ogawa, S., and Inoue, S. 1997. A questionnaire investigation on erosion control works among neighboring residents. Bulletin of the Ehime University Forest. 35: 55-76. College of Agriculture, Ehime University, Matsuyama, Japan.

Tokoro, M. 1980. Studies on the forestry history in the rrecent times. Kobunkan, Tokyo, 858p. (in Japanese).

Totman, C. 1983. From exploitation to plantation forestry in early modern Japan. P. 270-283 *in* Steen, H. K., ed., History of Sustained Yield Forestry: A Symposium. Western Forestry Center, Portland, OR. Oct. 18-19. Forest History Society and the International Union of Forestry Research Organizations, 270-283.

Williams, M. 1989. Americans and Their Forests: A Historical Geography. Cambridge University Press, New York. 599p.

PART 2

CONTRIBUTION OF MODELING EFFORTS TO FOREST SECTOR ANALYSIS

Forest Sector Modeling

- Current state and promise for the future -

Darius M. Adams and Richard W. Haynes

Key words: Supply-demand relations, forest sector modeling, forest economics

Abstract: To understand current trends in forest sector modeling and to speculate on potential future developments, we consider the form and nature of today's forest policy debates. Expansion in the breadth of system attributes to be considered and the need to understand the interconnectedness of these attributes in policy decisions are major concerns for sector modeling. This paper reviews forest sector market models classified by: (1) number of endogenous market levels, (2) treatment of spatial market relations, and (3) assumptions about producer and consumer knowledge of the market and the future. Models are considered in terms of their characteristics, decisions in developing their structure, how the structure affects the things that can be made endogenous, and the implications of these features for conducting policy studies. We conclude that future efforts should: (a) develop spatial, intertemporal optimization models with multiple market levels; (b) examine constraints on adjustment flexibility in the intertemporal welfare maximization format; (c) develop methods that allow joint simulation of market and resource models retaining their respective detail; (d) directly embrace the "messiness" of real world market-resource conditions; and (e) be explicit about the uncertainty inherent in policy scenarios and projections.

1. Introduction

Unlike times in the past, the needs of policy analysis are presently driving the development of forest sector models. To understand current trends in forest sector modeling and to speculate on potential future developments, one needs to consider the form and nature of today's forest policy debates. From the sector modeler's standpoint, expansion in the *breadth of system attributes* to be considered and the need to understand the *interconnectedness* of these attributes in policy decisions are aspects of major concern. Around the world, countries are coming to recognize a broader array of values in, and outputs from, the forest (beyond commodity production) and the complexity of the biophysical relations that support this wider spectrum of forest characteristics. Forest policy is being supplanted by more inclusive environmental policies, and policy mechanisms are being considered and adopted that act to regulate both actions and outcomes in far more specific ways than in the past.

In this milieu, forest sector models have also become more comprehensive and complex to remain useful to policy makers. This has generally meant that models have been modified or developed to:

(1) account for a broader array of *market and resource attributes* and show information that will provide a clearer picture as to *market and resource conditions*. For example, models of the resource base may be expanded to give estimates of wildlife habitat area and quality (attributes) by projecting details of forest type and age or size class distributions (conditions), rather than total timber volume or area alone. Additional market attributes that may be important to some policy studies might include multiple market levels (stumpage, log, product[1]) or multiple products. Descriptions of market conditions might be expanded to include more complete accounting for producer profitability (beyond price and quantity sold).

(2) account for *spatial-geographic relations* within the system. This could involve spatial patterns of habitat for wildlife population analysis or interactions between trading partners in an analysis of transboundary policy effects.

Today's policy environment is also characterized by fluidity: rapidly changing political conditions (e.g., shifts in governments and boundaries due to the combination or dissolution of political units) and changes in market and institutional conditions (e.g., mergers and spinoffs in corporations or "privatization" of previously public forest land ownerships) that can shift the analytical context dramatically. This propensity for change requires models that are highly flexible in their basic structure. The modeling issue here is not the ability to predict such changes but to be able to provide credible simulations of policy scenarios that involve such dramatic shifts from historical conditions.

Accompanying these policy-related needs for model development has been a confounding trend, in some countries, related to the availability and quality of data with which to construct and calibrate models. As both public and private sectors have become increasingly conscious of the costs of management and as institutions have changed, data collection and support of decision analysis have generally declined. In some cases data on critical elements of forest sector activity are no longer collected or no longer published. In instances where alternative sources are available, the data are often of uncertain quality or incompatible with past data. Further, some institutions that have provided important ancillary input to forest sector modeling in the past (such as long-range macroeconomic projections) no longer support these services.

[1] Stumpage hereinafter is used to refer to standing timber; logs to felled and bucked timber either in the woods or at some point between woods and mill depending on the application. Product, unless otherwise noted, refers to some good processed from logs such as lumber, plywood, pulp, panels, etc.

How have forest sector models adapted to these changing conditions for policy and policy analysis? The next part of this paper attempts an overview and classification of one large group of forest sector models. The discussion considers the attributes of models that influence their use in policy studies. As the extremely valuable summary by Nabuurs and Paivenen (1996) amply illustrates, there are a great many types of forest sector models that address many aspects of forest sector policy. To keep the comments in this section manageable, to allow some specificity, and to stay within the authors' area of expertise, attention is limited to the set of forest sector models often termed "market models." These models embrace both production and consumption behavior and explain the establishment of price (or the rationing of output to potential users). In this latter aspect they differ from the computationally simpler (and older) timber trend models[2]. The large class of "resource use" or "harvest scheduling" models is not considered, though the boundary between these groups and market models is at times indistinct, and there are instances where market and resource use models can be employed jointly to examine policy questions.

A final section offers some thoughts on the potential for future development of forest sector models.

2. Existing Market Models: Structures, Functions and Policy Applications

Past surveys of forest sector (market) models have classified studies in the literature based primarily on analytical method (e.g., Buongiorno [1996] considers econometric, linear programming and systems dynamics categories). Given the focus of the present discussion, however, an alternative approach linking model attributes to policy was needed. As described below, the proposed taxonomy employs three dimensions: (1) the number of market levels endogenous to the model (one-multiple), (2) the treatment of spatial market relations (spatial-non-spatial) and (3) assumptions made about producer and consumer knowledge of the market system and the future (limited-myopic--perfect-intertemporal optimizer).

Using this classification, models are considered in terms of their basic characteristics (as these are related to market levels, spatial trade flows and future knowledge), some of the key decisions involved in developing their structure, how the structure affects the things that can be made endogenous in (explained by) the model, and some of the implications of these features

[2] Timber trade models extrapolate volumes of production and consumption separately under assumptions of constant real prices. Inferences can be drawn about the likely direction of equilibrium price movements from the base period given the relative positions of consumption and production, but nothing can be said about the prospective future levels of prices. See for example, USDA Forest Service (1982).

for conducting policy studies. The focus of this paper is on the more recently developed models; a full historical review of the forest sector market model literature was not attempted. Similarly, in light of the virtually unlimited array of possible policies that might be relevant in any circumstance, observations on policy application are intended to be indicative and not exhaustive. As an organizing reference, Figure 1 lays out the proposed three-way classification and provides examples of models from the literature.

LIMITED KNOWLEDGE / MYOPIC		
	SPATIAL	NON-SPATIAL
MULTIPLE MARKET LEVELS	Adams (1985) CGTM (Cardellichio et al. 1989) FOHOW (Schwarzbauer 1992) FORSIM (Veltkamp et al. 1983) GTM (Kallio et al. 1987) Lönnstedt (1986) NTM (Trømborg and Solberg 1995) PELPS (Zhang et al. 1993) TAMM (Adams and Haynes 1980,1996) Yukutake (1985)	Adams (1974) McKillop (1967) Robinson (1974)
SINGLE MARKET LEVEL	Adams et al.(1986)	McKillop (1969) Newman (1987) Yukutake et al. (1994)
PERFECT KNOWLEDGE / INTERTEMPORAL OPTIMIZATION		
	SPATIAL	NON-SPATIAL
MULTIPLE MARKET LEVELS	----	----
SINGLE MARKET LEVEL	----	Berck (1979) TSM (Sedjo and Lyon 1990) FASOM (Adams et al. 1996a) Sohngen and Mendelsohn (in press)

Figure 1. Examples of forest sector market models classified by number of market levels considered, treatment of spatial relations, and assumptions about future knowledge of market agents (Notes: ---- indicates no examples known to authors)

2.1 Market level

Market level in a forest sector model refers to one of the several processing steps in the continuum of product flow from forest to final consumer. At each level there are (derived) demand and supply relations that determine a set of prices at producing and consuming points. A good at one ("lower") market level is a factor in the production of the good at the next "higher" market level. Thus there is a chain of product and factor markets. For example, a single market level model might address just the market for softwood lumber. A multi-level model might include softwood lumber and the markets for logs and stumpage (moving "down" the processing chain) or new construction (moving "up" the chain) depending on the focus of the application.

This model attribute is relevant to policy applications in at least two ways. In current policy deliberations, there is often uncertainty over whether some objective can be more readily achieved through actions aimed at one or several market levels. For example, in an analysis of the need for a countervailing duty on an imported forest product based on subsidization of timber costs by the exporter, alternative policy actions might include a direct tariff on the imported product (an action in the product market) or shifts in the exporter's policies to increase the costs of timber (an action in the timber market) or some combination. These same questions arise in the debate over appropriate policies to mitigate global warming through actions in the forest sector, viz., subsidies for forest plantations versus taxes on short-lived forest products that release carbon during decomposition after disposal.

Levels of markets also correspond to different groups of consumers and producers. Thus single or multi-level market models give different options for assessing the distributional impacts of alternative policies. In a single level model of a competitive market, welfare impacts measured, for example, on producers include quasi-rent impacts on producers both in that market and all lower (factor) markets (see Just et al. 1982). Thus in a model of the lumber market, welfare changes for producers include changes for log and stumpage producers as well. If it is important to explicitly identify welfare effects across specific populations, a multi-level market model may be more useful.

As suggested in Figure 1, the largest number of recent forest sector models have been developed to address more than one market level. Most commonly this involves one or more horizontally linked product markets (such as lumber and panels or various grades of paper and paperboard) and the log or stumpage level. In a very few instances, three levels are included--as in the NAPAP application of the PELPS modeling system where pulpwood, market pulp, and paper and paperboard markets are included (Zhang et al. 1993, Ince 1994).

But while many studies include the stumpage or log level together with a product level, treatment of the production linkage of the log-stumpage factor to the product level is far from uniform. Since logs (and not stumpage directly) are the input to processing of both solidwood and fiber products even in fully integrated operations, there is potentially some need to differentiate between the two goods.

On the supply side, it is uncommon to find both log and stumpage supply treated separately in a model. Log supply relations are often specified in the same way as stumpage supply relations by simply adding the cost of logging and hauling to a price-dependent stumpage supply equation and shifting stumpage volume to log volume (see, for example, Cardellichio et al. [1989] and Nagata and Tachibana [1994]). This may or may not be an adequate procedure depending in part on whether "logging and hauling services" (that convert stumpage to logs) exist in perfectly elastic supply and use stumpage in fixed proportions with other factors to produce logs. And even in cases where stumpage supply alone is estimated, delivered log price is most frequently found by simply adding log and haul cost (see, for example, Adams and Haynes 1996).

On the demand side (i.e., the derived demand for logs or stumpage by product producers), three distinct approaches to modeling are available.

(1) Some studies have estimated separate relationships for product output and log demand assuming variable proportions in the use of logs and other factors (e.g., Adams 1974, Yukutake 1989, Nagata and Tachibana 1994). In most instances these relations (product supply and factor demand) have not been restricted by the usual conditions of symmetry and reciprocity (see Chambers 1988). Studies using this approach have generally been simultaneous equation models with equilibrium found by simultaneous solution of the system.

(2) The largest group of studies assumes a fixed proportions/Leontief production relation between product and logs. In some models this relation is assumed to hold for all factors and the entire product supply process is built on this basis (see, for example, Kallio et al. 1987, Zhang et al. 1993). In other cases, fixed proportions hold only for the linkage between product and logs, while other factors may be used in variable proportions (though in most cases the prices of these factors are assumed fixed). Studies using this approach may modify the entire supply representation to recognize this special input-output relation (e.g., Lönnstedt 1986 and Adams and Haynes 1996) or simply impose it on the wood factor without modifying supply (e.g., Adams and Haynes 1980). Equilibrium solutions for models in this group are usually found via some form of net market surplus[3] maximization. If stumpage or log supply is price sensitive, the fixed proportions assumption for the wood input allows computation of the area under the product supply curve as the sum of the area under the wood (stumpage or log) supply curve weighted by the input-output coefficient plus the appropriately weighted area under the marginal cost relations for other factors.

(3) A further alternative would involve explicit representation of variable proportions in the combination of factors and at least one factor price (presumably logs or stumpage) varying with quantity consumed. In the net surplus approach to equilibrium solution, this generalization would

[3] This would be the sum of producers' and consumers' surplus, usually computed as the consumers' willingness to pay (the areas under the demand curves) less producers' total variable costs (the areas under the supply curves).

require modifications in the usual surplus objective function and the inclusion of the production function (or some approximation) as a constraint in the problem. To date there appears to have been no use of this approach in forestry. Önal et al. (1991) describe an application in a fishery using variable input proportions, but assume that factor prices are fixed.

2.2 A spatial model

A spatial model, for the purposes of the present discussion, is one that employs at least two supply regions and two demand regions, so that bilateral flows between pairs of producers and consumers can be identified. There are many examples of forest sector models that include multiple demand or supply regions but not both. These models provide spatial information on the location of production or consumption, but because flows between pairs of trading partners are not explained, they are considered to be nonspatial models by the foregoing definition (for reasons related to policy analysis applications as noted below). Models that include multiple market levels need not be spatial at all levels. For example, the TAMM model (Adams and Haynes 1996) is spatial at the product level (lumber, plywood, OSB) but not at the stumpage or log level, while the NAPAP model (Ince 1994) is spatial at the wood fiber (pulpwood and chipped residue), intermediate product (market pulp), and product (paper and paperboard) levels. In the case of the TAMM model, domestic trade in sawlogs is ignored in light of the limited interregional movement of sawlogs in the U.S.

Spatial and nonspatial models differ in several ways that may be important for purposes of policy analysis. There are, of course, two obvious limitations to nonspatial models: (1) regional detail on policy impacts is only available on the side of the market that is spatially disaggregated (if demand regions are aggregated, the demand region effects of a supply policy change can not be directly estimated); and (2) changes in costs or conditions affecting trade flows between specific trading pairs (such as trade restrictions and transportation costs or constraints) can not be directly simulated. In a similar fashion, simulating the impacts of policy changes originating on the aggregated side of the market may require some exogenous (extra-model) analysis if the policy change is not uniform across all regions in the aggregate.

Some forms of nonspatial models may also entail fairly strong assumptions about interregional price relations. For example, Adams and Haynes (1987) identify a "multi-region, nonspatial price equilibrium" model form that has been employed in several forest sector models (e.g., Binkley and Cardellichio 1983, Sedjo and Lyon 1990, Adams et al. 1996b). As most commonly applied, there are several supply-producing regions and a single demand region. Prices in the producing regions are linked to price in the aggregate demand region by transport costs or price adjustments that may be fixed or variable over time but are, in either case, exogenous. In equilibrium, price differentials between producing regions are fixed by the differences in these transport costs. In effect there is only one price in this model form. As a result, a policy change initiated in one supply region will produce price changes in all regions, but the price differentials between regions (in any given period) are always the same as the differences in transport costs. There may be circumstances, however, when one might anticipate that the extent of price change would vary between the region of the initial policy shift and other regions.[4]

A further distinction between spatial and nonspatial models is the ability of the former to project the opening and closing of trade and specific trade routes in response to policy shifts based on relative costs and prices. This may be a useful capability in analysis of obviously extreme policy measures. But even modest changes can alter trade patterns or the desirability of trade.

Though the tabulation of studies in Figure 1 does not purport to be complete, it is fair to say that the largest number of recently developed forest sector models within the myopic group have employed some spatial format. These vary widely in scope and complexity. TAMM, GTM, CGTM, and PELPS applications have all used multiple demand and supply regions and employ a direct "net social surplus" maximization to arrive at an equilibrium. Others, such as Lönnstedt (1986) and FOHOW (Schwarzbauer 1992) typically employ only two regions. In this latter case, the second region is typically the "rest of the world", though in principle more regions could be used.

Intertemporal optimization models, in contrast, have generally employed nonspatial forms. Berck (1979) considered only a single region. Trade was not considered since the model dealt with the stumpage market level. The TSM (Sedjo and Lyon 1990), FASOM (Adams et al. 1996a) and Sohngen and Mendelsohn (in press) models deal with logs (so trade is possible), but all use the multi-region, nonspatial price equilibrium form discussed above. Given the computational difficulties in finding equilibrium solutions in these latter models, there is strong motivation to simplify model structure (to reduce the number of state variables or activities). As a partial consequence of these concerns, they consider only the log market level. But in either the global case (for the TSM) or the contiguous United States case (for FASOM and Sohngen and Mendelsohn), there is relatively limited trade in logs across

[4] This model form is attractive, particularly in intertemporal optimization models, because of its simplicity while still allowing the interregional propagation of the market impacts of policy changes. Adams et al. (1996b) offer one rationale for its use, but the fixity of interregional price differences remains a constraint.

the regional boundaries used in the models. Market impacts of policy changes are transmitted across regions at higher market levels (lumber, plywood, etc.) as a result of cost impacts on producers of these products, not through direct interactions in log markets. The multi-region, nonspatial price equilibrium form allows some pathway for transborder market impacts while considering only the log market and keeping the model simple.

2.3 Producer and consumer knowledge of the market system and the future

Producer and consumer knowledge of the market system and the future refers to the basic decision-making context of the model. With perfect knowledge, market agents can predict with certainty the influence of a production or consumption decision on prices and quantities in the market at any future time. With imperfect knowledge, nothing is known with certainty beyond the current time period. Decisions must be based on available (past) information on markets and on *expectations* of the future and about the future outcomes of decisions. We know that imperfect knowledge is the true state of the world. At the same time, we believe that economic agents do know much about their market contexts and capabilities and that they may be able to generate perceptive expectations that recognize interactions among markets agents (responses to their decisions). Thus projections with models built under the assumption of perfect knowledge may provide useful insights into actual market responses.

Specific examples may help to clarify the differences between the two model forms. A highly simplified myopic or imperfect foresight forest sector model dealing only with the timber sector might appear as follows.

[1] $d(t) = d_t(P_s(t), P_F(t), K(t))$ DEMAND

[2] $s(t) = s_t(P_s(t), I(t))$ SUPPLY

[3] $d(t) = s(t)$ MARKET BALANCE

[4] $I(t) = \sum_{i=1}^{n} A_i(t)V_i,$

[5] *CUT OLDEST AGE CLASSES FIRST* $[n, n-1, \ldots, etc.]$ *UNTIL*

[6] $s(t) = \sum_{i=lo}^{n} C_i(t)V_i$ INVENTORY MODEL

[7] *so that* $C_i(t) \leq A_i(t)$

where $d(t)$ is the quantity of timber demanded in period t, $s(t)$ is the quantity of timber harvested in period t, $P_s(t)$ is the timber price, $P_F(t)$ is the price of products using timber, $K(t)$ is capacity in the timber using industry, $I(t)$ is inventory volume at the start of period t, $A_i(t)$ is the area in age class i in period t, n is the oldest age class, V_i is the volume per unit area in age class i

(the yield function), $C_i(t)$ is the area cut in age class i in period t, lo is the smallest age class cut so that the area harvested produces just the volume needed to meet $s(t)$, and d_t, s_t are timber demand and supply functions, respectively. At the end of each period, the area harvested is replanted (placed in the youngest age class) and the residual area in all pre-existing age classes is advanced one age class.

Solution of this model is obtained by simply equating demand and supply in each period and solving for prices and then quantities. The demand and supply equations might contain expectational variables (such as future prices) but these expectations are taken as functions (distributed lags) of current and past levels of other variables in the model. Thus the solution advances one period at a time, with no link to future market developments

A corresponding perfect foresight model (borrowing from Johnson and Scheurman [1977]) might appears as:

[8] [*OBJECTIVE*]

$$\max \sum_{t=0}^{T} (1+r)^{-t} [\int d_t^{-1}(s(t), P_F(t), K(t)) ds(t) - F(X(c,t), N(p,t))]$$

SUBJECT TO:

[9] [*HARVEST*] $s(t) - \sum_{c} X(c,t) V_{c+t} - \sum_{p=0}^{t} N(p,t) V_{t-p} \leq 0$

[10] [*EXISTING STANDS*] $\sum_{t} X(c,t) \leq A_F \, \forall c$

[11] [*NEW STANDS*] $\sum_{f=1}^{T-t} N(t,t+f) - \sum_{c} X(c,t) - \sum_{p=0}^{t-1} N(p,t) \leq 0 \forall t$

where r is the discount rate, $d^{-1}(\bullet)$ is the inverse demand equation, $s(t)$ is supply or harvest volume, $X(c,t)$ and $N(p,t)$ are areas harvested in period t in existing stands that were c years old at the start of the projection (X) and areas of new stands created in year p after the start of the projection (N). Appropriate summations of the X and N would yield the sum of the $C_i(t)$ in the myopic problem and the total forest area (the sum of the $A_i(t)$). $F(\bullet,\bullet)$ gives the costs of current period harvest and regeneration of as a function of the areas cut and replanted from X and N. V_i is the harvest volume per unit area as in the myopic problem. A_F is a vector of initial areas by age class for the stands that existed at the start of the projection.

There are two sets of constraints controlling existing and new stands: (existing) the sum of the areas cut over the projection from each age class existing at the start of the projection can not exceed the initial area in that age class; (new) in each period no more area can be planted to new stands than has been cut from existing stands or mature new stands created since the start of the projection.

With perfect knowledge the producer becomes an intertemporal optimizer, maximizing wealth over the projection period (to time T). Buyer-seller interactions in a competitive market would maximize the sum of producers' plus consumers' surpluses; consumers' total willingness to pay less the costs of timber production (area under the "supply" curve) discounted over all periods, as shown in the objective expression.

One of the key changes in model structure associated with the shift to a perfect knowledge format is that production decisions can be represented in their most elemental form.[5] In the example, the supply curve of the myopic model has been replaced by its equivalent in the intertemporal context: a set of activities (X and N) that describe the areas harvested in existing and planted in new stands at each period in the projection and the costs (F) associated with those decisions. If the myopic model had contained an investment function that explained the cost or intensity of regeneration activity in planted stands[6], this same process could be represented in the intertemporal model. We would add a new activity, say, $M(t)$, to denote the investment in management at time t; the functions F would be augmented to include the costs of alternative levels of planting intensity; areas in N would have different planting intensities over time; and the yields in the harvest constraint would be modified to reflect the impacts of varying management intensity.

In principle one should be able to examine the same types of policies with either the myopic or perfect foresight models. With appropriate addition of equations and variables, neither form should be more flexible in this respect than the other. As a practical matter, however, the computational requirements associated with the perfect foresight model do raise some limits. As previously noted, solutions of the perfect foresight model become burdensome with large numbers of activities or state variables. Reducing the number forces simplification and aggregation, which can impact the model's ability to simulate some types of policies in as much detail as the user may wish. Common examples here are the need to aggregate geographic units (viz., grouping counties into states or prefectures into regions) and the inability to include a full set of silvicultural alternatives when the model embraces the timber sector. In this latter instance it is common to limit the timing of silvicultural decisions to the point of regeneration (a one time choice when the stand is planted) rather than allow a sequence of independent decisions over the life of the stand.

5 This would be true for consumption (factor use) decisions as well if appropriate to the modeling situation.

6 Newman and Ware (1993) provide an example of such a function estimated from historical data.

A further critical issue is whether the myopic and perfect foresight forms give different indications as to the impacts of identical policy changes. There are a few instances in the literature where there is an opportunity to compare the results of the two forms. For example, Adams et al. (1996b) examine the impacts of reductions in public timber harvest on U.S. timber markets using a perfect foresight model. Their results are qualitatively similar to earlier analysis with a myopic model in Montgomery et al. (1994). More recently, however, Sohngen and Sedjo (1998) consider this question in a carefully structured comparative study. They consider a simple, contrived stumpage sector like the examples in the preceding paragraphs. Both types of models are developed from the same inventory database. They are then used to simulate six example policy or exogenous shifts.

Sohngen and Sedjo's (1998) results show clear differences in some of the cases but not in others. Since the demand relations are fixed and imposed in both models, differences derive only from producer actions. Given the strong investment efficiency focus of producer behavior in the perfect foresight model, any exogenous change that presents an opportunity for shifts in investment decisions is likely to lead to behavioral differences. Beyond the specific cases, however, it is difficult to make useful generalizations as to when divergences will arise.

Some additional observations about the behavior of perfect foresight models may be useful at this point, based in part on work with the FASOM model (Adams et al. 1996a). Given the optimizing structure of perfect foresight models, shifts in exogenous conditions will be met by behavioral adjustments to reduce the impact on the net surplus objective. In the Sohngen and Sedjo (1998) experiments, the timing of timber harvest was the only behavioral attribute that could be adjusted and the market impacts were in many cases large and long-lived. As additional aspects of the market environment are made endogenous in the analysis, however, the opportunities for behavioral adjustment increase and the extent and duration of impacts may decline. For example, the FASOM model (a linked model of the U.S. forest and agricultural sectors) was employed to examine the effects of an afforestation subsidy for agricultural lands on the net area of U.S. forest land. In the FASOM model lands of suitable qualities can move in both directions between sectors. The afforestation subsidy had the anticipated effect of inducing land to move from agriculture to forestry. At the same time this shift altered the relative levels of product and land prices between the sectors, inducing a movement of forest land to agriculture. The result was almost no net forest land gain. Without a link to the agricultural sector and endogenous land use decisions, the model would have given quite a different picture of the impact of the subsidy.

As noted above, many aspects of production decisions can be explicitly represented in a perfect foresight model as a direct producer selection process. Forest management investment is a particularly critical example. Land owner decisions can be represented as a chcice from among a set of management investment options in each period of the projection. Given the final equilibrium prices from the model solution (and the underlying investment costs and yields), the optimality conditions for the intertemporal problem guarantee that the investment decisions will be present net worth maximizing. In light of the paucity of information on private forest management investment decisions and the general absence of econometric models describing this behavior, this capacity of perfect foresight models is a significant addition to our analytical tools. At the same time, some caution in interpreting results from perfect foresight models with endogenous forest management investment may be in order (examples include, Sedjo and Lyon [1990], Adams et al. [1996a], Sohngen and Mendelsohn [in press]). In most past cases, these models have assumed perfect capital markets (constant interest rates). Investment is restricted only by yield, cost and market characteristics, investable capital is unlimited.

The results, as discussed in Adams et al. (1998) can be levels of investment well beyond historically observed ranges. As Stavins (in press) has argued in a slightly different but analogous context, investment models of this sort may fail to capture other motivations or limitations of land owners in their investment decisions. Concerns such as capital limitations (liquidity constraints), irreversibility's and option values, riskiness, or other than pecuniary yields from the land may serve to reduce the extent of investment. And since endogenous investment adds a further element of "behavioral flexibility" to perfect foresight models, simulated policy impacts may be damped as a result.

Finally, it is in fact too rigid to consider the policy analyst's choice as between a myopic model or a perfect foresight model. Hybrids are possible. Consider an example in the context of forest management investment. Adams et al. (1982) describe a process in which a straightforward present worth maximizing investment selection process is grafted onto a myopic forest sector model. Given an initial guess of future prices, a sequence of investments is selected that promise to return positive present values at some fixed interest rate. The investments are implemented in the market model over time and the price projection revised. The investment selection is revised accordingly and the projection redone. This process is continued until the prices used in the investment selection are (perfectly seen as) those that result from the market projection (within some tolerance). In this case we exploit the flexibility of the perfect foresight approach to model investment (for which there is little other information) while retaining the market model and its myopic harvest decision or stumpage supply components.

3. Market Models of the Future

To no one's surprise, the foregoing survey has exposed a wide diversity of forms and functions in current market modeling efforts. In some policy applications one of these model forms may well be superior to another; though one suspects that models are as often selected because of limitations in available data as for known or supposed superiority in projections or analytical capabilities. What is more the search for the "best" model may not be well directed. As Sohngen and Sedjo (1998) and Adams et al. (1996b) suggest, differences between model projections when considered in light of the differences in model structures may provide useful information to analysts about the ranges of possible outcomes under alternative behavioral assumptions. Even so, more will likely be needed to meet the burgeoning demand for forest sector analysis than simply comparing results from two or more model forms, even if one can afford to build multiple models. There are gaps in our model "gallery", and it may be very useful to fill them.

The empty cells in the Figure 1 are a reasonable place to start. To date there have been few attempts to develop a spatial, intertemporal optimization model with multiple market levels. Not discounting the computational difficulties associated with such a model, it would provide an extremely valuable tool. Most existing models focus on the log sector. But it is difficult to convincingly simulate the impacts of multi-regional or global scale developments such as climate change without the interaction of trading partners through the product market. In the United States, for example, models focusing on log markets alone force us to ignore Canada in considering the impacts of forest type shifts due to climate warming--an unacceptable omission. Of course, analysts can devise approximations and adjustments that purport to overcome these limitations. But these adaptations are generally not very robust across types of policy or exogenous shifts.

As we have noted above, one of the powerful aspects of perfect foresight models lies in their ability to represent intertemporal production activities endogenously as a process of optimal selection among alternatives. At the same time, these representations may imply too much flexibility in the decision process. Paraphrasing Stavins (in press), they may not fully reflect the actual preferences of decision makers. An alternative, while still taking advantage of the intertemporal welfare maximization format, might be to constrain aspects of the decision process. For example, Adams et al. (1998) impose liquidity constraints on nonindustrial land owners (upper bounds on investable funds in a given time period) to reduce flexibility in forest management investment. The results of this approach, of course, are likely to be behavioral projections that are more akin to myopic models, or what Stavins calls models based on the revealed preferences of decision makers.

Current debate on environmental policy issues and regulatory activities is increasingly sensitive to the complexity of interactions in the natural systems being protected and the social systems that use these resources. Thus policy analysis requires representations of social and biophysical processes that are ever more sophisticated and detailed. These bioeconomic models are a means to help clarify information. Models built by social scientists with highly simplified representations of natural systems, and models by biologists with minimal links to human use, may generate much disciplinary interest but are of limited value to policy analysts who need to consider both biophysical as well as socioeconomic variables in decisions. The challenge is to develop methods that allow merger or joint simulation of market and resource models while retaining their respective detail. Recent work by Sohngen and Mendelsohn (in press) on the effects of global warming is instructive in this regard. An approach was found to link complex global climate and biogeographic projections with a perfect foresight model of timber harvest and management investment. The result, as Sohngen and Mendelsohn note, is to exploit the wealth of detail available in the biophysical models while recognizing how humans will respond to and modify climate-induced vegetational changes.

Meeting the three research challenges noted to this point will, of course, lead to some very large and very complex models. In the past, modelers have been less than aggressive in taking up such efforts, in part because of the daunting computational burdens. The rapid development of desktop computing capabilities alone (not to mention mainframe and supercomputing options), however, suggests that this should no longer be a major concern. As a result, it should be possible to take a very broad view of the systems and decisions of interest when initially framing the questions to be answered and the model(s) to be built, rather than trying from the outset to "shoehorn" a problem into some "manageable" format. What is manageable in computing changes daily, even on a given hardware platform. This is not to be construed as a critique of simplification. Simplicity certainly has its virtues, not the least of which are compact model code and short simulation times. Yet, as we are faced with increasingly complex modeling problems, it seems more useful to directly embrace the "messiness" of these situations and at least not reject out of hand a large and messy model response to a messy problem.

Finally, we believe that the growing complexity of the environmental policy debate makes it ever more important for modelers to be explicit about the uncertainty inherent in policy scenarios and projections. As in many other aspects of the policy analysis process, the analyst has the opportunity to affect the mode of decision making. It is incumbent upon us to at least attempt to inform policy makers of the uncertainty attendant on the outcomes of many of their policy decisions. As has been argued elsewhere, there is a need to embrace risk as a fundamental characteristic of the management and policy environment (Adams 1993). While this notion may be *passé* to some involved in modeling, there seems to be little evidence that its recognition is widespread among policy makers. It is probably true, as Binkley (1993) argues, that producing confidence intervals or showing ever-widening bands of variation around projections will do little to aid decision making directly. The issue we raise, however, is that of gaining recognition that resource policy outcomes are uncertain. Appropriate coping devices, such as adaptive management as suggested by Binkley or some other approach, are a step beyond.

4. References

Adams, D.M. 1974. Forest products prices and national forest timber supply in the Douglas-fir Region. Forest Science. 20(3):243-259.

Adams, D.M. 1985. A spatial model of African-European trade in tropical logs and sawn wood. Forest Ecology and Management. 13:265-287.

Adams, D.M. 1993. Comments on papers by Shugart, et al. and Gardner et al., *in* LeMaster, D.C., and Sedjo, R.A. (eds.), Modeling sustainable forest ecosystems, Proceedings of a Workshop in Washington, DC. Forest Policy Center, American Forests, Washington, DC.

Adams, D.M., and Haynes, R.W. 1980. The 1980 RPA softwood timber assessment market model: structure, projections, and policy simulations. Forest Science Monograph 22.

Adams, D.M., and Haynes, R.W. 1987. Interregional modeling, *in* Dykstra, D., Kallio, M., and Binkley, C. (eds.), The global forest sector: An analytical perspective, John Wiley & Sons, Ltd.

Adams, D.M., and Haynes, R.W. 1996. The 1993 timber assessment market model: structure, projections and policy simulations. U.S. Department of Agriculture, Forest Service, Pacific Northwest Research Station, General Technical Report PNW-GTR-358, Portland, OR. 58p.

Adams, D.M., Alig, R.J., Callaway, J.M., McCarl, B.A., and Winnett, S.M. 1996a. The forest and agriculture sector optimization model (FASOM): Model structure and policy applications. U.S. Department of Agriculture, Forest Service, Pacific Northwest Research Station, General Technical Report PNW-RP-495, Portland, OR.

Adams, D.M., Alig, R.J., McCarl, B.A., Callaway, J.M., and Winnett, S.M. 1996b. An analysis of the impacts of public timber harvest policies on private forest management in the U.S. Forest Science. 42(3): 343-358.

Adams, D.M., Alig, R.J., McCarl, B.A., Winnett, S.M., and Callaway, J.M. 1998. The effects of factor supply assumptions on intertemporal timber supply behavior: The cases of investable funds and land. Canadian Journal of Forest Research. 28:239-247.

Adams, D.M., Haynes, R.W., Dutrow, G.F., Barber, R.L., and Vasievich, J.M. 1982. Private investment in forest management and the long-term supply of timber. American Journal of Agricultural Economics, 64(2):232-241.

Berck, P. 1979. The Economics of Timber: A renewable resource in the long run. Bell Journal of Economics 10: 447-462.

Binkley, C.S. 1993. Integrating economic and ecological models for forest sector analysis, *in* LeMaster, D. C., and R. A. Sedjo (eds.), Modeling sustainable forest ecosystems, Proceedings of a Workshop in Washington, DC. Forest Policy Center, American Forests, Washington, DC.

Binkley, C.S., and Cardellichio, P.A. 1983. Modeling the hardwood sector: Issues and prospects, *in* R. Seppälä, C. Row, and A. Morgan. (eds.), Forest sector models. AB Academic Press, Berkhamsted, UK.

Buongiorno, J. 1996. Forest sector modeling: A synthesis of econometrics, mathematical programming and system dynamics methods. International Journal of Forecasting. 12(3): 329-343.

Cardellichio, P.A., Youn, Y-C., Adams, D.M., Joo, R-W., and Chmelik, T. 1989. A preliminary analysis of timber and timber products production, consumption, trade, and prices in the Pacific Rim until 2000. CINTRAFOR, Univ. of Washington, Working Paper WP-22, Seattle, WA.

Chambers, R.G. 1988. Applied production analysis: A dual approach. Cambridge University Press, Cambridge, UK.

Ince, P.J. 1994. Recycling and long-range timber outlook. U.S. Department of Agriculture, Forest Service, Rocky Mountain Forest and Range Experiment Station, General Technical Report, RM-GTR-242, Ft. Collins, CO.

Johnson, K.N., and Scheurman, H.L. 1977. Techniques for prescribing optimal timber harvests and investment under different objectives. Forest Science Monograph 18.

Just, R.E., Hueth, D.L., and Schmitz, A. 1982. Applied welfare economics and public policy. Prentice-Hall, Englewood Cliffs.

Kallio, M., Dykstra, D., and Binkley, C. 1987. The global forest sector: An analytical perspective, John Wiley & Sons, Ltd., New York. 703p.

Lönnstedt, L. 1986. A dynamic forest sector model with a Swedish case. Forest Science. 32(2):377-397.

McKillop, W. 1967. Supply and demand for forest products: An econometric study. Hilgardia 38(1):1-32.

McKillop, W. 1969. An econometric model of the market for redwood lumber. Forest Science. 15(2): 159-170.

Montgomery, C., Brown, G. Jr., and Adams, D.M. 1994. The marginal cost of species preservation: The northern spotted owl. Journal of Environmental Economics and Management 26: 111-128.

Nabuurs, G. J., and Päivenen, R. 1996. Large scale forestry scenario models: A compilation and review. European Forest Institute, Working Paper No. 10, Joensuu, Finland.

Nagata, S., and Tachibana, S. 1994. Modeling the timber trade structure of japan with the USA and Canada. Paper presented at the SATT Symposium, Miyazaki, Japan.

Newman, D. H. 1987. An econometric analysis of the southern softwood stumpage market: 1950-1980. Forest Science. 33(4): 922-945.

Newman, D.H., and Ware, D.N. 1993. Production economics of private forestry: A comparison of industrial and nonindustrial forest owners. American Journal of Agricultural Economics 75(3):674-684.

Önal, H., McCarl, B.A., Griffin, W.L., Matlock, G., and Clark, J. 1991. A bioeconomic analysis of the texas shrimp fishery and its optimal management. American Journal of Agricultural Economics 73(4):1161-1170.

Robinson, V.L. 1974. An econometric model of softwood lumber and stumpage markets, 1947-1967. Forest Science. 20:171-179.

Schwarzbauer, P. 1992. Potential impacts of forest decline on the Austrian forest products markets: An analysis using a forest sector simulation model, *in* Adams, D., Haynes R., Lippke B., and Perez-Garcia J. (compilers), Forest sector, trade and environmental impact models: Theory and applications, Proceedings of An International Symposium April 30 - May 1, 1992, University of Washington, Center for International Trade in Forest Products, Seattle, WA.

Sedjo, R., and Lyon, K. 1990. The Long-term adequacy of world timber supply. Resources for the Future, Washington, DC.

Sohngen, B., and Sedjo, R. 1998. A comparison of timber market models: Static simulation and optimal control approaches. Forest Science. 44(1):24-26.

Sohngen, B., and Mendelsohn, R. In press. Valuing the impact of large sale ecological change in a market: The effect of climate change on US timber. American Economic Review.

Stavins, R.N. In press. The costs of carbon sequestration: A revealed-preference approach. American Economic Review.

Trømborg, E., and Solberg, B. 1995. Description of a partial equilibrium model applied to the project 'Modeling the Norwegian forest sector'. Report 14/95. Norwegian Forest Research Institute, Ås, Norway.

U.S. Department of Agriculture, Forest Service. 1982. An analysis of the timber situation in the United States 1952-2030. Forest Resource Report No. 23. U.S. Government Printing Office. Washington, DC. 272p. [plus appendix].

Veltkamp, J.J., Young, R., and Berg, R. 1983. The Data Resources, Inc., Approach to modeling demand in the softwood lumber and plywood and pulp and paper industries, *in* Seppälä R., Row C., and Morgan A. (eds.), Forest sector models. AB Academic Press, Berkhamsted, UK.

Yukutake K. 1989. Simulation analysis of demand/supply relationship of forest. The Current State of Japanese Forestry (VI). The Japanese Forest Economic Society, 23-38.

Yukutake, K., Saikai, T., and Yoshimoto, A. 1994. Modeling the Japanese timber markets. Paper presented at the SATT Symposium, Miyazaki, Japan.

Zhang, D., Buongiorno, J., and Ince, P. 1993. PELPS III a microcomputer price endogenous linear programming system for economic modeling. USDA, Forest Service, Forest Products Laboratory, Research Paper FPL-RP-526, Madison, WI.

Modeling the Global and Asia-Pacific Forest Sectors

- Experiences and prospects -

J. Buongiorno, S. Zhu and D. Tomberlin

Key words: Global forest resources, forest products trade

Abstract: Experience with a hierarchical world model of production, consumption, and trade Global Forest Product Model (GFPM), with Asia-Pacific as a subregion, confirms that equilibrium theory is a powerful way of getting coherent forecasts of many interacting forest economies. The GFPM model has produced forecasts at country and commodity level that are useful for Food and Agriculture Organization of the United Nations (FAO) global and regional outlook studies. However, several areas of improvement exist. The modeling work has revealed many shortcomings of the basic national statistics. The regional submodels are currently being replaced with a single model considering all countries simultaneously. The econometric demand equations can still be improved, especially their price elasticities. International econometric wood supply equations are almost nonexistent, and alternative models of wood supply need to be developed. The description of technology by activity analysis is powerful, but the comparative cost of production in different countries can be improved. The dynamics of trade flows and capacity expansion still require adjustments and perhaps new theory. Still, operational regional and global forest sector models are now available to individual analysts, giving them the means to produce their own forecasts, based on alternative scenarios of global growth and forest policies.

1. Introduction

Forest products are vital commodities for the sustenance and growth of every country in the world. The FAO is responsible for producing long-term projections of future consumption, production, and trade of forest products, by country. The objective of this study was to revamp and make operational the FAO system for global forest product outlook projections.

The approach was to develop a dynamic spatial equilibrium model of the world forest sector. Then, the model was applied to project forest product consumption, production, and trade under the most likely scenario of economic growth and timber supply constraints. As in past FAO projections, the assumed rates of economic growth as measured by Growth Domestic Product (GDP) are critical in determining final demand. A new feature is the use of constraints defining the timber supply potential of each country. The projected demand for final products in the world cannot exceed the global wood growing potential, after allowing for trade in raw materials or finished products between countries. Depending on market conditions, less wood may be cut than can potentially be produced, but not more. Therefore, the assumptions regarding timber supply potential are also critical.

The first part of the paper introduces the methodology of the Global Outlook Product Model of 1998 (GFPM), describes the countries or regions and the products included in the model, and discusses the demand equations and the supply constraint assumptions. Part two summarizes projections for 1995 to 2010 of consumption, production, and trade in the main regions of the world.

2. The Global Forest Product Model (GFPM)

The GFPM model, was developed with the Price Endogenous Linear Programming System (PELPS III) (Zhang et al. 1993). PELPS is a general modeling tool for sector analysis. It finds a spatial, competitive equilibrium by maximizing the sum of producer and consumer surplus subject to material balance and capacity constraints in each country, for each year. Because material flows throughout the system must balance, it helps check the consistency of data, and insures the coherence of projections.

In each projection year, for each country and commodity, supply (domestic production plus imports) is equal to demand (final consumption, plus input in other processes, plus exports). Demand for wood or intermediate products is derived from the demand for final products through input-output coefficients that describe the technology in each country. The supply of raw wood in each country is constrained by the wood supply potential. The supply of recycled paper is constrained by the waste paper supply, which depends on paper consumption and recycling rates. Imports and exports are balanced, at world level and in each country. And, the projected prices for each commodity are such that they clear markets at those prices demands are equal to supplies in each country.

From one year to the next, demand changes in each country owing to assumed changes in income. The wood supply potential changes according to the chosen scenario. Technology also affects demand, especially the amount of recycled fiber used in making paper and paperboard. Capacity increases or decreases according to new investments that depend on past production and the profitability of production in different countries, as revealed by the shadow price of capacity. Then, a new equilibrium is computed subject to the new demand, supply conditions, new technology, new capacity, and inertia constraints that limit the change in exports and imports from year to year. The general principle of the PELPS system is, then, that markets optimize the allocation of resources in the short run (within one year). Long run resource allocation is partly governed by market forces, as in capacity expansion and trade, and also by other forces such as the wood supply potential determined by forest policy, the waste paper recovery rates determined by environmental policy, the trade inertia that depend in part on international agreements, and the techniques of production determined by exogenous progress.

This system has been used to develop the North American pulp and paper model (NAPAP), the North American solid wood model (NASAW), and by the International Tropical Timber Organization (ITTO) to develop the Asia-Pacific Tropical Timber Trade model. The Asia-Pacific Forest Products Model of Zhang et al. (1996), is also based on the PELPS system. The GFPM model is a natural extension of past FAO methodology. The demand for forest products continues to depend critically on income. The additions are a detailed description of supply linked to demand so that production and consumption projections are fully consistent, and prices that clear markets, as they must in the real world. The mathematical formulation of the model is given below.

2.1 Hierarchical structure

The GFPM model has a hierarchical structure. It consists of one main model called WORLD, and one sub-model for each region: AFRICA, AMERICAS, ASIA-OCEANIA and EUROPE-FORMER USSR, each one solved with trade flows that are consistent with the WORLD model.

Each region of the WORLD model has: (1) demand (consumption) of fuelwood, other industrial roundwood, sawnwood, veneer and plywood, particle board, fiberboard, newsprint, printing and writing paper and other paper and paperboard; (2) production (supply) of fuelwood, industrial roundwood, other industrial roundwood, other fiber pulp, waste paper, sawnwood, veneer and plywood, particle board, fiberboard, mechanical pulp, chemical pulp, newsprint, printing and writing paper and other paper and paperboard. Also, each region imports from and exports to the "world market."

A similar structure applies to the regional sub-models. For example, in the AMERICAS sub-model, there are 35 countries, each with a demand and supply of the same products as in the WORLD model. Each country has imports from and exports to the "world market." However, the total imports and exports of all countries in, say, the AMERICAS are constrained to be equal to the total imports and exports of the AMERICAS in the WORLD model.

2.2 Countries and products

The countries in each sub-model are those recognized in the FAO Yearbook of Forest Products. The AFRICA sub-model has 50 countries, the AMERICAS has 35, ASIA-OCEANIA has 50, and EUROPE-FORMER USSR has 45 countries. In addition to country detail, seven regions and two economic classes are used in reporting the results.

GFPM has final demand equations for fuelwood and charcoal, other industrial roundwood, sawnwood, veneer sheets and plywood, particle board, fiberboard, newsprint, printing and writing paper, and other paper and paperboard. The supply equations for fuelwood and charcoal, other industrial roundwood, and other fiber pulp are horizontal, up to the maximum sustainable production. Waste paper supply is constrained by the quantity of paper recovered, a function of paper consumption. Industrial roundwood supply is price elastic, and shifts exogenously, as described in section 2.5.

2.3 Final product demand

The demand equations of final products in GFPM assume that national demand for forest products is determined by economic activity (GDP) and real prices. Let q_{it} (of price p_{it}) be the amount of a forest product consumed or imported by a country i during year t; with it an index of the price of all other goods and services used in combination with this product to produce an output designated by the variable y_{it}, here the real gross domestic product. Then, assuming cost minimization under a Cobb-Douglas technology with constant returns to scale, coupled with a partial adjustment of actual towards desired consumption, leads to the following derived demand (Chou and Buongiorno 1984, Baudin and Lundberg 1987):

$$[1] \quad \ln(q_{it}) = \alpha_i + \beta \ln(\frac{p_{it}}{i_{it}}) + \gamma \ln(y_{it}) + \delta \ln(q_{it-1}) + u_{it}$$

For estimation, all countries in the world were divided into high, medium, and low income countries. The data on production, imports, and exports were obtained from (FAO 1996). Gross Domestic Products (GDP) were measured in constant U.S. dollars with data from World Bank (1995). The parameters were estimated by covariance analysis, one example is in Table 1.

Table 1. Example of elasticities used in the GFPM model

				Long-term				
	Price	GDP	Lag	Price	GDP	R^2	DW	DF
Sawnwood								
High income	-0.10	0.14	0.54	-0.22	0.31	0.99	2.17	871
	(0.03)*	(0.03)*	(0.03)*	(0.07)*	(0.06)*			
Low income	-0.08	0.09	0.67	-0.23	0.28	0.95	1.99	1893
	(0.03)*	(0.03)*	(0.02)*	(0.09)*	(0.08)*			
Printing and Writing Paper								
High income	-0.27	0.54	0.55	-0.61	1.19	0.99	1.95	833
	(0.04)*	(0.05)*	(0.03)*	(0.09)*	(0.07)*			
Low income	-0.30	0.17	0.55	-0.66	0.37	0.95	2.14	1589
	(0.03)*	(0.03)*	(0.02)*	(0.07)*	(0.06)*			

* Significant at 5% level

In the GFPM model, prices are endogenous (the equilibrium between demand and supply leads to the prices). Only national GDPs shift demands, and are strictly exogenous. In the assumed scenario, Africa grew at 3.9% per year from 1995 to 2000 and then approached 4% by 2010. For North and Central America, GDP grew at 2.5% per year from 1995 to 2010. In South America, growth rate was 3.6% from 1995 to 2000 and 4.3% from 2000 to 2010. Asia continued to grow fast in the projections, at more than 4.5% to 2010. Oceania was projected to grow at 3.4% from 1995 to 2010. In Europe, growth that started at 2.4% in 1995 was assumed to increase to 2.7% by 2001 and continue until 2010. The former USSR projections assumed a strong recovery, with a 5.7% GDP growth rate from 2001 to 2010.

2.4 Manufacturing activities

The supply of final products and the demand and supply of pulp are represented by activity analysis. Input-output coefficients represent the technical relation, e.g., between fiber and paper, and a manufacturing cost indicates the cost of all other inputs, not accounted for endogenously in the model (labor, energy, capital). Table 2 shows an example for the paper industry in Italy. The input output coefficients were calibrated to reflect national statistics, while remaining technologically plausible. The manufacturing cost is the world price of the product minus the cost of fiber inputs.

Table 2. Subset of GFPM input-output and manufacturing cost coefficients for Italy

Input	Output		
	Newsprint	Printing and writing paper	Other paper and paperboard
Mechanical pulp	0.05	0.01	0.01
Chemical pulp	0.25	0.45	0.31
Other fiber pulp	0.03	0.02	0.02
Waste paper	0.70	0.56	0.70
Manufacturing cost ($/t)	246	536	486

2.5 Wood supply constraints

In the current version of GFPM, the supply of industrial roundwood in each country is a function of the price of industrial roundwood. The price elasticity of supply was set at 0.8, except for Asia, where, according to expert opinion, the elasticity varied by country. In addition, the supply equations were assumed to shift over time, to simulate changes in supply independent of prices. The rates of shifts were estimates of how much timber supply would change without any change in price. These rates of shift vary by country, and are based on a variety of information regarding past production, forest area and stock, growth rates, extent of plantations, and policies of each country. In some countries of Asia, upper bounds were set on the timber supply, regardless of price, to reflect perceived limits on the potential sustainable timber production of the country. The rate of shifts relied mostly on the data of the Global Fiber Supply Study (GFSS, FAO 1998), the rate of shift being equal to the annual percentage change in the commercially available wood supply projected by the GFSS. The resulting aggregate average rates of shift of industrial roundwood supply, by region, were 0.7% in America, 0.6% in Europe, 1.1% in Africa, and 1.4% in Asia.

2.6 Other assumptions

Other assumptions governing the projections concerned changes in capacity, fiber input mix, and trade inertia constraints. In GFPM, capacity is governed by past production and by relative profitability. Capacity data were estimated from the most recent production data. For most products the simplest capacity projection allowed by PELPS was used. It consisted of distributed-lag functions linking capacity changes to changes in past production at the global level, followed by allocating capacity growth to different countries according to the shadow price (the marginal value of one unit of additional capacity) revealed by the equilibrium computations (See Appendix, and Gilless and Buongiorno 1987, Zhang et al. 1993).

The technology, defined by input-output coefficients was held constant at the 1994 level, except for paper and paperboard. In the latest GFPM projections (Zhu et al. 1998), the input-output coefficients were changed substantially relative to those used in the Provisional Outlook (FAO 19-7), in that fewer constraints were put on the input-output coefficients to satisfy the national statistics. Instead, the coefficients varied little across country, while the industrial roundwood production data and the pulp production data were adjusted to fit the other production and trade data. For paper and paperboard, it was assumed that the fiber input mix for newsprint, printing and writing paper and paper and paperboard would change substantially between 1994 and 2010.

The levels and trends of trade in forest products are determined by a wide range of factors, of which only the main elements were recognized: demand, cost of wood, and other input costs. With the implementation of GATT-WTO underway, the member countries are projected to reduce or eliminate trade tariffs and non-trade barriers in the near future. Thus, it was assumed that forest products would be traded more freely worldwide between 1994 and 2010 in response to market forces. However the adjustment of trade flows between countries or regions is bound to take time, thus inertia constraints were introduced to limit trade in a given year to be within a specified fraction of the previous year's trade (Buongiorno and Gilless 1-84, Kallio et al. 1987, Zhang et al. 1997).

3. Global Outlook to 2010

The GFPM model has been applied to project production, consumption, exports, and imports from 1995 to 2010 under each scenario. The detailed tables by country and product are in FAO (1997), updated in Zhu et al. (1998). The projections indicate that the global demand for forest products will continue to rise, pulled by the demand in rapidly growing economies, especially in Asia. The total roundwood consumption and production would reach 4.1 billion m^3 by the year 2010, for an average growth rate of 1.5% per year, but faster in Oceania, Asia, and the former USSR (Table 3).

Table 3. Roundwood consumption (10^6 m^3)

	Actual			Projection		% yr^{-1}	
	1970	1990	1994	2000	2010	1970-1990	1994-2010
Africa	257	438	490	535	626	2.7%	1.5%
N America	471	724	727	736	765	2.2%	0.3%
S America	164	294	320	344	392	3.0%	1.3%
Asia	751	1097	1168	1349	1666	1.9%	2.2%
Oceania	24	31	30	37	47	1.4%	2.7%
Europe	351	402	352	387	419	0.7%	1.1%
Former USSR	370	375	123	185	175	0.1%	2.2%
World	2387	3362	3210	3574	4090	1.7%	1.5%

Of this, about 2.2 billion m^3 would be fuelwood consumed largely in developing countries. World sawnwood consumption, at near 498 million m^3 in 2010 would be lower than in 1990, owing to the sharp decline in the former USSR from 1990 to 1994, not fully recovered by 2010. In the rest of the world, sawnwood consumption would be slightly higher than in 1990 (Table 4). Wood-based panel consumption would total 179 million m^3 worldwide by 2010, growing at an average of 2.2% per year. Paper consumption is expected to grow the fastest, averaging 2.4% annual growth, leading to a world consumption of 391 million metric tons in 2010 (Table 5).

Table 4. Sawnwood consumption (10^6 m^3)

	Actual			Projection		% yr^{-1}	
	1970	1990	1994	2000	2010	1970-1990	1994-2010
Africa	6	10	9	10	11	3.0%	1.1%
N America	107	153	149	156	168	1.8%	0.8%
S America	12	24	24	25	28	3.4%	1.1%
Asia	79	112	112	123	143	1.8%	1.5%
Oceania	6.2	6.5	6.7	7	8	0.2%	0.9%
Europe	92	103	87	92	101	0.6%	0.9%
Former USSR	113	99	24	34	40	-0.6%	3.3%
World	413	508	411	447	498	1.0%	1.2%

In 2010, Asia is projected to be the largest net importer, for all products. The former USSR and South America would become the main net exporters of industrial roundwood (Figure 1). North-Central America and the former

USSR would be the main exporters of sawnwood, while for panels, Europe and Oceania would take that role. Europe and North-Central America would be the main net exporters of paper and paperboard (Figure 2).

Table 5. Paper and paperboard consumption (10^6 t)

	Actual			Projection		% yr^{-1}	
	1970	1990	1994	2000	2010	1970-1990	1994-2010
Africa	1.6	3.7	3.6	4	5	4.3%	2.3%
N America	55	88	96	110	132	2.4%	2.0%
S America	3.5	8	10	11	13	4.0%	1.7%
Asia	19	62	81	97	130	6.1%	3.0%
Oceania	1.9	3.3	4	4	6	3.0%	2.8%
Europe	38	65	70	79	99	2.7%	2.2%
Former USSR	6.5	10	2	4	6	2.4%	6.6%
World	126	240	266	308	391	3.3%	2.4%

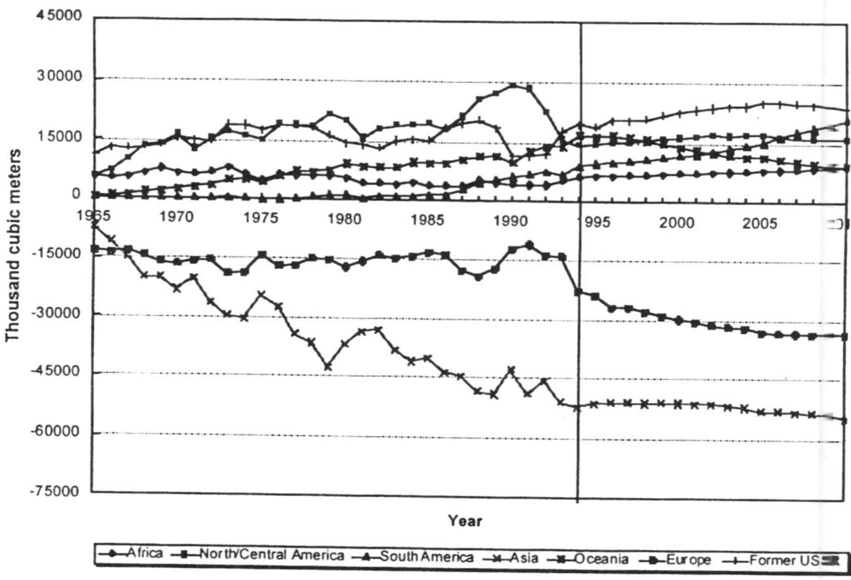

Figure 1. Industrial roundwood net trade

Under the assumed scenario, the real prices of industrial roundwood and sawnwood are projected to change little until the year 2010 (Figure 3). The real prices of wood-based panels, paper, and paperboard would also generally remain constant, owing in part to the growth in world waste paper recovery (Figure 4).

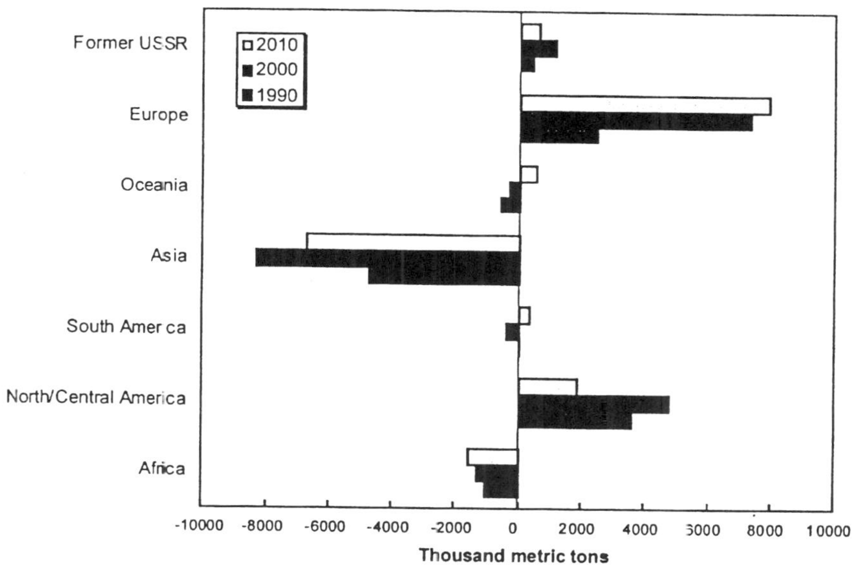

Figure 2. Paper and paperboard net trade

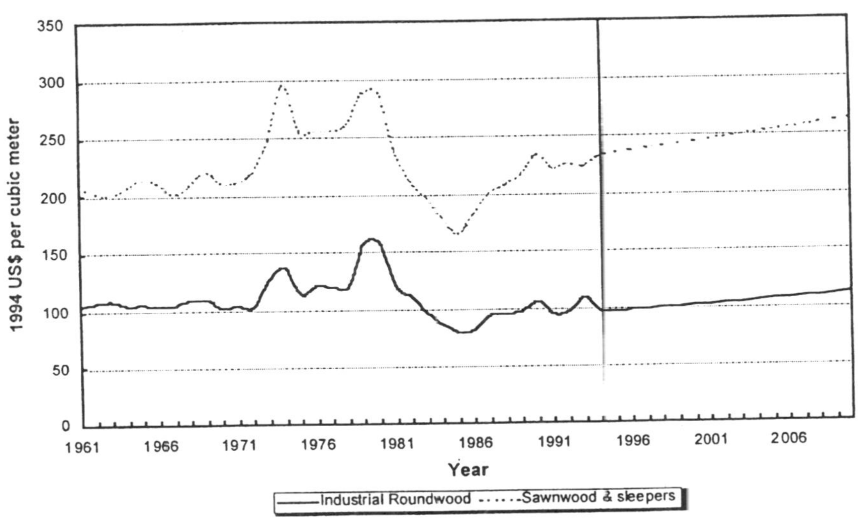

Figure 3. Real world price of solid wood products

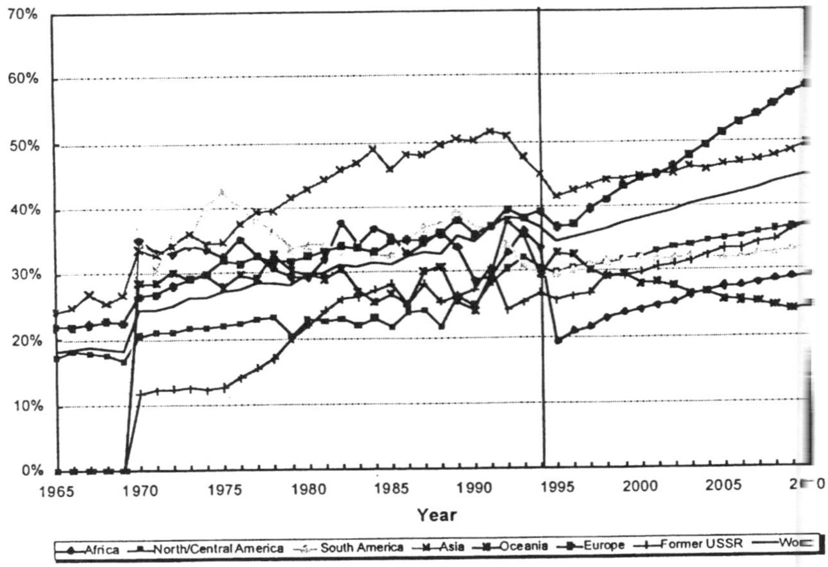

Figure 4. Recycled fiber utilization rates

4. Conclusion

The GFPM structure has been applied repeatedly for global projections of production, consumption, trade, and prices of forest products, globally and for the Asia-Pacific subregion. The hierarchical structure of the model facilitates such regional analyses, reflecting the full interaction of a region with the rest of the world. The GFPM is anchored on equilibrium theory. As a result, it insures full coherence between all subsectors, and it produces prices that are market clearing, worldwide. The ability to balance material flows at all stages of manufacturing is a powerful feature of the method. The GFPM has proved feasible with fine level of product disaggregation, while recognizing each individual country. This level of detail is essential to obtain good user feedback. Expertise exists mostly at commodity and country levels. The GFPM model has produced forecasts that have been useful for FAO's global and regional outlooks, and in major policy papers like the 1997 State of World Forests Report.

However, as is usual in this field, several improvements are possible. First and foremost, the modeling work, especially the establishment of wood and fiber accounts, has revealed many shortcomings of the basic national statistics. In several countries, wood pulp consumption statistics were incompatible with those of paper production. In the most flagrant cases, we computed our own pulp production statistics, based on paper statistics and pulp trade statistics. The same occurred for industrial roundwood data.

The decomposition of the large global equilibrium model into a world model with four regions, followed by a regional solution with each country is an imperfect way of attacking the model size issue. We found, for example, that when constraints on timber supply were imposed at country level, prices could differ from those of the world model. This was minimized by building the model "bottom up," from country level to regional level data. But there is no perfect solution to this problem, except building a very large model where all the countries are considered simultaneously in all their detail.

The econometric demand equations can still be improved, especially their price elasticities. In some regions, such as the European Union, tests have shown that it is appropriate to assume that elasticities are the same across countries. But this may not be true in general. How to use national and international data to arrive at the best possible estimate of elasticity is still a research topic. International econometric wood supply equations are almost inexistent, and better models of wood supply need to be developed to make the shift of supply endogenous, instead of relying on expert opinion.

The description of technology by activity analysis is very attractive, but a systematic method is needed to reconcile a-priori technical knowledge with national statistics, and to arrive at good estimates of the comparative cost of production in different countries. How these relative costs translate into different shadow prices and capacity expansion in different countries is theoretically sound, but the magnitude of the attendant parameters is still uncertain. Last, the dynamics of trade flows that now use inertia constraints need to be made more flexible, and here a new theoretical construct may be needed to model realistically the progressive response of trade to price signals.

Despite these imperfections, the current version of the GFPM is an operational, fully documented global forest sector model, available to a wide number of individuals working at global, regional, or national level. Instead of a single periodic study at long time intervals, it opens the real possibility of producing almost continuously many alternative projections as new data become available, or new scenarios for future economic, demographic, and environmental change are formulated.

6. References

Baudin, A., and Lundberg, L. 1987. A world model of the demand for paper and paperboard. Forest 33(1): 185-196.

Buongiorno, J., and Gilless, J.K. 1984. A model of international trade of forest products, with an application to newsprint. Journal of World Forest Resource Management 1: 65-80.

Chou, J., and Buongiorno, J. 1984. Demand functions for United States forest product exports to the European Community. Wood and Fiber Science 16(2):158-168.

FAO. 1996. FAOSTAT•PC. FAO, Rome.

FAO. 1997. FAO provisional outlook for global forest products consumption, production and trade to 2010. Forestry Policy and Planning Division, Forestry Department, FAO, Rome.

FAO. 1998. Global fiber supply study. Forest Products Division, Forestry Department, FAO, Rome.

Gilless J. K., and Buongiorno, J. 1987. PAPYRUS: A model of the North American pulp and paper industry. Forest Science Monograph 28.

Tomberlin, D., Zhu, S., and Buongiornom J. 1998. Using the GFPM global forest products model of consumption, production, and trade. Forestry Policy and Planning Division, Forestry Department, FAO, Rome.

World Bank. 1995. World tables. The Johns Hopkins University Press. Baltimore.

Zhang, D., Buongiorno, J., and Ince, P. 1993. PELPS III: A microcomputer price-endogenous linear programming system for economic modeling. Research Paper FPL-RP-526. USDA Forest Service, Forest Products Laboratory, Madison, WI.

Zhang, D., Buongiorno, J., and Zhu, S. 1997. Trend and outlook for forest products consumption, production and trade in the Asia-Pacific region. Working Paper No:APFSOS/WP/12. Forestry Policy and Planning Division, Rome.

Zhu, S., Tomberlin, D., and Buongiorno, J. 1998. Global forest products consumption, production, trade, and prices: GFPM model projections to 2010. Forestry Policy and Planning Division, Forestry Department, FAO, Rome.

Acknowledgements

The research leading to this paper was supported in parts by the FAO by (Cooperative State Research Educational and Extension Service (CREES) grants 95-37400-1851 and 98-35400-6110, by McIntire-Stennis grant D946, by the USDA Forest Service, Pacific Northwest Research Station, and by the School of Natural Resources, University of Wisconsin-Madison. We thank Michael Martin and David Brooks for their help in setting the study objectives, and Adrian Whiteman, Chris Brown, and Gary Bull for helping us evaluate the data, parameters, and results. The programming of Chia-Sheng Liu and econometric help of Wen-Shiuan Chao and Bintang Simangunsong are also gratefully acknowledged. This paper would not have been possible without the contribution of Dali Zhang, Antti Rytkonen, and Yibing Zhang to previous versions of the GFPM and Asia-Pacific models.

Appendix: Mathematical Formulation of GFPM

GFPM was built with the general PELPS III structure (Zhang et al. 1993). It simulates a dynamic market equilibrium for the global forest sector. Every year, demand, supply, trade and prices are computed that clear markets for all products and in all regions (static phase). Then, the model parameters are updated to reflect exogenous and endogenous changes from one year to the next (dynamic phase). Yearly exogenous changes include economic growth, technical change, potential timber supply and trade inertia, they are assumptions. Yearly endogenous changes include capacity growth and availability of recycled fibers, they are determined by the model. The model then computes the quantity-price equilibrium next year as shaped by the intervening changes. It reiterates the static and dynamic phases for every year until the end of the projection.

Static phase

The equilibrium for each year of the projection is obtained with an optimization model which finds the production, consumption and trade that maximizes the total value of consumption minus the total cost of production for all products in all countries, in a given year. All the variables refer to that year.

Objective function:

[A1] $$\max Z = \sum_i \sum_k \int_0^{D_{ik}} P_{ik}(D_{ik})dD_{ik} - \sum_i \sum_k \int_0^{S_{ik}} P_{ik}(S_{ik})dS_{ik} - \sum_i \sum_k Y_{ik} m_{ik}$$

where i,k = country, commodity, P = price in US dollars, D = final product demand, S = raw material supply, Y = quantity manufactured, m = cost of manufacture.

Demand for final products:

[A2] $$D_{ik} = a_{ik} P_{ik}^{\sigma_{ik}} X_i^{\alpha_{ik}} D_{ik,-1}^{\eta_{ik}}$$

where D_{-1} = demand in the previous year, X = gross domestic product, in real US dollars, σ, α, η = elasticities with respect to price, GDP, and past demand.

Wood supply:

[A3] $$P_{ik} = P_{ik}^0 \, with S_{ik} \leq S_{ik}^U$$

where P^0 = cost of production, S^U = potential supply.

Material balance:

[A4] $$\sum_j T_{jik} + S_{ik} + Y_{ik} - D_{ik} - \sum_n a_{ikn} Y_{in} - \sum_j T_{ijk} = 0 \; \forall i,k$$
$$T_{ijk}^L \leq T_{ijk} \leq T_{ijk}^u \; \forall i,j,k$$

where a_{ikn} = input of commodity k per unit of n, T = trade flow. Each country exports to and imports from a "world market", T^u, T^L = upper and lower bounds on imports and exports.

Manufacturing capacity:

[A5] $Y_{ik} \leq K_{ik} \; \forall i,k$ where K = current capacity of production.

Waste paper recovery:

[A6] $S_{ir} \leq S_{ir}^U \; \forall i,r$ where r = recycled paper, S^U = upper bound on recycled paper supply.

Dynamic phase:

Yearly changes in the market equilibrium constraints are brought about by:

1) shifts of the demand curves, due to changes in X at the projected GDP growth rate,

2) shifts of the upper bounds on wood supply, *S*, at the assumed rate of growth of potential supply,
3) changes of manufacturing coefficients m due to reflect technical change (increasing use of recycled paper),
4) changes of capacity, determined as follows: At global level, capacity change is a function of changes in production during the past three years:

[A7] $\Delta K_k = b_{1k}\Delta Y_{k,-1} + b_{2k}\Delta Y_{k,-2} + b_{3k}\Delta Y_{k,-3}$

then, global capacity change is allocated to each country in proportion of its production level and the marginal value of capacity revealed by the shadow price of capacity in the static phase.

[A8] $$\Delta K_{ik} = \frac{Y_{ik}^{\theta}\pi^{\varpi}}{\sum_j Y_{ik}^{\theta}\pi^{\varpi}}\Delta K_k$$

where Δ = yearly change, π = shadow price of capacity (endogenous), b, θ, ϖ = parameters.

5) Changes in waste paper recovery:

[A9] $S_{ir}^{U} = \sum_k w_{ikr} D_{ik,-1}$

where w_{ikr} = maximum possible recovery rate (exogenous).

6) Trade inertia:

[A10] $T_{ijk}^{u} = (1+\varphi)T_{ijk,-1} and T_{ijk}^{uL} = (1-\varphi)T_{ijk,-1}$

where φ = upper bound on relative change in trade flow (exogenous).

Japanese Forest Sector Modeling

- Possibility of increasing domestic timber production -

Atsushi Yoshimoto, Junji Kajita and Kiyoshi Yukutake

Key words: Forest economics, timber trade model, econometrics, forest sector analysis

Abstract: In this paper we present a spatial and temporal market equilibrium model for Japanese lumber markets in conjunction with a Japanese lumber import model. The proposed model is based on the Samuelson type market equilibrium model with applying a linear function to lumber demand and supply functions in each region as well as a lumber import function. These functions are derived by using price elasticity of lumber demand and supply, which is estimated by using annual regional lumber production and consumption data from 1974 to 1995. As for the import model, data from 1970 to 1996 are used to estimate the price elasticity. Japanese lumber markets are divided into eight regional markets, i.e., Tohoku, Kanto, Hokuriku, Chubu, Kinki, Chugoku, Shikoku, and Kyushu. We consider three types of lumber in the model; Japanese domestic lumber, lumber processed in Japan from logs imported from the United States, and imported lumber. We investigate possibilities of increasing lumber production in Japan using different assumptions in the model specification and market situation. Our simulation analysis indicates that reduction of U.S. log imports does not have much influence on domestic lumber production, but instead accelerates lumber imports to Japan, which can be observed mainly in Kanto.

1. Introduction

During the 1950s, owing to the rapid economic growth in Japan, the first housing boom stimulated a large demand for lumber. This resulted in a shortage of forest resources and a large increase in lumber prices. In the 1960s, the Japanese Forestry Agency conducted a large-scale forest plantation project over the entire nation in order to prepare for the future demand for lumber. The total amount of plantations was estimated to be 10.4 million ha, of which sugi (*Cryptomeria japonica*) occupied 43.8 % and hinoki (*Chamaecypress obtusa*) covered 24.4% (Rinyacho 1995).

The shortage of timber supply after World War II had been remedied by not only non-wood products but also imported timber products (logs, lumber, and plywood) mainly from Asian nations, the United States and Canada. Importing timber products had been accelerated by removing the foreign exchange control from almost all logs, some lumber, and plywood in 1960, and replacing it with an automatic approval system. Before this period, foreign exchange control had played an important role on regulating the amount of imported products, and tariff on raw materials was almost eliminated in 1951. In 1964, the foreign exchange control on these timber products was completely removed, and the free market for timber products was introduced (Nomura 1978).

The amount of imported timber products in 1960 was estimated at 7.[illegible]4 million m^3, which covered about 13.3% of total timber products consumption in Japan, as opposed to 2.48 million m^3 in 1955 (5.5% of the total consumption). In 1973, owing to the record increase in housing starts with 1.9 million new houses, imported timber products occupied more than 60% (75.4 million m^3) of total domestic consumption (117.5 million m^3) for the first time. In 1979, market share of imported timber products reached its peak, 69.2% (76 million $m^{3)}$. Since then, because of economic recession and decrease in housing starts, domestic demand for timber products has shrunk. This resulted in a small decrease in the market share of imported domestic products to 64.3% (58 million m^3 out of 90.2 million m^3) in 1982, and 64.[illegible]% (58.5 million m^3 out of 91.4 million m^3) in 1984, respectively. Logs had been the main imported timber products. In 1986, tariffs on most timber products such as lumber, particleboard, plywood and fiberboard, was removed under the MOSS (Market Oriented Sector Selective). These imported timber products then gradually increased with recovery of the Japanese economy and housing activity. Recent statistics shows 89.02 million m^3 of imported timber products in 1995 and 89.84 million m^3 in 1996 (Rinyacho 1989 and 1997).

Because of the shortage of forest resources in Japan and a large amount of imported timber products as well as the use of non-wood products as a substitute, the structure of Japanese timber markets has changed drastically during recent decades. There are several econometric models to analyze Japanese timber markets (McKillop 1973, Gallagher 1980, Yukutake 1984 and 1985, Mori 1991, Vincent et al. 1991, Furuido et al. 1991, Nagata et al. 1992, Tachibana 1994, and Flora and Lane 1994). These mainly focused on estimating timber demand and supply functions and timber export and import functions in order to analyze characteristics of Japanese markets as a whole. None of the above attempted to analyze regional markets within Japan.

While the structure of the timber market changed, most forests planted during the 1960s have reached harvest age. The Japanese Forest Agency is now trying to motivate forest owners to harvest. This has also stimulated the development of the first forest sector model in Japan called JAFSEM (*JA*panese *F*orest *SE*ctor *M*odel) which was developed by Yukutake et al. (1996). They linked the timber export and import model and the Japanese regional timber market model through the Koopmans-Hitchcock linear programming equilibrium model (Labys 1989). They found that an increase in merchantable forest resources had a positive effect on the domestic timber supply function. Using JAFSEM, the effect of imported logs and lumber from North America on the Japanese regional markets was analyzed by Yukutake and Yoshimoto (1996). Since JAFSEM only takes into account transportation costs to determine a final solution for the trade, any effect of shape on timber demand and supply functions cannot be revealed. In other words, although increase in merchantable forest resources has a positive effect on supplying timber, it cannot be considered in the optimization process directly. In reality the surrounding economic environments seem not to favor harvesting domestic forests and to have a much greater effect on harvesting than does an increase in forest resources. It is important to incorporate economic factors into the optimization process.

The objective of this paper is to seek a situation where there will be an increase in domestic timber production. First, we investigate the potential of harvestable volume from sugi (*Cryptomeria japonica*) and hinoki (*Chamecypress obtusa*) forests using the strata-based linear programming harvest scheduling model. Using lumber demand and supply functions, we propose a spatial and temporal market equilibrium model for Japanese lumber markets. Other forest products are not considered here. To reflect characteristics of markets, we divided the Japanese lumber markets into the following eight geographical regions, i.e., Tohoku, Kanto, Hokuriku, Chubu, Kinki, Chugoku, Shikoku, and Kyushu. All markets are interrelated with each other. Historical data show that Tohoku and Kyushu regions have extensively produced domestic lumber, while the Chugoku region mainly imports logs from the United States. The Kanto, Chubu, and Kinki regions are the areas of largest lumber consumption. Simulation analysis is conducted by shifting lumber supply curves in the markets.

2. Harvestable Volume from Domestic Forest Resources

Total lumber consumed in 1995 was estimated 34.8 million m^3. There are three different types of lumber in the Japanese markets, i.e., domestic lumber, imported lumber, and lumber processed in Japan from imported logs. The United States has been the largest log supplier to Japan, amounting to 9.2 million m^3. The rest of the lumber processed in Japan summed up to 4.5 million m^3. In contrast, the amount of imported lumber was 10.0 million m^3, while that of domestic lumber was 11.0 million m^3. In our analysis, we only consider imported lumber, domestic lumber and lumber processed in Japan from imported logs from the United States (US-JP lumber) for the analysis. Note that we excluded Hokkaido from the analysis because the market and forest resources in Hokkaido are quite different from others.

Change in the market share of lumber in Japan is shown in Figure 1 from 1974 to 1995. As can be observed, the domestic lumber share has shown a constant decrease with about a 30% reduction, while imported lumber has shown an increasing share of the market to more than twice as much. The US-JP lumber fluctuated throughout the period. Currently, a diminishing trend of this segment can be predicted owing to difficulty of importing logs from the United States.

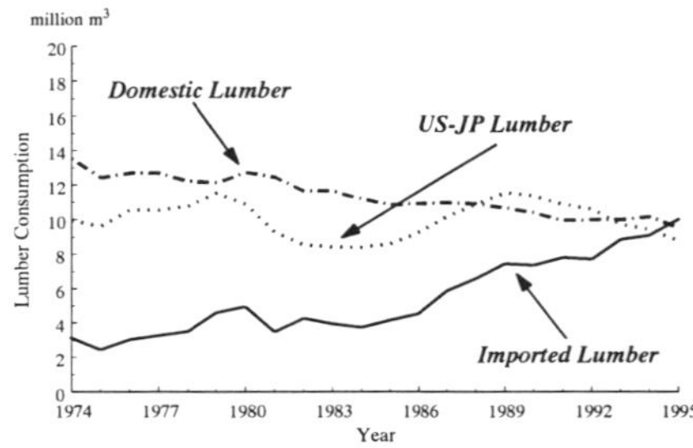

Figure 1. Dynamics of market share in Japan

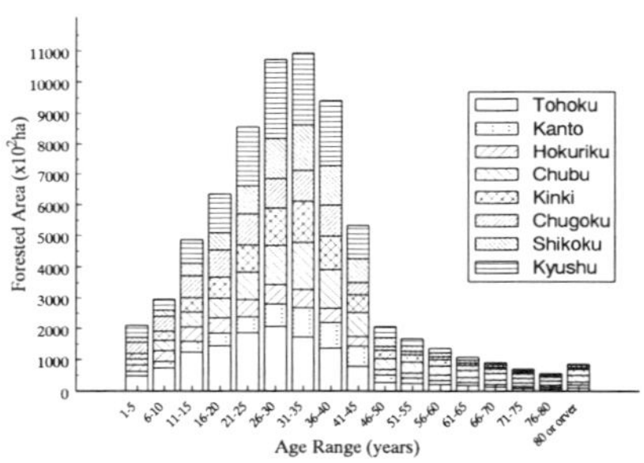

Figure 2. Age distribution of forest resources in 1995

Artificial forest resources of sugi and hinoki in Japan can be characterized by their age distribution (Figure 2). Effect of the large-scale plantation projects appears in age classes from 20 to 40 years. About 60% of forested area is categorized in these classes with 86.48 million m^3 forest stock. This implies that there could be a large lumber supply from domestic forest resources. Given this age distribution and growth projection for sugi and hinoki forests, we formulated a harvest scheduling problem to see how much timber could be available in the coming decades.

Using the following growth function to project forest growth (Richards 1958),

[1] $w(t) = M(1 - e^{-a \cdot t})^b$

where $w(t)$ represents volume per hectare at age t, M, a, and b are coefficients, we estimated coefficients of the function for each region with forest inventory data (Rinyacho 1995). Resultant growth projection for sugi and hinoki is given in Figure 3-a and b, respectively. Owing to a lack of timber quality data, the growth projection for hinoki for all regions except for Tohoku region showed early attainment of its asymptotic level--after 50 years or older.

To calculate harvestable volume from these sugi and hinoki forests, the following strata-based harvest scheduling problem was used (Johnson and Stuart 1987). The objective was to maximize harvestable volume over the planning time horizon, which was set to 100 years or 20 periods of 5 years each.

[2] $J = \max \sum_{i=1}^{n} c_i \cdot x_i$

where $\{x_i\}$ is a set of decision variables, representing harvest treatments, e.g., "harvesting 20% of the 5-th strata at period 3," and $\{c_i\}$ is a coefficient set for the corresponding decision variables, representing harvestable volume by implementing the i-th treatment over the planning time horizon. An n is the number of the decision variables, and is equal to the number of treatments. We use two types of constraints. One is the land accounting constraint [3], regulating land availability for harvesting. We assumed no new forest land created. The other type of constraint is the harvest flow constraint [4]. Considering sustainability of forest management, we set harvest flow from each period to be equal over the planning time horizon in each region.

[3] $\underline{L} \cdot \underline{X} \leq \underline{1}$

[4] $\sum_{k \in A_j} v_{i,k} \cdot x_k = \sum_{k \in A_j} v_{i+1,k} \cdot x_k \quad \text{for} \quad i = 1,2,,,T-1, \; j = 1,2,,,m$

[5] $\underline{X} \geq \underline{0}$

where $\underline{L}$ is a land accounting matrix with an (m x n) dimension, with m equal to the number of regions, $\underline{X}=(x_1, x_2, x_3,,, x_n)^t$ with t indicating transpose, and A_j is an index set for the j-th region. If the k-th treatment, x_k, relates to the j-th region, then an index k becomes an element of A_j. Harvestable volume from the k-th treatment at the i-th period is represented by $v_{i,k}$. Thus the constraint [4] implies that harvestable volume at the i-th period from the j-th region is equal to that at the next period.

Solving the above problem with minimum rotation period set to 7 periods, we had a result given in Table 1. The total harvestable volume over one period (5 years) became 272.66 (or 54.53 per year) million m^3, which is more than double that of the current harvest level, 117.18 (or 23.45 per year) million m^3. Among eight regions, Kyushu showed the highest harvestable volume, 60.83 million m^3, followed by Tohoku, 48.10 million m^3. The current harvest level and harvestable volume derived from the above problem in each region were used to estimate potential of increase in harvest for each region. As can be seen, possible increase in harvest level is the highest in Kyushu, then Kinki, Tohoku, Shikoku, Chugoku, Chubu, Kanto, and Hokuriku.

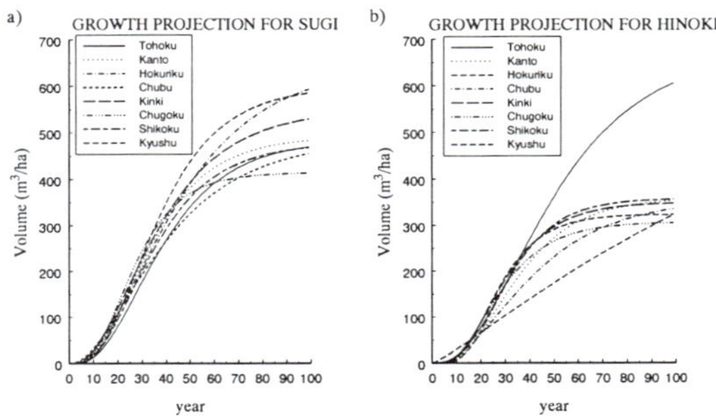

Figure 3. Growth projection for sugi and hinoki

Table 1. Estimated harvestable volume over 5 year period (million m^3)

	1991-1995	Estimated	Potential
Tohoku	27.26*	48.10	20.83
Kanto	9.48	22.99	13.51
Hokuriku	7.44	20.83	13.39
Chubu	13.97	29.06	15.09
Kinki	8.96	33.09	24.13
Chugoku	10.71	26.92	16.21
Shikoku	11.48	30.84	19.36
Kyushu	27.88	60.84	32.96
Total	117.18	272.67	155.48

* The value was calculated from merchantable volume using 65% conversion rate from raw materials.

3. A Spatial and Temporal Market Equilibrium Model

In this paper, we used a spatial and temporal market equilibrium model introduced by Samuelson (1952). Unlike the Koopmans-Hitchcock type equilibrium model, the Samuelson model takes into account demand and supply functions in order to search for an equilibrium, so that an effect of change in those functions can be revealed through the model.

As stated by Samuelson (1952), an equilibrium solution in several markets is determined by maximizing net social profit (NSP) involved in the markets. NSP is defined by the sum of consumer surplus and producer surplus minus transportation costs to deliver products from producer markets to consumer markets. Given demand and supply functions for m consumer markets and n producer markets, the proposed problem is formulated by

[6] $$\underset{[\{d_i\},\{s_j\},\{q_{i,j}\}]}{\text{maximize}}\ NSP = \sum_{i=1}^{m}\int_0^{d_i} PD_i(x)dx - \sum_{j=1}^{n}\int_0^{s_j} PS_j(x)dx - \sum_{i=1}^{m}\sum_{j=1}^{n} T_{i,j}\cdot q_{i,j}$$

where $PD_i(\mathrm{x})$ is a demand function of quantity for the i-th consumer market, and $PS_j(\mathrm{x})$ is a supply function of quantity for the j-th producer market. Transportation cost per unit quantity is expressed by $T_{i,j}$ from the j-th producer market to the i-th consumer market. Variables, d_i, s_j, and $q_{i,j}$ are the total quantity delivered to the i-th consumer market, the total quantity delivered from the j-th producer market, and the quantity delivered from the j-th producer market to the i-th consumer market, respectively. An optimal solution for them is searched by maximizing NSP subject to the relation among them,

[7] $$d_i = \sum_{j=1}^{n} q_{i,j} \qquad \forall i,$$

[8] $$s_j = \sum_{i=1}^{m} q_{i,j} \qquad \forall j$$

[9] $$\{d_i\},\{s_j\},\{q_{i,j}\} \geq 0$$

Constraint [7] is to equate quantity demanded in the i-th consumer market to the sum of quantities delivered from all producer markets. Constraint [8] is for quantity supplied from the j-th producer market equated to the sum of quantities delivered to all consumer markets. The last constraint [9] is to force variables to be a non-negative real value.

4. Japanese Forest Sector Model

Figure 4 shows the framework of the proposed Japanese forest sector model. We used three different types of lumber, i.e., domestic lumber produced from domestic forest resources, imported lumber, and US-JP lumber. The other lumber processed in Japan from imported logs was not considered here.

In the proposed model, we applied a linear function for demand and supply functions, which are derived by using price elasticity with one set of price and quantity. Provided price elasticity for demand function in the i-th consumer market, d_i, the demand function, $PD_i(\mathrm{x})$, becomes

[10] $$PD_i(x) = (1+\frac{1}{\varepsilon_i})\bar{p} - \frac{\bar{p}}{\bar{d}}\cdot\frac{1}{\varepsilon_i}\cdot x$$

where $(\bar{p},\bar{d})$ is one set of price and quantity. We used average value of price and quantity over the last three years. Likewise, supply function for the j-th producer market with ε_j price elasticity, is obtained by

[11] $PS_j(x) = (1 - \frac{1}{\varepsilon_j})\bar{p} + \frac{\bar{p}}{\bar{d}} \cdot \frac{1}{\varepsilon_j} \cdot x$

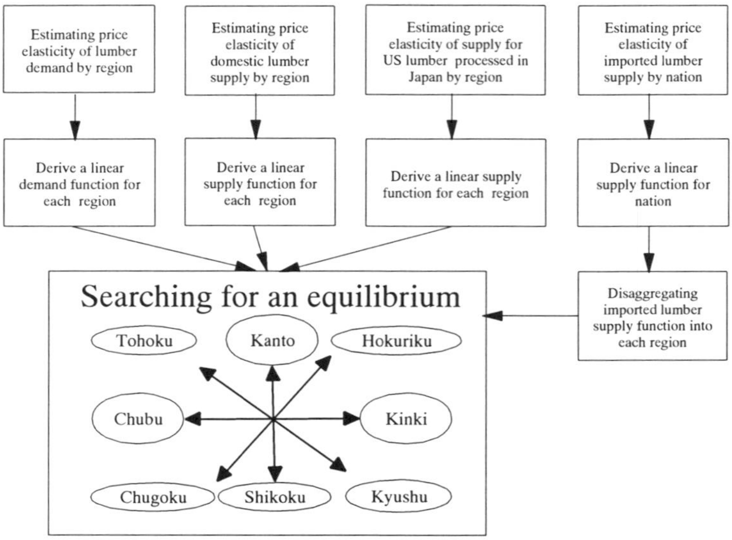

Figure 4. Framework of the proposed model

As a result, the proposed optimization problem becomes a quadratic programming problem. We used MINOS (Murtagh and Saunders 1993) to solve the problem.

Estimation of price elasticity requires regression analysis on demand and supply for lumber. We used the following logarithmic functions to estimate the elasticity. Price elasticity of the total domestic lumber demand function is estimated by

[12] $\log(Q_i^D) = a_1 + a_2 \cdot \log(P_i^W) + a_3 \cdot \log(H_i)$

The coefficient, a_2, becomes price elasticity of the total domestic lumber demand function. Likewise, because for price elasticity of the domestic lumber supply function we used

[13] $\log(Q_i^{DS}) = b_1 + b_2 \cdot \log(P_i^D) + b_3 \cdot \log(C_i^H)$

then b_2 is price elasticity. For price elasticity of the supply function for the US-JP lumber,

[14] $\log(Q_i^{US}) = c_1 + c_2 \cdot \log(P_i^U) + c_3 \cdot \log(Q_i^L) + c_4 \cdot \log(W_i^P)$.

is used, and

[15] $\log(Q^{IS}) = d_1 + d_2 \cdot \log(P^I) + d_3 \cdot \log(W) + d_4 \cdot \log(E)$

for price elasticity of imported lumber supply function. The second coefficient on the right-hand side of all equations is regarded as price elasticity, and it is calculated for each region except the last one. Variables are defined as follows.

Q_i^D : Total demand for lumber ($10^3 m^3$) in the i-th region[1)]
P_i^W : Weighted average price of lumber (10^3Yen) in the i-th region[1)]
H_i : Wood based construction starts ($10^3 m^2$) in the i-th region[2)]
Q^{IS} : Imported lumber supply (m^3) to Japan[3)]
P^I : Imported lumber price (US$) [3)]
W : Wage index[4)]
E : Foreign exchange rate (Yen/US$) [4)]
Q_i^{US} : US-JP lumber supply ($10^3 m^3$) in the i-th region[1)]
P_i^U : US-JP lumber price (10^3Yen) in the i-th region[1)]
Q_i^L : Imported log supply ($10^3 m^3$) in the i-th region[1)]
W_i^P : Wage for manufacturing in Japan (10^3Yen) in the i-th region[1)]
Q_i^{DS} : Domestic lumber supply ($10^3 m^3$) in the i-th region[1)]
P_i^D : Domestic lumber price (10^3yen) in the i-th region[1)]
C_i^H : Harvesting wage (Yen/m^3) in the i-th region[5)]

Note that 1) indicates data from "Forest Product Demand & Supply Report" (Rinyacho, 1975-1996), 2) from "Annual Statistical Report of Construction" (Ministry of Construction, 1975-1996), 3) from "FAOSTAT Database" provided by FAO on their Internet homepage, and 5) from "Annual Statistics Survey of Forest and Forestry (Ringyo Tokei Yoran)" (Rinyacho, 1976-1997). Wage index in 4) is an average wage in US$ of the following 16 countries, Australia, Canada, India, Japan, South Korea, Mexico, New Zealand, United States, Austria, Finland, Germany, Italy, Norway, Sweden, and United Kingdom. The data in 4) are from "International Financial Statistics Yearbook" (International Monetary Fund, 1997). Annual data from 1974 to 1995 were used for estimating coefficients of equation [12], [13] and [14]. As for the imported lumber, annual data from 1970 to 1996 were used.

Equations [10] and [11] were used for deriving the total lumber demand function, the domestic lumber supply function, and the US-JP lumber supply function in each region. Regarding the imported lumber supply function, owing to a lack of data in each region, only price elasticity for the nation was estimated. Thus, the imported lumber supply function for each region was determined with the assumption that each supply function is a portion of the derived imported lumber supply function for the nation. In other words, assume $Q_i^{IS}(\cdot)$ be the supply function of price for the i-th region. Since the supply function for the nation, $Q^{IS}(\cdot)$, is aggregation of those for eight regions, thus we have

$$[16] \quad Q^{IS}(\cdot) = \sum_{i=1}^{8} Q_i^{IS}(\cdot)$$

Further assuming that each supply function is a portion of that for the nation, we set

$$[17] \quad Q_i^{IS}(\cdot) = r_i \cdot Q^{IS}(\cdot) \qquad where \sum_i r_i = 1$$

where r_j is the proportion of imported lumber in the j-th region of total imported lumber. Using the current data on the imported lumber in each region in 1995, we set 7.9%, 40.9%, 2.4%, 14.4%, 22.0%, 0.7%, 1.5%, 6.0% for the corresponding proportion, $\{r_i\}$, in Tohoku, Kanto, Hokuriku, Chubu, Kinki, Chugoku, Shikoku, and Kyushu, respectively. Applying equation [17] to equation [11] for deriving the imported lumber supply function in the j-th region, we have

$$[18] \quad PS_j(x) = (1-\frac{1}{\varepsilon})\overline{p} + \frac{1}{r_j} \cdot \frac{\overline{p}}{\overline{d}} \cdot \frac{1}{\varepsilon} \cdot x$$

where ε is price elasticity estimated from equation [15].

Table 2. Results of estimation of price elasticity for lumber demand and supply

		Region							
		Tohoku	Kanto	Hokuriku	Chubu	Kinki	Chugoku	Shikoku	Kyushu
Total Lumber Demand	a_1	-8.9060	-5.5540	-4.1843	-3.6394	-2.7664	-4.9957	-4.2370	0.6[illegible]31
		(-2.528)	(-1.644)	(-1.431)	(-1.134)	(-0.763)	(-1.855)	(-2.196)	(0.3[illegible])
	a_2	-0.5522	-0.8640	-0.5287	-0.5924	-0.3429	-0.8386	-0.4757	-0.0[illegible]27
		(-1.629)	(-2.367)	(-1.658)	(-2.196)	(-1.043)	(-2.481)	(-2.172)	(-0.0[illegible]8)
	a_3	1.6928	1.5833	1.2588	1.2790	1.1097	1.4415	1.1984	0.6[illegible]38
		(4.307)	(4.016)	(3.774)	(3.631)	(2.774)	(4.554)	(5.388)	(3.3[illegible]0)
	R^2	0.5243	0.4350	0.4240	0.3653	0.2844	0.4976	0.6258	0.6[illegible]55
	DW	0.5562	0.5535	0.4525	0.5241	0.3275	0.9155	0.9920	0.4[illegible]92
Domestic Lumber	b_1	9.0409	8.6823	8.2146	8.5387	8.4500	8.7673	6.5597	6.5[illegible]37
		(29.357)	(18.844)	(21.772)	(31.366)	(26.880)	(25.581)	(29.051)	(33.7[illegible]4)
	b_2	0.0965	0.0417	0.0425	0.0583	0.1024	-0.1633	0.1254	0.1[illegible]35
		(1.263)	(0.382)	(0.485)	(0.971)	(1.629)	(-2.155)	(2.711)	(4.[illegible]9)
	b_3	-0.7485	-0.8181	-0.8545	-0.6137	-0.8796	-0.5776	-0.0983	0.2[illegible]26
		(-10.459)	(-11.847)	(-15.591)	(-12.217)	(-15.865)	(-10.400)	(-2.510)	(7.0[illegible]6)
	R^2	0.8384	0.8692	0.9199	0.8760	0.9317	0.8384	0.4308	0.7[illegible]90
	DW	1.3145	0.8858	0.9451	0.8704	1.0718	1.1951	1.3027	2.2[illegible]43
US-JP Lumber Supply	c_1	-2.5332	2.8669	2.5618	4.1585	2.6640	2.4878	3.3594	0.0[illegible]34
		(-1.567)	(2.927)	(2.538)	(4.126)	(0.756)	(1.828)	(3.149)	(0.[illegible]4)
	c_2	0.5281	0.5522	0.0948	0.2148	0.4529	0.7567	0.4969	0.1[illegible]61
		(1.513)	(3.994)	(0.329)	(2.190)	(1.075)	(1.434)	(2.303)	(0.[illegible]7)
	c_3	0.5490	0.4042	0.3593	0.4854	0.6068	0.5458	0.2885	0.6[illegible]51
		(2.722)	(6.156)	(2.413)	(5.958)	(2.353)	(1.826)	(2.301)	(2.[illegible]4)
	c_4	0.7247	-0.1657	0.2630	-0.2008	-0.2572	-0.3794	-0.0723	0.3[illegible]29
		(4.685)	(-2.195)	(1.538)	(-2.741)	(-1.063)	(-1.597)	(-0.691)	(0.[illegible]0)
	R^2	0.7397	0.8219	0.6053	0.8845	0.3739	0.3299	0.3204	0.1[illegible]14
	DW	1.0236	1.6083	2.4837	1.5336	2.2478	2.2567	1.5362	2.1[illegible]41

Notes: value in ( ) indicates t-statistics. DW: Durbin-Watson Statistics

Table 3. Results of estimation of price elasticity for imported lumber supply

	d_1	d_2	d_3	d_4	R-square	DW
Estimates	17.5240	0.4777	-0.4547	-0.5354	0.8927	2.0958
	(6.928)	(1.924)	(-2.448)	(-1.941)		

Notes: value in ( ) indicates t-statistics. DW: Durbin-Watson Statistics

Table 2 shows results of coefficient estimation for equations [12], [13], and [14], and Table 3 shows results for equation [15]. As was observed in Table 2, the total domestic lumber demand in Kanto and Chugoku became highly elastic relative to price, while in Kyushu it became inelastic. The domestic lumber supply became inelastic in all regions, with even a negative value for Chugoku. For this region, we assumed a vertical supply curve for the analysis. As for the US-JP lumber supply, it became elastic relative to price in Chugoku, and inelastic in Hokuriku, Chubu, and Kyushu.

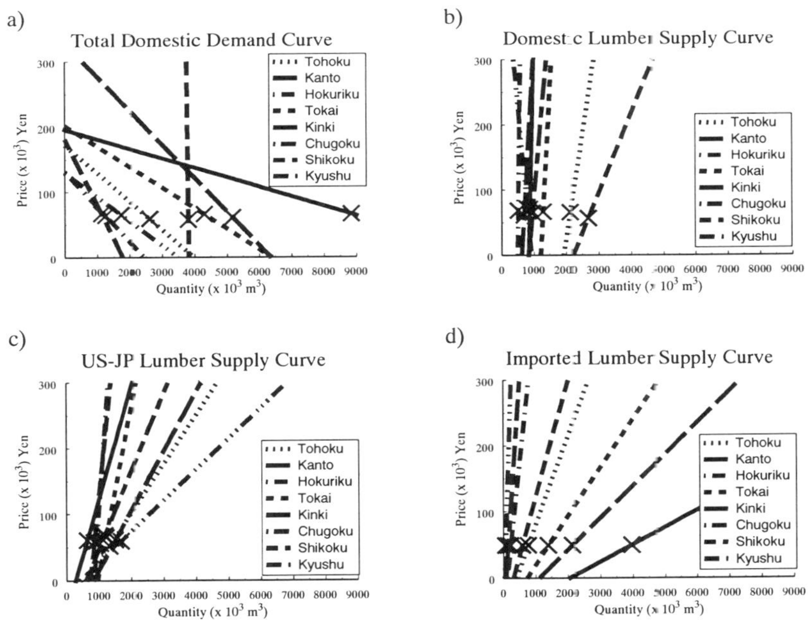

Figure 5. Derived demand and supply curves

Based on the results in Table 2, the resultant demand and supply curves are given in Figure 5. Since all supply functions derived here had a negative value for an intersect to the y-axis, the objective function [6] needs to be modified. This was done by converting integration of supply function with respect to quantity into that with respect to price as follows.

[6]'

$$\underset{[\{d_i\},\{s_j\},\{q_{i,j}\}]}{\text{maximize}}\ NSP = \sum_{i=1}^{m}\{\int_0^{d_i} PD_i(x)dx - d_i PD_i(d_i)\} + \sum_{j=1}^{n}\int_0^{p_j} QS_j(y)dy - \sum_{i=1}^{m}\sum_{j=1}^{n} T_{i,j}q_{i,j}$$

where

[19] $QS_j(y) = \bar{a} + \bar{b}\cdot y$

and coefficients of equation [19] are calculated by

[20] $$\bar{a} = \begin{cases} -r_j\cdot\bar{d}(\varepsilon-1) & \text{for imported lumber supply function} \\ -\bar{d}_j(\varepsilon_j-1) & \text{for other supply function} \end{cases}$$

[21] $$\bar{b} = \begin{cases} r_j\cdot\varepsilon\cdot\dfrac{\bar{d}}{\bar{p}} & \text{for imported lumber supply function} \\ \varepsilon_j\cdot\dfrac{\bar{d}}{\bar{p}} & \text{for other supply functions} \end{cases}$$

Once an optimal value of p_j is sought, it is transformed into quantity of supply by

[22] $$s_j = \begin{cases} \bar{a}+\bar{b}\cdot p_j & \text{if } p_j > 0 \\ 0 & \text{if } p_j = 0 \end{cases}$$

in order to search for an equilibrium solution for the markets.

5. Simulation Analysis

Solving the proposed problem given in the last section along with the transportation cost data in Table 4, an equilibrium solution was sought (Table 5). The total amount of domestic lumber demand was estimated at 29.25 million m^3, which was satisfied by 9.59 million m^3 from domestic lumber, 9.32 million m^3 from US-JP lumber, and 10.34 million m^3 from imported lumber. Regions with excess lumber supply were Tohoku, Hokuriku, Chugoku, Shikoku, and Kyushu, among which Tohoku had the largest excess, followed by Shikoku, Chugoku, Hokuriku, and Kyushu. Excess lumber demand was observed for the other regions, Kanto, Chubu, and Kinki. About 50% of the lumber demand in Kanto was dependent upon other regions, while about 10% of the demand in Chubu and Kinki was from outside of the region.

Table 4. Transportation costs among regions (Yen/m^3)

	Tohoku	Kanto	Hokuriku	Chubu	Kinki	Chugoku	Shikoku	Kyushu
Tohoku	2000	3500	4000	4500	7000	8000	8000	10000
Kanto	-	2000	3500	3000	3500	3500	4000	7000
Hokuriku	-	-	2000	3000	2800	4000	4000	7000
Chubu	-	-	-	2000	3000	4000	4000	7000
Kinki	-	-	-	-	2000	3000	3000	4000
Chugoku	-	-	-	-	-	2000	2000	4000
Shikoku	-	-	-	-	-	-	2000	4000
Kyushu	-	-	-	-	-	-	-	2000

Simulation analysis was conducted to investigate the possibility of increase in the domestic lumber supply. The first analysis was implemented by reducing transportation costs only for domestic lumber. This is to provide a price advantage for domestic lumber suppliers. Change was carried out by reducing the costs by 0% to 100% from the current level in Table 4. Figure 6 shows effects of reduction in transportation costs for domestic lumber delivery. As was observed from these figures, there was almost no effect on each lumber supply. This could result from the fact that most domestic lumber supply functions are inelastic to price change. Reduction of transportation costs would not be able to increase the domestic lumber supply.

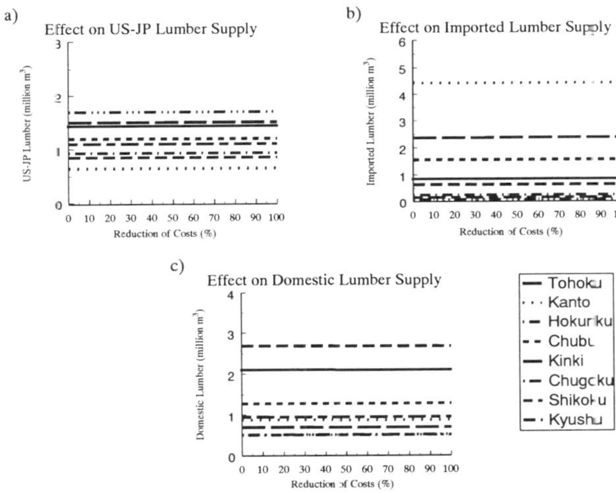

Figure 6. Effect of reduction of transportation costs

The second analysis was to see how reduction of the United States log imports or the US-JP lumber supply influences the amount of domestic lumber supply. Since there is a tendency of reducing the U.S. log exports owing to the environmental movement of regulating exporting raw materials, it is important to investigate its effect. The analysis was carried out by increasing the value of the intersect of the US-JP lumber supply function in the following manner:

$$[23]\quad \bar{a}_{new} = \bar{a} + \frac{i}{100} \cdot Q_{eq} \cdot \bar{b}$$

where $\bar{a}_{new}$ is the new intersect and Q_{eq} is an amount in the equilibrium in Table 5. Increase was specified by changing i from 0% to 100%. Since increase in the intersect shifts the US-JP lumber supply curve toward the left, the amount traded in the markets of the US-JP lumber in each region is expected to decrease. Results are depicted in Figure 7. Changing the value of the intersect by 0% to 100% resulted in reduction of the US-JP lumber supply by 67% to 96%, varying in regions. The reduction of the US-JP lumber supply in each region was covered mostly by imported lumber. Through all regions, about 20 to 25% increase in the imported lumber supply was observed, while increase in the domestic lumber supply was less than 5% for all regions except Kyushu with 8.7%.

Table 5. An equilibrium solution for the markets (x10^3 m^3)

Supply Origin				Demand Destination							
				Tohoku	Kanto	Hokuriku	Chubu	Kinki	Chugoku	Shikoku	Kyushu
				**2,553	9,040	1,296	4,401	5,141	1,837	1,180	3,800
Tohoku	*4,375	DL1)		279	1,822	0	0	0	0	0	0
		US2)		1,428	0	0	0	0	0	0	0
		IL3)		846	0	0	0	0	0	0	0
Kanto	5,943	DL		0	872	0	0	0	0	0	0
		US		0	644	0	0	0	0	0	0
		IL		0	4,427	0	0	0	0	0	0
Hokurik	1,682	DL		0	0	484	0	19	0	0	0
		US		0	0	812	115	0	0	0	0
		IL		0	0	0	252	0	0	0	0
Chubu	4,034	DL		0	0	0	1,281	0	0	0	0
		US		0	0	0	1,195	0	0	0	0
		IL		0	0	0	1,557	0	0	0	0
Kinki	4,562	DL		0	0	0	0	691	0	0	0
		US		0	0	0	0	1,496	0	0	0
		IL		0	0	0	0	2,376	0	0	0
Chugok	2,270	DL		0	0	0	0	0	507	0	0
		US		0	1,275	0	0	22	0	389	0
		IL		0	0	0	0	0	77	0	0
Shikoku	2,204	DL		0	0	0	0	0	947	0	0
		US		0	0	0	0	0	306	790	0
		IL		0	0	0	0	161	0	0	0
Kyushu	4,173	DL		0	0	0	0	376	0	0	2,312
		US		0	0	0	0	0	0	0	847
		IL		0	0	0	0	0	0	0	640

* Total lumber supply, ** Total lumber demand, 1) Domestic lumber supply, 2) Supply of the US lumber processed in Japan, 3) Imported lumber supply

The last simulation analysis was conducted by increasing the domestic lumber supply in the same way but with a curve shift direction opposite that for the US-JP lumber supply change. Figure 8 shows the change in each lumber supply. With a 95% increase in the domestic lumber traded in the markets, a 20% reduction of the imported lumber supply was observed. For the US-JP lumber supply, a reduction was also observed, but it was less than 20%. This implies that change in the domestic lumber supply has more influence on the imported lumber supply than on the US-JP lumber supply.

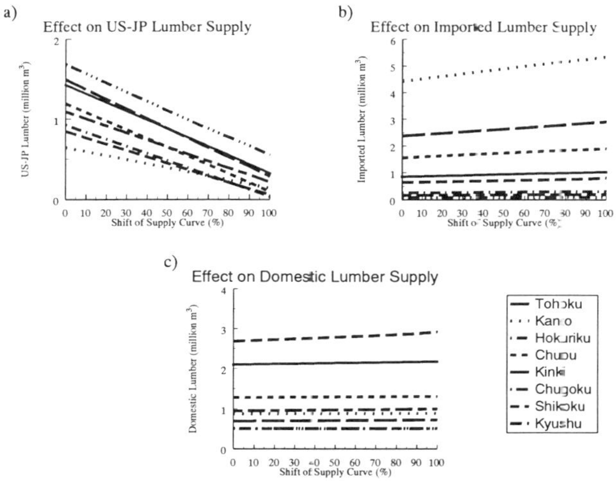

Figure 7. Effect of shifting US-JP lumber supply curve

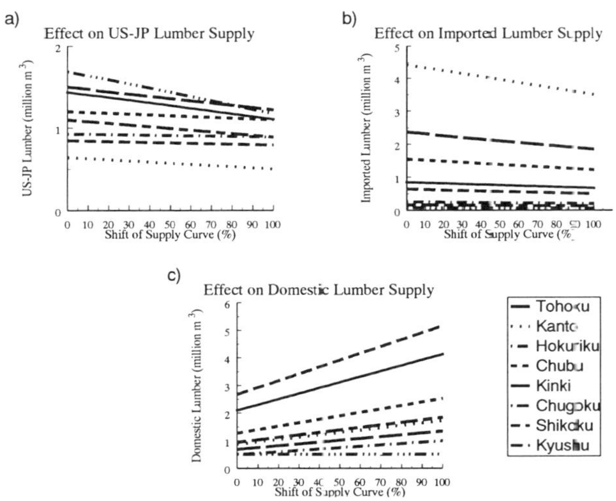

Figure 8. Effect of shifting domestic lumber supply curve

6. Conclusions

In the last several decades, it was often thought that the domestic lumber supply could increase along with an increase in merchantable domestic forest resources in Japan, which were planted during the 1960s. However, the surrounding economic environment and internationalization of timber markets has become less favorable to domestic lumber production. In this paper, we investigated a situation using a spatial and temporal market equilibrium model for the Japanese lumber markets, which could result in increase in the domestic lumber production.

We considered three cases for the analysis. In the first case, we investigated the effect of transportation costs. By reducing the transportation costs to favor Japanese lumber, almost no change in the market share was observed. Because the domestic lumber supply functions in all regions were found to be inelastic to price change, the effect of reducing transportation costs turned out almost negligible. This led to no effect on the imported lumber supply as well as the US-JP lumber supply. The second case was to see if a decrease in U.S. log imports could lead to an increase in domestic lumber supply, while the third case was completed by increasing the domestic lumber supply manually. Our analysis showed that there was almost no increase in the amount of the domestic lumber traded in the markets, rather an increase in imported lumber. Increases in the domestic lumber supply pushes away some portion of the imported lumber from markets more than it did the US-JP lumber. Through the analysis, all results may stem from the fact that the imported lumber supply functions are elastic to price change while the domestic lumber supply functions are not. Inelastic domestic lumber supply functions could have resulted from the nature of the Japanese forest production planning system. Since the production plan is often made for the next decade or so under a top-down decision process from the Forestry Agency to the local governments, change in production level from the target amount in response to price change appears unlikely.

It is also true that as long as the Samuelson-type market equilibrium model is used, inelastic functions have less contribution to the objective function than do elastic ones in terms of small change in quantity or price. This implies that an optimal solution tends to be the one that increases more surplus created by elastic functions. This raises a point that estimation of elasticity is one of the important tasks for this kind of forest sector analysis. For this research, further analysis should be carried out to see any effect of changing price elasticity of the domestic lumber supply in order to seek the potential of an increase in the domestic lumber traded in the markets.

Solving the strata-based harvest scheduling problem for Japanese plantation sugi and hinoki forests, the result indicated a potential increase in the domestic lumber production to more than twice the current level. If the Japanese lumber market environment becomes favorable to domestic lumber, it is possible for domestic lumber suppliers to dominate the market. It should be noted however that even if it becomes possible, sustainability of timber production has to be taken into consideration. Because of non-uniform age distribution of Japanese forest resources, careful scheduling has to be implemented in order to achieve sustainable forest resource management.

7. References

Flora, D., and Lane, C. 1994. Equilibrium modeling of pacific rim trade in logs and lumber. Journal of Forest Economics. 126: 9-11.

Furuido, H., Kato, T., Inoue, M., Nagata, S., and Oka, H. 1991. Econometric analysis of forest products trade in the pacific region (1) Softwood lumber exports from USA to Japan. Transactions of the Annual Meeting of the Japanese Forestry Society. 102: 33-35.

Gallagher, P. 1980. An analysis of the softwood log trade between the United States and Japan. Technical Bulletin 330 Forestry Series. 34: 2-19, Agricultural Experiment Station, University of Minnesota.

International Monetary Fund. 1997. International financial statistics yearbook. International Monetary Fund. Washington, DC. 827p.

Johnson, K.N., and Stuart, T.W. 1987. FORPLAN Version 2: Mathematical programmer's guide. USDA For. Serv., Land Management Planning Systems Section, Washington, DC.

Labys, W.C. 1989. Spatial and temporal price and allocation models of mineral and energy markets. P.17-47 *in* Labys, W.C., Takayama, T., Uri, N.D. (eds.), Quantative methods for market-oriented economic analysis over space and time. Avebury, Vermont.

McKillop, W.L.M. 1973. Structural analysis of Japanese-North America trade in forest products, Forest Science. 19: 63-74.

Ministry of Construction, 1975-96. Annual statistical report of construction. Ministry of Construction. Kensetu Bukka Kyokai, Tokyo.

Mori, Y. 1991. Timber market in Japan: An econometric analysis. Memoirs of the College of Agriculture, Kyoto Univ., 179-191.

Murtagh, B.A., and Saunders, M.A. 1993. MINOS 5.4 user's guide (preliminary). Department of Operations Research, Stanford University, California. 135p.

Nagata, S., Furuido, H., Kato, T., and Oka, H. 1992. Econometric analysis of forest products trade in the pacific region (2) Making of an econometric model for coniferous wood trade between Japan and North America. Transactions of the Annual Meeting of the Japanese Forestry Society. 103: 29-31.

Nomura, I. 1978. Shingaizaidokuhon, Ringyoshinbunsha, 419p.

Richards, F.J. 1958. A flexible growth function to empirical use. Journal of Experimental Botany. 10: 290-300.

Rinyacho. 1975-96. Annual report. Report on Supply and Demand of Timber. Ministry of Agriculture, Forestry and Fisheries of Japan. Norin Toukei Kyokai, Tokyo.

Rinyacho. 1976-97. Annual statistics survey of forest and forestry. Japanese Forestry Agency. Ringyo Tokei Yoran. Rinya Kousai Kai, Tokyo.

Rinyacho. 1995. Current states of forest resources, Forestry Agency, Tokyo, 90p.

Rinyacho. 1989. Mokuzaijukyuu-to-mokuzaikougyou-no-genkyo (Current states of demand and ssupply of forest products and forest industries), Forestry Agency, Tokyo, 496p.

Rinyacho. 1997. Mokuzaijukyuu-to-mokuzaikougyou-no-genkyo (Current states of demand and supply of forest products and forest industries), Forestry Agency, Tokyo, 560p.

Samuelson, P. A. 1952. Spatial price equilibrium and linear programming, The American Economic Review. 42: 283-303.

Tachibana, S. 1994. An econometric analysis of timber trade between Japan and North America: Relationship between the environmental conservation movement and the timber trade, Journal of Forest Economics. 125: 36-41.

Vincent, J. R., Brooks, D. J., and Gandapur, A. K. 1991. Substitution between tropical and temperate sawlogs. Forest Science. 37: 1484-1491.

Yukutake, K. 1984. Econometric analysis of South Sea and USSR log market in Japan. Journal of Agricultural Research Quarterly. 17: 269-279.

Yukutake, K. 1985. Simulation analysis of log and lumber: Econometric model for Japan. The Current State of Japanese Forestry. 4: 1-14.

Yukutake, K., and Yoshimoto, A. 1996. Effects of timber export from North America to Japan on regional timber production: A quantitative analysis using the Koopmans-Hitchcock type market equilibrium model, Transactions of the Japanese Forestry Society. 107: 413-414.

Yukutake, K., Yoshimoto, A., Nagata, S., and Tachibana, S. 1996. Forest sector modeling in Japan. P. 111-134 *in* Proc. of Project Group P.6.11 FORESEA Meetings at the 20th IUFRO World Congress, Univ. of Washington.

Acknowledgements

This research was supported by a Grant-in-Aid for International Scientific Research (No.08041066) from the Ministry of Education, Science, Sports and Culture of Japan.

Modeling the North American Demand for Solid Wood Products with an Emphasis on Canada

Susan Phelps

Key words: Forest sector modeling, linear programming, PELPS III, NASAW

Abstract: A spatial equilibrium model of the North American solid wood sector has been developed jointly by the Canadian Forest Service and the Forest Products Laboratory of the USDA Forest Service. The model, NASAW, is constructed using the Price Endogenous Linear Programming System, PELPS III, and provides 20-year forecasts of prices, quantities and manufacturing capacities for six North American regions: representing Canada (British Columbia Coast, British Columbia Interior, and the rest of Canada) and the United States (North, South and West). Two broad offshore demand regions (Atlantic and Pacific) are also included for each country. Demand and supply are described by equations that define quantity as a function of price. Manufacturing technologies are represented by processes that convert raw materials into primary products (lumber, plywood, OSB-waferboard and particleboard-MDF) and primary products into secondary products (engineered lumber, trusses, foam core panels). Both primary and secondary products are used in single and multi-family house construction. The model tracks shifts in the type of construction system used as a result of changes in relative prices. Raw materials, primary and secondary products, and by-products can all be exported. Forecasts are made for housing demand, manufactured product consumption and production (lumber, structural and non-structural panels), and fiber supply (hardwood and softwood). The model provides a useful framework for analyzing a broad range of trade-related and product substitution issues. Preliminary results show a continued increase in the importance of the United States South and eastern Canada as major suppliers of lumber and structural panels to the North American market.

1. Introduction

Several economic studies of the North American forest sector have been undertaken in recent years by researchers at the USDA Forest Service and the Canadian Forest Service (CFS). These studies have focused primarily on the pulp and paper sector and have used the method of price-endogenous linear programming to project how the sector may evolve over time under various economic conditions. In their investigations, the studies by Ince (1994), Jacques (1992) and Jacques et al. (1993) used a model based on the general price endogenous linear programming system PELPS III developed by Zhang et al. (1993).

PELPS III is a general system for combining regional information on supply and demand curves, manufacturing technologies and transportation costs into spatial sector models and was originally applied by the authors to the North American pulp and paper sector. This model was extended to become NAPAP (the North American Pulp and Paper model) developed jointly in 1991 by researchers at the USDA Forest Service Forest Products Laboratory in Madison, Wisconsin, and the Canadian Forest Service in Ottawa, Ontario. NAPAP is described in detail by Ince (1994) and is currently used by the USDA Forest Service in the RPA Timber Assessment (see Haynes et al. 1995), along with several other forest sector models reported in the literature (i.e., TAMM, ATLAS, FASOM see Adams and Haynes paper for definitions). The RPA Timber Assessment is required periodically to assess the long-term supply and demand situation of the United States forest sector.

In 1994, a second joint project was undertaken by the two research agencies to develop a model of the North American solid wood sector (NASAW) also based on price endogenous linear programming.[1] This model, NASAW, facilitates the analysis of sectoral trade issues which are of interest to Canada and provides insight into the dynamics of the North American solid wood sector, especially the substitution of building products in the house construction industry. It is expected that NASAW will complement the suite of models currently used in the RPA Timber Assessment process.

The purpose of this paper is to present an overview of the structure of the NASAW model with some of the data estimated for Canada, and to provide some examples of the types of output that can be generated and analyses that can be undertaken using the model.

2. Structure of the Model

PELPS III requires first defining regions, commodities and manufacturing processes for the sector under study. Once determined, a database is created by entering information for each of these into a series of templates provided with PELPS III as a Lotus 1-2-3 spreadsheet file, which is then input to the matrix generating component of the system. The matrix generator of PELPS III is actually a series of computer programs, written in Pascal, which interact with the input data to generate an LP matrix solved by LINDO (LINDO Systems Inc. 1996).

NASAW is an annual model calibrated for 1986 (also the base year of NAPAP). All Canadian prices and costs have been converted to US$ and deflated to 1986 values.[2] The model is metric; all products are expressed in

[1] Participants include Henry Spelter, Denise Ingram and Irene Durbak of the USDA Forest Service and, in addition to the author, Darcie Booth and Tony Lemprière of the Canadian Forest Service.

[2] To convert to 1997 $US, multiply by 1.37

m^3 for volume, m^2 for area, or kg for weight. Most equations were estimated statistically from historical data using ordinary least squares (OLS) or extensions of OLS, although in some cases published estimates from other sources were used. Generally, the historical information available from public sources such as Statistics Canada provided over 30 years of data beginning about 1960. The following sections generally describe the structure of NASAW in the context of the PELPS III data requirements and include some of the data making up the Canadian component of the model.

2.1 Regions

The United States is both an important producer and consumer of solid wood products. However, the areas of production and consumption within the country are not homogeneously distributed. For example, the major timber producing regions of the United States are the Pacific Northwest and the South, while the largest wood product consuming region is in the northeast section of the country. The United States is represented in NASAW by three regions, which are both producers and consumers: the North, the South, and the West. Canada, on the other hand, with its relatively small population is a much less important consumer of forest products, but is a significant producer. Consequently, Canada is represented in NASAW as a single consuming region and as three producing regions: the British Columbia Coast, British Columbia Interior, and the rest of Canada (including the three prairie provinces, plus Ontario, Québec and the four Atlantic provinces).

In addition, two broad export regions have been defined: Atlantic and Pacific. Although the specific country make-up of these two regions is a function of the primary destination of the majority of exports of each product from each country, the Atlantic region generally includes countries in the European Union, while the Pacific region usually includes Japan and Korea.

2.2 Manufactured commodities

All lumber, structural panel and non-structural panel activities that make up most of the economic activity in the Canadian and United States solid wood sector have been included in NASAW. These are currently represented by three primary products, two by-products and four secondary manufactured products, noted as (P), (B), and (S) below. Another important secondary product, laminated veneer lumber (LVL), is currently not included in the model. The nine products are:

- softwood lumber (P)
- softwood plywood (P)
- oriented strandboard-waferboard (OSB) (P)
- chippable residues (for the pulp and paper sector) (B)
- non-chippable residues (for use by particleboard-MDF industries) (B)
- particleboard-medium density fiberboard (MDF) (S)

- wood I-joists (S)
- trusses (S)
- foam core panels (S)

2.3 End-use demand

The demand for final products is represented in NASAW by traditional economic demand curves, where quantity is estimated as a function of price. These require a starting price and quantity, a price elasticity, elasticities for up to three shifter variables plus a lagged term, and an optional lower bound.

Most solid wood products consumed in North America are used in building construction and for each of the four demand regions a demand curve has been estimated to represent each of the two major types of buildings constructed: single family and multi-family houses. The demand for housing has been estimated in millions of square meters (m^2) and is calculated as the product of the number of units (starts) and the average unit size (m^2) of each dwelling type. The regional demand for wood products used in mobile homes, nonresidential construction, industrial buildings, and repairs and renovations has been included in NASAW by demand curves that represent the demand for the individual wood products used in these applications.

Demand curves have also been estimated for the three types of wood fiber required by the pulp and paper sector (hardwood and softwood pulpwood and softwood pulp chips), but it is anticipated that eventually NASAW and NAPAP (Ince 1994) will be merged, eliminating the need for these equations. Table 1 summarizes the domestic demand equations estimated for NASAW. With the exception of softwood pulpwood, all equations in Table 1 represent Canadian demand. Softwood pulpwood is estimated for two sub-regions (western and eastern Canada, noted in the table as WC and EC, respectively).

To complete the demand picture, export equations have been estimated for solid wood products shipped to the Atlantic and Pacific regions. Table 2 summarizes these estimates for Canadian exports.

The demand equation shifter variables shown in Tables 1 and 2 vary in definition from product to product. For example, the 1st and 2nd shifter variables of the demand curve for Canadian single family housing shown in Table 1 represent real household disposable income and the 5-year mortgage rate, respectively. The 1st and 2nd shifter variables of the demand curve for exports of softwood lumber to the Atlantic region in Table 2 represent respectively, the aggregated real Gross Domestic Product (GDP) of the major importing countries in the region and an indicator of the recent concerns within the European Union regarding the potential introduction of pinewood nematode from imports of Canadian softwood lumber.

Table 1. Canadian demand elasticities and base period assumptions used in NASAW

Product	Base Period Price	Base Period Quantity	Own Price Elasticity	Elasticity of 1st Shifter	Elasticity of 2nd Shifter	Elasticity of 3rd Shifter
Single Family Houses (m^2)	597.3	17.47	-0.26	0.87	-0.46	
Multi-family Houses (m^2)	507.9	8.45	-0.64	1.29	-0.80	
Hardwood Pulpwood (m^3)	19.4	6.22	-0.10	1.00		
Softwood Pulp Chips (m^3)	19.0	52.27	-0.10	1.00		
Softwood Lumber (m^3)	98.7	11.30	-0.26	-2.06	-0.25	1.58
Structural Panels (m^3)	256.5	1.54	-0.26	-0.30	-0.32	-0.19
Engineered Wood (m^3)	212.0	0.02	-0.40	0.10		
Particleboard/MDF (m^3)	122.4	1.07	-0.37	1.72		
Softwood Pulpwood (WC) (m^3)	20.5	6.33	-0.10	1.00		
Softwood Pulpwood (EC) (m^3)	28.6	30.93	-0.10	1.00		

Table 2. Canadian export demand elasticities and base period assumptions used in NASAW

Region and Product	Base Period Price	Base Period Quantity	Own Price Elasticity	Elasticity of 1st Shifter	Elasticity of 2nd Shifter	Partial Adjustment Coefficient
PACIFIC DEMAND:						
Softwood Pulp Chips (m^3)	17.6	1.10	-0.72			1.14
Softwood Sawlogs (m^3)	52.4	2.31	-0.52	-0.16		0.88
Softwood Lumber (m^3)	98.7	2.61	-0.06	0.46		0.51
Structural Panels (m^3)	172.6	0.02	-0.58	1.89		
Particleboard/MDF (m^3)	122.4	0.05	-0.41	0.41		
ATLANTIC DEMAND:						
Softwood Lumber (m^3)	98.7	2.37	-0.16	0.58	-1.45	
Structural Panels (m^3)	172.6	0.25	-0.35	-0.74		
Particleboard/MDF (m^3)	122.4	0.04	-4.03	-1.23		-1.04

2.4 Fiber supply

Fiber supplies are represented in NASAW by traditional economic supply curves, where quantity is estimated as a function of price. As for demand, these require a starting price and quantity, a price elasticity, elasticities for up to three shifter variables plus a lagged term, and an optional upper bound. For the United States and Canada, input supply is distinguished between pulplogs and sawlogs (including veneer logs), species type (softwood and hardwood) and type of owner (public or private). With the exception of hardwood sawlogs, which are not currently included in the model, a supply curve has been estimated for each of these categories for each of the six supply regions. In addition, for each Canadian region the upper bound on supply has been used to approximate the annual allowable cut (AAC). This bound represents the maximum volume of timber type that can be supplied by each region per time period. Table 3 summarizes these Canadian supply curves.

2.5 Manufacturing processes

The production of a commodity in NASAW is modeled as a manufacturing process, described by activity analysis, which is a term introduced by Takayama and Judge (1964). Each process has a cost and a limited capacity and represents one combination of inputs or input mixes that can be used to manufacture a particular output. In NASAW, outputs may be either primary or secondary manufactured products or housing and inputs may be either raw materials, or primary or secondary products. Manufacturing activities may also result in the production of by-products such as chip residues, sawdust and shavings, which are then available for other uses.

Table 3. Canadian supply elasticities and base period assumptions used in NASAW

Region and Product	Base Period Price	Base Period Quantity	Own Price Elasticity	Base Period Upper Bound
B.C. COAST:				
Hardwood Pulpwood (m^3)	17.49	0.000	1.00	0.00
Softwood Pulpwood (m^3)	18.46	6.138	0.225	6.79
Softwood Sawtimber (m^3)	23.35	20.241	0.56	29.00
B.C. INTERIOR:				
Hardwood Pulpwood (m^3)	17.49	0.028	1.00	2.40
Softwood Pulpwood (m^3)	18.46	0.190	0.225	0.21
Softwood Sawtimber (m^3)	19.18	50.671	0.37	57.00
REST OF CANADA:				
Hardwood Pulpwood (m^3)	17.49	9.206	1.00	58.10
Softwood Pulpwood (m^3)	27.25	30.927	0.248	54.00
Softwood Sawtimber (m^3)	23.2	42.503	0.32	60.00

In NASAW, input-output coefficients (recovery factors) and manufacturing costs have been obtained or estimated for a number of processes used to manufacture the products outlined in section 2.2 and the two housing demands described in section 2.3. Additionally, several dummy processes have been created to overcome some of the restrictions of PELPS III and to model special situations, such as the Canada-United States softwood lumber agreement.

2.5.1 Solid wood products

Softwood lumber and plywood in NASAW may be manufactured completely from softwood sawlogs or from some combination of sawlogs and pulplogs. Canadian producers are assumed to use softwood sawlogs only. Depending on relative prices, it may also be profitable from time to time for these processes to operate in an overtime situation to produce a limited amount of additional product. Two processes are used to represent this incremental production at a slightly higher cost than the two regular capacity processes.

OSB-waferboard is produced in NASAW by only one process, but with two input mixes, allowing the model to choose between the two types of fiber used in North American OSB-waferboard production: hardwood pulpwood (mostly aspen) and softwood pulpwood (mostly pine). Non-structural panels, represented in NASAW by particleboard-MDF, are produced by a single process using residues generated as a by-product of lumber and plywood manufacturing. Only residues that are not sent to the pulp and paper industry as wood chips are available to the particleboard-MDF industry in NASAW.

Wood I-joists (engineered lumber), made by gluing a lumber flange to a structural panel web, are manufactured in NASAW by a single process, but with two input mixes. These represent the combination of flange and web materials used in wood I-joist construction. For both input mixes, the flange component is softwood lumber, but the web component varies between OSB-waferboard and softwood plywood.

Trusses are manufactured in NASAW from softwood lumber using only one process. One process has also been defined for foam core panel (FCP) production, but two input mixes have been identified to represent the two types of sheathing used in FCP construction: softwood plywood and OSB-waferboard.

Table 4 summarizes the Canadian base period input-output coefficients (volume of input required per m^3 of output produced) and manufacturing costs (1986 US$/$m^3$) for these solid wood processes. Table 5 shows the base period by-product factors (volume of by-product generated per m^3 of output produced) for the lumber and plywood manufacturing processes.

Table 4. Base period input coefficients and costs for Canadian solid wood product manufacturing processes used in NASAW

Region and Input	Softwood Lumber	Softwood Plywood	OSB/ Waferboard	Particleboard/ MDF	Wood I-Joists
B.C. COAST:					
Hardwood Pulpwood (m^3)	–	–	1.920	–	–
Softwood Pulpwood (m^3)	–	–	1.920	–	–
Softwood Sawtimber (m^3)	1.851	3.122	–	–	–
Softwood Residues ('000 kg)	–	–	–	0.798	–
Softwood Lumber (m^3)	–	–	–	–	0.500
Softwood Plywood (m^3)	–	–	–	–	0.156
OSB/Waferboard (m^3)	–	–	–	–	0.156
Manufacturing Costs (1986 US$)	41.89	96.62	60.22	60.40	175.87
B.C. INTERIOR:					
Hardwood Pulpwood (m^3)	–	–	1.920	–	–
Softwood Pulpwood (m^3)	–	–	1.920	–	–
Softwood Sawtimber (m^3)	1.926	2.354	–	–	–
Softwood Residues ('000 kg)	–	–	–	0.798	–
Softwood Lumber (m^3)	–	–	–	–	0.500
Softwood Plywood (m^3)	–	–	–	–	0.156
OSB/Waferboard (m^3)	–	–	–	–	0.156
Manufacturing Costs (1986 US$)	35.61	102.55	76.71	74.77	175.87
REST OF CANADA:					
Hardwood Pulpwood (m^3)	–	–	1.920	–	–
Softwood Pulpwood (m^3)	–	–	1.920	–	–
Softwood Sawtimber (m^3)	1.883	2.354	–	–	–
Softwood Residues ('000 kg)	–	–	–	0.727	–
Softwood Lumber (m^3)	–	–	–	–	0.500
Softwood Plywood (m^3)	–	–	–	–	0.156
OSB/Waferboard (m^3)	–	–	–	–	0.156
Manufacturing Costs (1986 US$)	45.60	101.29	72.36	79.85	175.87

Table 5. Base period by-product coefficients for Canadian lumber and plywood manufacturing processes used in NASAW

Region and By-Product	Softwood Lumber	Softwood Plywood
B.C. COAST:		
Chip Residues (m^3)	0.573	1.341
Sawdust and Other Residues ('000 kg)	0.116	
B.C. INTERIOR:		
Chip Residues (m^3)	0.841	0.765
Sawdust and Other Residues ('000 kg)	0.161	
REST OF CANADA:		
Chip Residues (m^3)	0.901	0.765
Sawdust and Other Residues ('000 kg)	0.107	

2.5.2 Housing

The demand for wood products used in North American single family and multi-family house construction is a function of the type of framing and sheathing system used. Several factors influence the choice of house construction system, but the most obvious are cost and consumer preference. Henry Spelter at the USDA Forest Products Lab, has identified six types of house framing systems and two types of sheathing systems for NASAW. Spelter's framing systems are defined as lumber, truss, engineered lumber, foam core panel (FCP), steel, and concrete and the sheathing systems are plywood and OSB-waferboard. Depending on relative input prices and housing demand, the model selects the most appropriate combination of framing and sheathing system, which then determines the type and volume of wood products used.

Each housing system has been included in NASAW as a separate manufacturing process, with an estimated (non-wood) cost and a limiting capacity. Each process defines a particular set of wood inputs required to produce a particular combination of framing and sheathing system. Table 6 summarizes Spelter's base period input-output coefficients (volume of wood input required per m^2 of house constructed) and non-wood construction costs (1986 US\$/m^2) for each of the six single and multi-family house construction processes defined for Canada.

2.6 Manufacturing capacities

Capacity limits for each of the above manufacturing processes have been obtained from published production limits or from industrial capacity utilization statistics, or estimated from ad hoc procedures such as trends through peaks. Table 7 summarizes the base period capacities (in millions of m^3), depreciation rates, and replacement costs (1986 US\$/m^3) estimated for each of the Canadian solid wood manufacturing processes. Housing capacities have also been estimated, but are not shown here.

Table 6. Base period input coefficients and costs for Canadian single and multi-family house construction processes, by type of sheathing, used in NASAW

Type of Sheathing and Input	Lumber Framed	Truss Framed	Engineered Lumber Framed	Foam Core Panel Framed	Steel Framed	Concrete Framed
SINGLE FAMILY HOUSES						
Plywood Sheathing:						
Softwood Lumber (m^3)	0.2167	0.1560	0.1158	0.1499	0.0912	0.0762
Softwood Plywood (m^3)	0.0451	0.0413	0.0483	0.0255	0.0471	0.0333
OSB (m^3)	–	–	–	–	–	–
Wood I-joists (m^3)	–	–	0.0224	–	–	–
Truss (m^3)	–	1.7061	–	0.6234	1.7061	3.0184
Foam Core Panel (m^3)	–	–	–	2.0000	–	–
Manufacturing Costs (1986 US$)	513.9	508.9	507.6	499.4	523.1	473.6
OSB Sheathing:						
Softwood Lumber (m^3)	0.2167	0.1560	0.1158	0.1499	0.0912	0.0762
Softwood Plywood (m^3)	0.0021	0.0021	0.0032	0.0029	0.0021	0.0083
OSB (m^3)	0.0412	0.0388	0.0444	0.0231	0.0442	0.0278
Wood I-joists (m^3)	–	–	0.0224	–	–	–
Truss (m^3)	–	1.7061	–	0.6234	1.7061	3.0184
Foam Core Panel (m^3)	–	–	–	2.0000	–	–
Manufacturing Costs (1986 US$)	513.9	508.9	507.6	499.4	523.1	473.6
MULTI-FAMILY HOUSES						
Plywood Sheathing:						
Softwood Lumber (m^3)	0.1318	0.0668	0.0714	0.0653	0.0312	0.0551
Softwood Plywood (m^3)	0.0288	0.0317	0.0311	0.0207	0.0310	0.0170
OSB (m^3)	–	–	–	–	–	–
Wood I-joists (m^3)	–	–	0.0279	–	–	–
Truss (m^3)	–	1.5419	–	0.9842	0.5577	1.5419
Foam Core Panel (m^3)	–	–	–	1.3800	–	–
Manufacturing Costs (1986 US$)	457.3	454.8	455.6	447.8	473.7	476.1
OSB Sheathing:						
Softwood Lumber (m^3)	0.1318	0.0668	0.0714	0.0653	0.0312	0.0551
Softwood Plywood (m^3)	0.0018	0.0030	0.0017	0.0034	0.0018	0.0030
OSB (m^3)	0.0251	0.0276	0.0276	0.0167	0.0276	0.0138
Wood I-joists (m^3)	–	–	0.0279	–	–	–
Truss (m^3)	–	1.5419	–	0.9842	0.5577	1.5419
Foam Core Panel (m^3)	–	–	–	1.3800	–	–
Manufacturing Costs (1986 US$)	457.3	454.8	455.6	447.8	473.7	476.1

2.7 Transportation costs

All raw materials and manufactured products may flow between the six producing and six consuming regions in NASAW subject to a transportation cost. The existence of a transportation cost between a source and destination region does not guarantee that in the equilibrium the particular product will be transported across that arc, but it is required by PELPS III if there is trading potential between the two regions. The transportation cost matrix of PELPS III also allows for the inclusion of duties or tariffs that may exist on products shipped between the various regions. Table 8 shows the base period transportation costs per m^3 estimated for softwood lumber.

Within region, or intra-regional, transportation costs have also been estimated and are included as a component of manufacturing cost. Intra-regional costs capture the average cost of shipping a product from a harvesting or mill site to the region's main distribution center. These intra-regional costs have an impact on NASAW results, as the area of each region is large and a significant additional transportation cost may be incurred in moving products to the main distribution center of the region from mills which are often located far from this center. Excluding these may therefore understate actual production costs.

Table 7. Base period capacities, depreciation rates and replacement costs for Canadian solid wood manufacturing processes used in NASAW

Region and By-Product	Base Period Capacity (millions)	Capacity Depreciation Rate	Capacity Replacement Cost (1986 US$/m^3)
B.C. COAST:			
Softwood Lumber (m^3)	9.67	0.06	62.2[illegible]
Softwood Plywood (m^3)	0.96	0.00	192.1[illegible]
OSB/Waferboard (m^3)	0.00	0.09	232.7[illegible]
Particleboard/MDF (m^3)	0.095	0.00	158.7[illegible]
Engineered Wood (m^3)	0.024	0.04	127.1[illegible]
B.C. INTERIOR:			
Softwood Lumber (m^3)	24.27	0.06	62.2[illegible]
Softwood Plywood (m^3)	1.13	0.00	192.1[illegible]
OSB/Waferboard (m^3)	0.00	0.09	232.7[illegible]
Particleboard/MDF (m^3)	0.115	0.00	158.7[illegible]
Engineered Wood (m^3)	0.024	0.04	127.1[illegible]
REST OF CANADA:			
Softwood Lumber (m^3)	22.85	0.06	62.2[illegible]
Softwood Plywood (m^3)	0.30	0.00	192.1[illegible]
OSB/Waferboard (m^3)	1.593	0.09	232.7[illegible]
Particleboard/MDF (m^3)	1.108	0.00	158.7[illegible]
Engineered Wood (m^3)	0.083	0.04	127.1[illegible]

Table 8. Base period transportation costs for softwood lumber used in NASAW (1986 US$/m^3)

From: \ To:	U.S. North	U.S. South	U.S. West	Canada	Pacific	Atlantic
U.S. North	–	8.3	24.8	11.0	–	40.0
U.S. South	8.3	–	22.0	27.6	45.2	35.3
U.S. West	22.0	24.8	–	36.9	31.7	45.5
B.C. Coast	25.9	38.6	3.4	35.2	35.2	60.6
B.C. Interior	22.5	34.7	8.1	31.4	35.6	63.1
Rest of Canada	4.7	22.9	28.0	14.0	59.3	46.6

3. Projections

PELPS III consists of a static and a dynamic phase where the static phase is the solution to a medium-sized linear programming problem whose objective function is the maximization of the sum of producer and consumer surplus. The dynamic phase simulates partial long-run optimization behavior through a succession of static phases, one for each period in the forecast.

NASAW has been used to make 20-year projections of the supply and demand situation of the North American solid wood sector from 1986 to

2006. Changes in equilibrium from period to period are influenced by changes in the shape and position of the supply and demand curves (particularly changes in the shifter variables and upper and lower bounds) that are estimated to have occurred or predicted to occur over this 20-year period. These changes are input exogenously as are any observed and predicted changes to house construction costs, product manufacturing costs, input-output coefficients, transportation costs and tariffs.

Information on the changes over time in variables such as the demand and supply curve shifters and bounds, recovery and by-product factors (input-output coefficients), manufacturing costs, transportation costs and tariff levels for the historical period 1986 to 1996 are based on historical data. Changes in these variables from 1996 to 2006 are generally forecast based on assumptions about growth rates of external variables. Assumptions about how these various shifter variables might change over time was obtained from third party analyses and incorporated into the exogenous changes component of NASAW.

The Canadian supply of sawlogs and pulplogs is largely based on historical harvest and delivered wood cost data. Upper bounds on supply approximate the AAC of a region and for the historical period 1986 to 1996 were estimated from harvest data reported by the individual provincial agencies and summarized by the CFS (CCFM 1997). The rate of change in these bounds between 1996-2006 incorporates assumptions of how regional AACs are expected to change over this period.

One of the strengths of the PELPS III system is that investment in new capacity can be modeled using a technique based on Tobin's (1969) suggested *q* theory of investment. Spelter (1995) has modified this model from its original application to the North American pulp and paper sector (see Zhang and Buongiorno 1993) to take into account the longer lags observed in the panel sector between investment response and price stimulation. Rates of change in the capacity limits of each NASAW manufacturing process are determined endogenously using Spelter's model, except in cases where capacity limits are provided exogenously.

4. Results

While validation and testing are still continuing with the base period data and the growth rate assumptions in the exogenous variables outlined above, the following sections illustrate some of the preliminary results from the model using the base case scenario.

4.1 Housing

Initial results from NASAW suggest that the demand for new single family houses in North America will rise to slightly more than 315 million m^2 by the year 2006 from a level of about 217 million m^2 in 1986 and 240 million m^2 in 1996, increases of just over 45 % and 30%, respectively (see Figure 1). This increase in the demand for new single family housing from 1986 to 2006 reflects changes in both United States single family house size and number of starts. In comparison, the demand for new multi-family houses in North America is forecast to be about 50 million m^2 by 2006, an increase of just over 7% from 1996 demand levels, but an 8% decrease from 1986 demand levels.

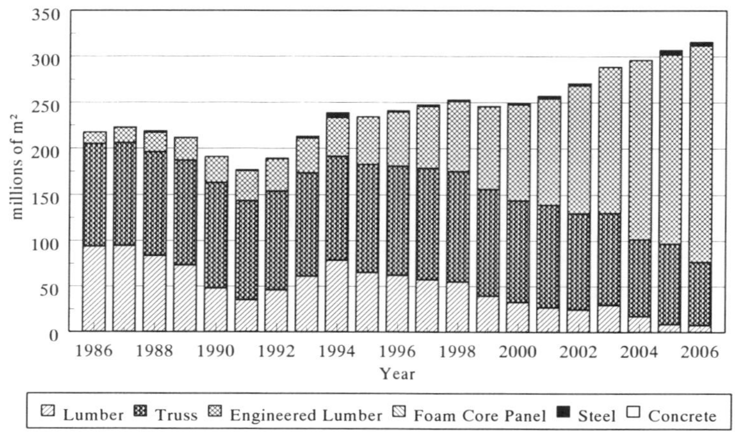

Figure 1. Forecasts of North American single family housing demand to 2006 by framing system

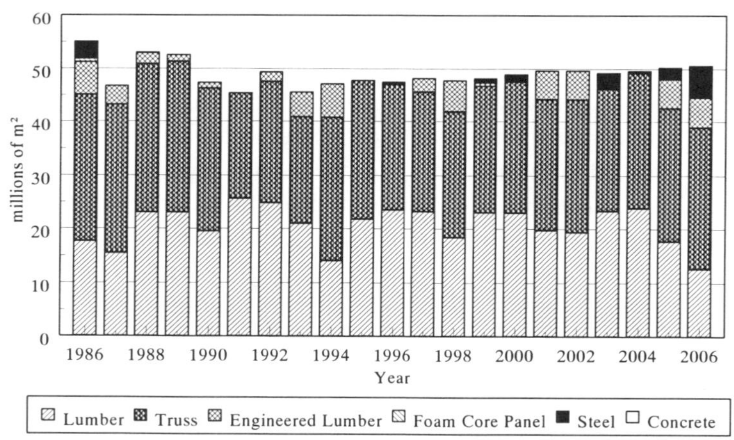

Figure 2. Forecasts of North American multi-family housing demand to 2006 by framing system

More interesting is the observed shift in the type of framing system chosen by the model for construction of single family homes. As can be seen in Figure 1, by the year 2006 engineered lumber is projected to be the most cost efficient type of house framing system in North America, displacing both lumber and truss-based systems. Over the projection period, only minimal use is made of steel and virtually no use is made of foam core panel or concrete framing in single family house construction. In contrast, multi-family house demand shown in Figure 2 displays a more consistent use of lumber and truss-based framing systems, with only small volumes of engineered lumber and steel-framing becoming cost effective from time to time. Single and multi-family house prices estimated by the model are highly stable after 1995, reflecting the assumption of no change in the non-wood component of house construction costs over this period. The implication of this assumption is that housing prices are affected by changes in wood cost alone.

4.2 Solid wood

North American production of softwood lumber is projected to reach 150 million m^3 by the year 2006, an increase of about 9.5% over production levels in 1986 and 4% above 1996 production. As illustrated in Figure 3, this modest increase in lumber production is expected to occur as a result of increased production in the United States South and Eastern Canada regions, which will offset projected declines in lumber production by the United States West. Softwood lumber production in British Columbia is forecast to remain relatively constant over the projection period. These results are consistent with current expectations regarding the pattern of growth anticipated in these regions.

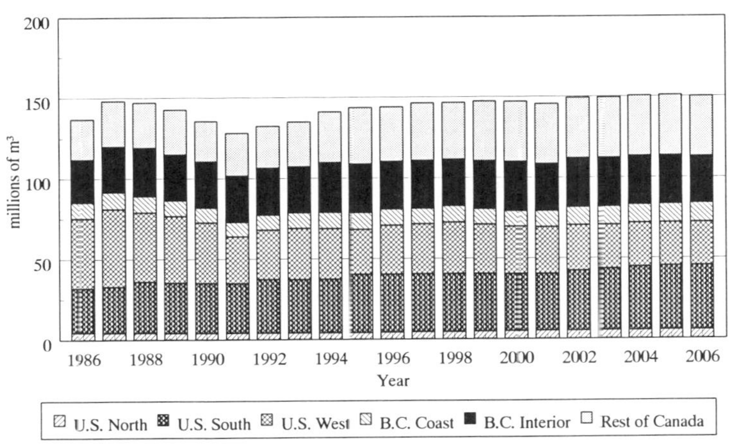

Figure 3. Forecasts of North American softwood lumber production to 2006, by region

Figure 4 shows that production of OSB-waferboard by North American manufacturers is forecast to increase dramatically over the projection period, eventually overtaking softwood plywood as the structural panel of choice in house construction. Total structural panel production is projected to increase from 27 million m^3 in 1986 to 36 million m^3 in 2006 and the share of OSB in total structural panel production is forecast to increase from 20% in 1986 to over 80% by 2006. This rapid increase in OSB-waferboard production can be mostly attributed to the change in the relative prices of the fiber resource used for panel production (hardwood and softwood pulpwood vs softwood sawlogs), but is also a function of the capacity change estimated by NASAW for both plywood and OSB-waferboard processes.

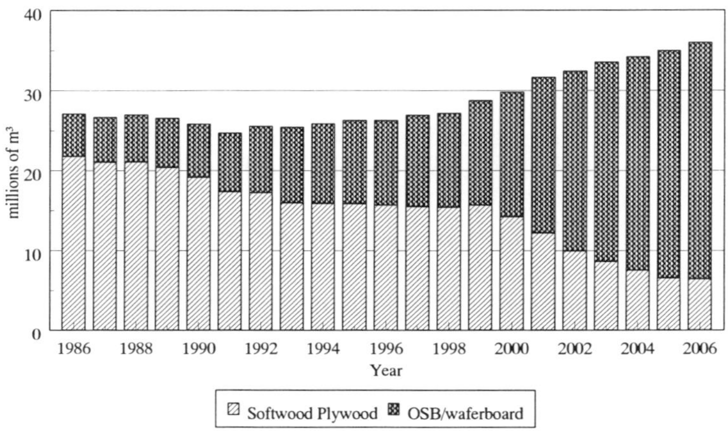

Figure 4. Forecasts of North American structural panel production to 2006

5. Summary

This paper has presented a brief overview of a model of the North American solid wood sector developed using the price endogenous linear programming system PELPS III. This model is called NASAW and has been structured so that the two key North American markets for solid wood products, single and multi-family housing, are represented in some detail. Each dwelling type can be constructed from different combinations of framing and sheathing systems, and the choice of system determines the demand for solid wood products. NASAW differs from other models of the North American solid wood sector in its attention to this house construction detail and allows investigators to analyze wood and non-wood substitution issues arising from the demand for North American housing. In addition, solid wood products used in house construction are manufactured by processes that are not only subject to the relative costs of labor and equipment, but also to the cost and availability of raw material, which in turn are subject to biological, economic and public policy constraints.

The model is an on-going project and the results and data reported here are not final. The purpose of this paper has been to present an overview of the structure of NASAW and to provide some examples of the types of output that can be generated and the types of analyses that can be performed with the model.

6. References

CCFM. 1997. Compendium of Canadian forestry statistics 1996: National forestry database. Canadian Council of Forest Ministers. Ottawa, ON.

Haynes, R.W., Adams, D.M., and Mills, J.R. 1995. The 1993 RPA timber assessment update. Gen. Tech. Rep. RM-GTR-259. Fort Collins, CO. USDA Forest Service, Rocky Mountain Forest and Range Experiment Station. 66p.

Ince, P.J. 1994. Recycling and long-range timber outlook: Background research report 1993 RPA assessment update. USDA Forest Service. Res. Pap. FPL-RP-534. Madison, WI. USDA Forest Service, Forest Products Laboratory. 110p.

Jacques, R. 1992. The economic impact of state legislation on mandatory recycled fiber content: An application of the North American pulp & paper model. Working Paper. Policy and Economics Directorate, Forestry Canada. Ottawa, ON.

Jacques, R., Booth, D., and Phelps, S. 1993. The economic impacts of U.S. recycled fiber content legislation on the Canadian Newsprint Industry. Working Paper. Policy and Economics Directorate, Forestry Canada. Ottawa, ON.

LINDO Systems, Inc. 1996. LINDO user's manual. Chicago, IL.

Takayama, T., and Judge, G.G. 1964. An interregional activity analysis model of the agricultural sector. Journal of Farm Economics. 49:349-365.

Tobin, J. 1969. A general equilibrium approach to monetary theory. J. Money Credit Banking. 1:15-29.

Spelter, H. 1995. Capacity changes in U.S. particleboard, southern pine plywood, and oriented strandboard industries. Canadian Journal of Forest Research. 25:614-620.

Zhang, D., and Buongiorno, J. 1993. Capacity changes in the United States paper and paperboard industries: Q theory and empirical models. Canadian Journal of Forest Research 23: 72-80.

Zhang, D., Buongiorno, J., and Ince, P.J. 1993. PELPS III: A microcomputer price endogenous linear programming system for economic modeling, Version 1.0. Res. Pap. FPL-RP-526. Madison, WI. USDA Forest Service, Forest Products Laboratory. 43p.

A Model for the Siberian Timber Products Market
- An engineering-economic approach -

Michael Obersteiner

Key words: Timber products market, forest economics, auction theory, simulation modeling

Abstract: The application of conventional economic models, e.g., partial equilibrium computation or LP models, do in many cases lack realism owing to a number of untested a priori assumptions. Computing power and information resources have dramatically improved in the recent past, which increasingly allows economists to substitute model assumptions with real data. The modeling approach taken in this paper was originally designed to model industrial restructuring and relocation of the Siberian forest sector during its transition from a command economy to a market-oriented economy. Based on the possibility of linking to an extensive GIS-database, the modeling strategy was to build an easy-to-understand and easy-to-compute economic model that makes maximum use of existing data and expert knowledge. The model is based on Vickery's Nobel Prize-winning auction theory (Vickery 1961) and simulates the possible future formation of Russia's forest industrial sector. In the model, I distinguish between a static phase and a dynamic phase. The static phase describes the auction mechanism involving a number of strategy sets, which is applied for each period, leading to a partial market clearing. In the dynamic phase, product prices adjust based on supply slack expectations, capital is invested or depreciates, contract partners revise contract policies and, finally, factor prices change exogenously reflecting the overall economic development. Demand equations are currently predicted using Bayesian econometrics of panel data. The Bayesian priors are estimated using consumption data of selected Western economies from 1960 to 1985. Such an agent-based model can be used for policy analysis of various scenarios at different levels of detail, by modeling the entire economic system on the basis of behavioral, resource, and technological characteristics of individual firms.

1. Introduction

Markets are central institutions of Western societies and are increasingly becoming the dominating institution in post-communist societies. However, even in Western economies we do not know much either about their functioning or about their emergence. Although, markets are so central to the functioning of our real economies and societies, economists make little use of the few insights into the real patterns and determinants of markets, when they model real economic phenomena and make predictions upon them. How are prices formed? By which ways do buyers and sellers interact and do business? How are the determinants of market functioning evolving? How embedded are markets in formal and informal institutions? All these questions make an analysis of real-world problems difficult and complex. However, it is indispensable for an analyst to take many of these phenomena into account if (s)he is to give an educated judgement of the status and future pattern of the market analyzed.

The purpose of this paper is certainly not to pretend answering all these questions and many others, but more modestly to try to incorporate some aspects of market functioning into a new economic model of production and trade based on simulated contract bidding of individual agents. The general theoretical model was specially tailored to model the Siberian forest industry on its transition to a 'market' system. For a more detailed description of the model the reader is referred to Obersteiner (1998).

Since the collapse of the socialist system in 1991, Russia has committed itself to pro-market reforms. The country is still undergoing a transition, which is characterized by the elimination of state subsidies, decentralization of decision making, privatization, and the abolition of price and wage controls. The former tradition of soft budgets, which guaranteed employment and low inflation, is abandoned.

Price liberalization and mass privatization meant at the enterprise level the elimination of non-competitive products and radical changes in the ownership structure. This, combined with the collapse of the COMECON and the Soviet Union markets, have caused a deep economic recession in all of Russia, and in the remote Siberian regions in particular, owing to a dramatic cut in transport subsidies and monopoly prizing. The entire economy and the forest sector have faced an economic decline unprecedented in modern history. The decline started in 1989 to 1990 and still continues with the exception of some export-oriented enterprises or product lines. Wood production has declined to a level one-quarter of that prevailing in the 1980s (Moiseyev et al. 1998).

The official rate of unemployment is rather low given the dramatic decline of production, but is in reality much higher. Hidden unemployment increasingly becomes apparent. Workers over forty are most affected by the transition. Pensioners barely survive owing to empty pension funds. The economic downturn has affected the most remote areas where most of the logging towns and forest- industry-dominated towns are located.

Decline in real incomes and construction activities adversely affect the consumption of forest products. Consumption will most likely be lower at the turning of the century than ten years ago. Adoption of a new economic culture securing an efficient functioning of markets will probably take generations. Well-established Mafia networks and unrealistic profit and rent seeking by most managers increase the concerns about slow adaptation. Economic problems in Russia are also connected with high political risks, which in the worst case may erupt in civil unrest owing to unequal distribution of wealth or nationality conflicts. Such unforeseeable events would, of course, destroy all scenario exercises.

The level of and geographic distribution of forest sector output will mainly depend on two factors: the first is competitiveness of domestic wood-working industries and the second is the trade pattern. After the deep economic slump there are few enterprises that are still competitive on domestic or international markets. Those who inherited decent working capital from the Soviet system are on the way to recovery. However, many producers had to close operations and many more will do so in the future.

Trade links with former Soviet satellite countries like Bulgaria, Hungary, East Germany, and the Baltic countries have dried up. Future delivery especially of roundwood from remote Siberian areas to European Russia will heavily depend on the future of transportation costs. Most of the trade is organized through barter contracts. However, there are promising prospects for timber and timber products trade with some former Soviet republics and countries with wood deficits in east and Southeastern Asia and the Middle East.

Future competitiveness of the Siberian forest sector centers now around the question of how the allocation and structure of production will change during the transition to a market economy. It is not only the change in prices that will drive this development. Today, economic agents in Siberia act according to different economic paradigms than they were used to under the previous regime. Different decision rules guide economic activities given a framework of changed constraints. Russia has also opened its borders, which allows more freedom of trade and capital flow. The interaction of these factors with many others will define the future path of the Siberian forest sector's battle for higher market shares both nationally and internationally.

It is the purpose of this paper to propose a new analytical tool that allows exploring future development of the Siberian forest sector given its geography and specifics in the transition to a market economy.

2.Model Formulation

2.1 Intuitive description

The first economic activity of business partners in the forest industry of Siberia is to establish trade contracts. Imagine now for simplicity a fictitious market hall at a trading point in Russia. This image can equally be treated as a traveling sales or purchase agent of the respective organization. A trading point in this sense is located either at one of the Siberian borders for export trade or located near a wood-working mill or by some consumer of finished products within Russia. Producers, depending on their competitive position and their available capacities, who are potentially willing to sell their products are invited to this market hall. In the market hall buyers and sellers meet. Some of them already know each other from joint business activities or even have established long-term contracts. Buyers learn about price trends and quality of the products producers try to sell. As a side effect of this discussion both buyers and sellers get a grasp of the reliability of their possible contract partners. In addition, buyers get to know the geographic location of producers; their productive capacities and other producer-related information. By this, buyers can make inferences about the total cost of delivery of each producer and the grades delivered, which they might use for the negotiation. In the model specification, which was used and adapted to model the Siberian lumber market, buyers believe, to some degree, that:

- Reliability and quality of services increase the profit margin each producer can potentially earn.
- Buying from a high-profit producer today will guarantee cheaper contracts in the future when competition might become stiffer and prices adjust.
- Buyers have the tendency to contract with large (small) firms to minimize transaction costs and risk.
- At a generally low price, producers are more likely to be uncertain contract partners whereas the reliability of buyers increases.
- Only producers who can profitably deliver their products can be reliable contact partners.

After all information is exchanged the auction can start. According to their preferences, buyers decide where to buy and give out their orders at the base price. Because not all producers are willing to sell their goods at the base price, only a portion of the total demanded volume is actually contracted at the base price. Depending on the bargaining power of buyers against sellers, quantities that are offered in each auction are exogenous. After buyers reveal their preferences to the individual producers, depending on their own bargaining power and the producer structure, producers decide to accept, modify or reject proposed offers. Producers, of course, will screen all proposals and will select among them. Most desirable (profitable) proposals are accepted immediately, whereas less desirable proposals will be partially accepted with the hope of coming back to a even better offer when tender prices increase, but with the risk of being out-competed by new entrants. Some relatively unprofitable proposals will never be accepted.

In the Russian business environment, ordered products might get lost or to some extent be of bad quality, so that not the entire contract volume as originally negotiated is actually realized. Strategically, producers have the tendency to deliver bad quality if prices are low. Buyers, on the other hand, will postpone payment or complain about delivered quality of the product with increasing prices.

It follows that after the first round of negotiation, buyers and sellers have to meet and the same auction procedure will start over again, but, with increased product prices and different tender volumes. If there is still demand for products in subsequent auctions, product prices rise continuously until either producers run out of capacity and resources or buyers have satisfied their total demand. Increasing prices cause new suppliers to enter the market leading to enhanced competition.

In the end, either demand is satisfied or producers run out of capacity or resources and no further auctions are conducted. All agents involved now analyze past market processes, update their information, and develop new strategies for investments and price and contract policies for the next period. Policy setting of each individual agent involves analysis of the past as well as expected market developments including prices, competition, total demand, overall investment climate, and uncertainty. Strategies, if feasible in the light of the financial and natural resource conditions, are implemented by each producer. Trading according to the tender mechanism starts again in the following period.

2.2 Formal formulation

2.2.1 Static phase

2.2.1.1 The buyers' problem

$$Y_{ijkt}^{B} = f\left(X_{ijkt}\right)D_{jt}a_{jkt}$$

$$for$$

[1] $$Y_{ijkt}^{B} > 0 \quad if \begin{cases} Y_{ijkt}^{B} \geq Y_{ij}^{\min} \\ C_{ijt} \leq \overline{P_{jkt}} \\ \sum_i \sum_k Y_{ijkt}^{B} \leq \overline{Y_{it}} \\ \sum_i \sum_k Y_{ijkt}^{B} \leq D_{jt} \\ \sum_i \sum_k \sum_e Y_{ijkte}^{B} \leq \overline{R_{et}} \end{cases}$$

$$Y_{ijkt}^{B} = 0 \quad if \quad else.$$

The buyers' initial proposal to the individual producers is computed according to,

Y_{ijkt}^{B} Contract offered by buyers of market j to producer i at time t in k^{th} tender.

X_{ijkt} Matrix of decision variables specific for buyers of market j purchasing from producer i in k^{th} tender at time t.

D_{jt} Total volume to be tendered from market j at time t.

a_{jkt} Parameter determining the size of tender volumes for the kth auction at market j at time t with $a_{jkt} \in]0,1]$.

$\overline{Y}_{it}$ Production capacity of producer i at time t.

$Y_{ij}^{\min}$ Minimum contract size. $Y_{ij}^{\min} \in \Re^{+}$.

C_{ijt} Unit cost (FOB or CIF) of producer i at time t delivering to market j.

$\overline{P_{jkt}}$ Apparent price at market j at time t in k^{th} auction.

$\overline{R_{te}}$ Biological resource constraint in Ecoregion e at time t (Annual Allowable Cut (AAC)).

In the current version of the model, the decision variables in X_{ijkt} are made up of measures like a technical cost approximation given certain technology options, the size of the firm, an indicator matrix of membership in a production network (domestic or international), current price, cost and contract uncertainty on both the buyer and seller side. Through transformation, premultiplication with the variance-covariance matrix, the decision variables are made orthogonal to each other, which allows the calculation of Euclidean distance measures. Each individual decision variable is assigned a 'shadow' value or parameter (v_{ijkt}), which allows the calculation of a contract affinity score of each buyer to each seller. The parameters can either be estimated from real data or be set by the analyst after some sensitivity analysis. The contract affinity score is computed such that $\sum_i f(X_{ijkt}, v_{ijkt}) = 1$ and is multiplied by a fraction, $D_{jt} a_{jtk}$, of the total demand in the kth tender. The parameter a_{jkt} and the price change within the simulated auction procedure $\overline{\Delta P_{jkt}} = \overline{P_{jkt}} - \overline{P_{jk-1t}}$ are used to mimic market power of the market participants. D_{jt} is predicted using Bayesian econometrics and panel data of the respective regions and macro-regions.

2.2.1.2 The sellers' problem

After buyers revealed their offers (Y_{ijkt}^{B}) in the kth tender to their potential set of producers each individually decides which offer he-she will accept, modify, or reject. In the version applied to the Siberian forest sector the seller will fully accept the offer if it comes from the buyer who guarantees sufficiently high returns. (S)He will modify, i.e., decrease the contract size if the contract will yield a profit below the benchmark profit and will reject if the offer would yield profits that would not justify delivery owing to the insufficient size of the contract. In the latter case the seller believes that she/he would be able to sell at a higher yield in a tender to come. This behavior is formalized by the following,

$$[2] \quad Y_{ijkt} \begin{cases} Y_{ijkt}^{B} \quad if \pi_{ijkt}^{'} > \pi_{ijkt}^{'*} \\ Y_{ijkt}^{B}\left(\dfrac{\pi_{ijkt}^{'}}{\pi_{ijkt}^{'*}}\right)r \quad if \pi_{ijkt} \qquad and \qquad Y_{ijkt}^{B}\left(\dfrac{\pi_{ijkt}^{'}}{\pi_{ijkt}^{'*}}\right)r \geq Y_{ij}^{\min} \\ 0 \quad if \pi_{ijkt}^{'} \leq \pi_{ijkt}^{'*} \qquad and \qquad Y_{ijkt}^{B}\left(\dfrac{\pi_{ijkt}^{'}}{\pi_{ijkt}^{'*}}\right)r \geq Y_{ij}^{\min} \end{cases}$$

The prime on $\pi_{ijkt}^{'}$ indicates that the seller selects offers from feasible contract partners only, as defined by the constraint set of the buyers' decision. $\pi_{ijkt}^{'*}$ indicates the benchmark profits. r is a logistic scaling factor for the modification of the revised contract size proposed by the seller. This simple one-dimensional model of quasi-profit maximization on the sellers' side can easily be substituted by any more complex goal function.

2.2.1.3 Optimal behavior

From a theoretical point of view the numerical calculation would converge to an optimal solution of the allocation pattern of production for the buyer if the buying-selling uncertainties would converge to zero, the parameter of profit maximization in the decision variable set of the buyers approach infinity (or a more complex goal function is optimized) and, all of the volume would be given out for tendering in the first tender, the price increase from one auction to the other is infinitely small but still positive, and the number of auctions within a period is allowed to be infinitely large such that all potential capacity could be used.

Producers are allowed to optimize only within an individual tender. They are not able to optimize dynamically over all tenders within a period and are thus also not able to optimize over periods. Currently there is work underway to prove that no analytical optimum can be computed in this setting.

Given the present uncertain economic environment and given the rather poor enterprise information, the parameter values of the model will not be in a close neighborhood of these optimal values, but will converge slowly towards an optimum. This type of model was constructed to allow agents to act according to an optimizing behavior, but only to a certain extent. Thus using this model the user can work with different degrees of optimizing behavior of the agents. In the model applied to the Siberian forest sector, I performed the calibration by educated judgements relying on personal experience or expert consultation. Another method of calibration would be through experiments with real economic agents or by the use of econometric techniques. The possibility of loosening the optimization 'constraint' is probably the most distinguishing feature of this approach in comparison to the standard models used in economics and operations research. This is the beauty but also the weakness of the modeling approach taken. The quality and the usefulness of the results hinge on the calibration of the model. We are truly opening Pandora's box allowing for a lot of arbitrary inputs featured by bounded and fuzzy rationale involving multidimensional pay-off functions. The trick is to open Pandora's box in a controlled manner, which then should justify the deviation from the standard model.

2.2.2 The dynamic phase

In the dynamic phase, changes in capacity are calculated following different investment depreciation regimes, prices are allowed to adapt according to expected demand and supply scenarios, maximum potential demand is forecast, and costs change according to input price and technology scenarios. Not all of the variables mentioned are fully internalized. The art of making the static model dynamic is determined by transition conditions for the internalized variables and finding the appropriate scenarios for exogenous variables.

2.2.2.1 Capacity changes

Investment is related to the current allocation of capacity, the current supply-demand pattern, the accumulated profits of individual enterprises or production networks, current and expected returns on investment given scenarios of the long-term future industry structure, strategic capital targeted to particular markets, and a number of behavioral patterns like the willingness of producers to reinvest in the same activities. Technological upgrading and capacity expansion are also indirectly dependent on the economic or technical depreciation of installed capacities.

In the current version of the model, three separate investment cycles have to be considered. There are two independent investment cycles and divestment. Capacity that will be available to each producer in the up-coming period, $\overline{Y}_{it+1}$, is made up of the sum of already existing capacities, $\overline{Y}_{it}$, plus capital additions determined in the dynamic phase, $\Delta\overline{Y_{it}^{i}}$, , minus capital assets $\Delta\overline{Y_{it}^{d}}$ that are depreciating owing to over-capacity of a particular producer. Capacity additions or investment are again the sum of two parts. The first is capital generated by internal profits. This is calculated from the discounted total accumulated profits targeted for a specific production technology. Profits can either be reinvested in the same activity (e.g., lumber production) or other activity within the forest industry or outside the forest industry. The second investment component is foreign (extra-enterprise or extra-network) venture capital. The total sum of venture capital is distributed among enterprises with market potential. Market potential is measured by a comparison of the market position at the time *T*, at a given market, and its current position. Technology is allowed to vary according to market, e.g., high-tech investment for expected deliveries to Japan and low-tech for deliveries to China.

$\overline{\Delta Y_{it}^{d}}$ defines the depreciated capital stock resulting from excess capacity or 'natural' depreciation. Excess capacity occurs if capacity that was inherited or built over time is not employed owing to the economic supply and demand scheme. Physical excess capacity depreciates at a different rate than economic depreciation and varies by technologies. In addition, capacities are bounded from below, which is motivated by the belief that an enterprise needs to be of a certain minimum size in order to stay in business.

2.2.2.2 The 'quasi-optimal' supply pattern at time *T*

The calculation of the quasi-optimal state follows the same procedure as discussed in the section describing the static phase. Parameters are adjusted such that the quasi-optimal state, at time *T*, are set according to the considerations discussed in the paragraph on optimal behavior. The quasi-optimal state is constantly updated. This causes some inertia in the evolution pattern of capacity building and is motivated by the fact that producers can base their decisions on information that is available at time *t*. Thus the effects of technological-price shocks and changes in the competitive position of enterprises can only to a limited extent be anticipated by the agents.

Overall, the set-up of the dynamics of the model is very similar to a standard inter-temporal stochastic optimization procedure solving time-continuous stochastic optimization problems. The distinguishing feature is that we do not necessarily arrive at an optimal solution in *T*; however, there is a persisting tendency to converge to this state. Perfect matching with the optimal state is not obtained owing to either insufficient resources, *inter alia* natural resources, investment capital, lack of demand, or to the stochastic behavior of contract building. In addition, owing to the model set-up, the optimal state is a moving target (within certain boundaries) making it difficult to converge.

2.2.2.3 Price determination

It is generally accepted among economists that under a certain market structure, prices need to adapt according to the interaction of current and future demand and supply. In fact, most mainstream economists believe that price rigidities explain departures from trend growth and equilibrium states. In the Russian case, prices are rather rigid owing to price control and information lags. Export prices are regularly fixed by semi-governmental organizations in Russia. The model tries to account for some of these imperfections and allows for altering policy strategies specially targeted at the sector or more general economy-wide policies.

In the model, changes in prices are endogenous. The price changes are directly determined by the model variables given some assumptions on the process of price determination. This is, from a computational and a model-building point of view, not a trivial task because the price change depends essentially on all other variables and coefficients of the model. Hence, it becomes difficult to make a correct and fast diagnosis of price fluctuations predicted by the model. In the model, price changes are modeled as a function of the expected supply slack, the expected total investments, and the current profitability of the industry. The expected supply slack is the difference between the current supply, which is the installed capacity, and the expected demand at time t of the product concerned at market j in period $t+t'$. The price changes are computed individually for each market. The logic here is that prices change according to the ratio of total expected capital requirements to total expected revenue calculated at time t in order the match the expected demand. Thus, prices adapt in such a way that there is a tendency to market clearing. If total predicted capital requirements for the upcoming period, with prediction period of t' years, is larger than today's total revenue then prices increase, and vice versa.

The ratio of total expected capital requirements to total expected revenue is used to calculate the change in the initial tender price in the next period ($\overline{P_{jk=1,t+1}}$) and the change in the price within the ascending price tender auction ($\overline{\Delta P_{jkt}}$). The ratio is adjusted by a logistic filter. The increase in prices in the ascending price tender auction depends thus crucially on the expected net supply slack. If the expected supply slack is small the initial price falls, and in addition the price change within the tender decreases. The latter affects both the variance and the mean of the prices achieved at the particular market in the particular year, which reflects degree of competition. This assumption, however, still lacks empirical testing. Implicitly, this modeling approach allows having investment and consequently prices be determined by the strategies of the agents being active on the concerned markets. That is to say that, e.g., raising investments in Siberia targeted for the Japanese market will lead to lower and less volatile prices for Japanese buyers in the subsequent periods.

2.2.2.4 Uncertainty

The model includes uncertainty on a number of processes active in the Siberian forest sector. These are uncertainties in the production costs, price setting, and contract reliability of Russian producers and traders. This issue is not further described in this paper and the reader is referred to (Obersteiner 1998).

3. Summary and Conclusions

Russian industry experienced a serious economic crisis involving also a major drop in output in the forest industry. This caused a dramatic decrease in real employment and income in forest and forest-industry-based communities. Especially in the more distant regions of Siberia, the economic downturn has been more severe than to the Russian average. In addition, Siberia's forest sector has recently gained considerable international interest for its environmental problems. So far, little is known about the possible future path of Siberia's forest sector and its role on the international forest products markets. I have developed a model that can be used as a decision-making tool for policy analysis of various scenarios and levels of detail analyzing the forest industries of Siberia. The different markets of the Siberian forest industry are modeled on the basis of the behavior and endowment of individual firms.

The model was designed to analyze changes in the production allocation patterns of individual producers in the Siberian forest sector during transition from a command economy to an economy based on market principles. Because of the implausibility of assumptions built in standard economic models combined with the possibility of linking to an extensive resource data base, the modeling strategy was to build an easy-to-understand and easy-to-compute economic model that makes sense and makes maximum use of existing data and expert knowledge. The model is based on Vickery's Nobel Prize-winning auction theory (Vickery 1961) in order to simulate the possible future formation of Russia's forest industrial sector. The model distinguishes between a static phase and a dynamic one. The static phase describes the auction mechanism, which is applied for each period, leading to a partial market clearing depending on producer, demand, and price variables. In the dynamic phase, product prices adjust, producers invest or depreciate capital, contract partners revise contract policies and, finally, prices change according to the overall economic development.

Enormous amounts of data from many different sources and expert knowledge have been compiled and have been woven into the model. The core model is flexible in the sense that new knowledge or data can rapidly be included. This is usually achieved by combining other models with the core model. So for example a cost module and a demand module are attached to the core model in the current version. The cost module calculates costs as a function of a number of variables starting from forest inventory information, forest management rules, to harvesting and processing technology. The demand module predicts demand on the basis of demand functions, which were estimated with the help of econometric tools. For the current version, I estimate econometric equations using panel data and Bayesian priors. Another very important feature of the model is that issues of international trade can be addressed. Trade, and especially international trade flows, not only depends on predicted supply and demand patterns of importing and exporting regions, but also factors like transportation costs, loading and reloading costs, tariffs and quotas, and network relations between economic agents. In addition, the model is capable of handling differences in the business approaches of different cultures to the trade pattern. As an illustration one can compare a Chinese trader, who is more concerned about prices, to Japanese markets, which demand high contract reliability and product quality.

It is worth pointing out that the distinguishing feature of the entire modeling approach is that exchange is not simulated by a very specific class of trade game (i.e., general equilibrium or oligopolistic competition), but by negotiations in an auction mechanism. Auctions are simulated using multiple decision criteria with different negotiation capabilities and market power of individual agents. However, optimizing behavior of any kind can be treated and implemented as special cases. In addition, I allow for heterogeneity in the cost structure and behavior of agents. Depending on the market power of the buyers or sellers, the algorithm either allows the buyers to purchase at a low price or the sellers sell at a high price. Either the producers or the buyers gain relatively more from the transaction. An increasing price auction with a reserve price is iteratively conducted until either producer or buyer constraints are violated. Buyers propose contacts to a selected subset of sellers, which after a screening of all received contracts decide to accept, modify or reject the proposals made by the buyers. Owing to the nature of the auction set-up, it is impossible (also theoretically impossible by backward induction) neither for the individual agents to compute their optimal strategy in the auction to compute the optimal strategy over periods using an inter-temporal optimization procedure. Nonetheless, the agents' behavior tends towards a quasi-optimal state in a distant future period.

Typically, output of the final product is given at an individual enterprise level. Trade flows between individual agents, prices negotiated by individual bargains, supply slacks, capital formation, profits, investment, and many other details can be reported on an individual mill or even contract level. Results indicate that the model is able to match reasonably well the pattern of output decline and relocation of the industry. Also price development, especially those for export prices, can be reconstructed almost realistically.

These are all very encouraging results. Nonetheless, the model will still have to be further extended and fine tuned. The most recent results from theory will also have to be continually incorporated in the model. Especially, there is a need to further strengthen the theoretic basis for the transition algorithms used in the dynamic phase. Also, experiments with real agents will have to be carried out in order to pin down negotiation behavior across different cultures.

At this stage it is impossible to foresee the future of the model. All depends on how this new modeling approach is accepted by the scientific community and possible users.

4.References

Moiseyev, A., Uusivuori, J., and Burdin N. 1998. Forest industries in the russian federation *in* Palo, M., and Uusinvuori, J. (eds.), World forests, society and environment (Vol 1). United Nations University / Institute for Advanced Studies, Tokyo, Japan.

Obersteiner, M. 1998. The pan siberian forest industry model (PSFIM): A theoretical concept for forest industry analysis. IIASA-Interim Report IR-98-033 (available through: http://www.iiasa.ac.at/Publications/Documents/IR-98-033.pdf).

Vickery, W. 1961. Counterspeculaation, auctions, and competitive sealed tenders. Journal of Finance. Vol. 16. March. 8-37.

An Economic and Environmental Assessment of Asian Forest Sectors

John Perez-Garcia, Yujiao Wang and Weihuan Xu

Key words: Global forest sector, assessment, scenario analyses, trade modeling

Abstract: Economic conditions in the Asian forest sector are rapidly changing owing to expanding economies, the planting of fast-growing plantation species, and restrictions on harvesting natural forests. These factors and their effect on forest sectors of the region are analyzed using the CINTRAFOR Global Trade Model (CGTM). The model, which has an emphasis on the Pacific Rim region, is used to help understand the implications of greater plantation harvests, lower mature timber supply, and expanding economies on the forest sectors of the region and the recent decline in Asian demand. Projections of production, consumption, prices, and trade for softwood and hardwood lumber and plywood are made using the CGTM. Scenario analysis is used to characterize who wins and who loses under alternative outlooks for supply and demand for timber. A discussion of the projections is presented with particular emphasis on environmental effects, capital investment flows, shifts in timber supply and timber demand, forest resource utilization, and economic and environmental policies.

1. Introduction

The last decade of the 20th century has brought with it a significant shift in the use of forest land from fiber for wood products to environmental amenities. Environmental concerns in the United States and abroad, building since the 1960s, have reduced the forest land base to supply hardwood and softwood fibers globally. The growth of urban populations and the shift in per capita income has placed increasing emphasis on non-timber attributes of forests, and, at the same time, demands on industrial wood products. There is little doubt that these environmental and population pressures on our land resources will continue into the 21st century.

What makes the shift in timber supplies important is their effects on market behavior. Forest products markets have responded to supply constraints with large fluctuations in product prices. These supply shocks have led to uncertainty in wood fiber availability. Uncertainty in supply will persist in the United States West as changes in forest land regulations will likely continue to constrain wood fiber supply. A key point to understand is that the dramatic price fluctuations that have occurred over the past several years are the result of supply limitations. Whereas more recent movements in prices have been the result of somewhat more predictable demand changes. The study illustrates several ways in which present fiber supply shortages are forcing markets to behave differently than historical observations.

Having noted the continued importance of supply constraints, we are currently witnessing a significant recession in Asia with large effects on the forest sector. Owing to lower Asian demand, prices for United States West coast logs have dropped considerably. Not only in the United States, but in other regions also, log prices for Asian-bound wood have collapsed from high peaks of the early to mid 1990s. In the United States, strong domestic demand has kept prices firm. As Asian appetite for wood continues to wane, there will be greater pressure on the United States to maintain healthy wood products markets.

With many effects from timber harvest restrictions already in place, the question becomes: What will forest product markets look like in the 21st century? Can we identify changes in these markets and important issues that the forest sector will face as we enter the next century? Given that environmental issues and land use regulations are not likely to disappear, how can forest sectors adjust to impacts from environmental and other policies? What might we expect from an eventual Asian economic recovery?

2. Study Outline

The study presents a near-term outlook for softwood lumber markets in North America and the world given a baseline condition. It also presents the result of alternative futures where timber harvests from the federal lands in the United States are reduced to zero and harvests from Chile and New Zealand reach only 75% of their projected 2010 levels. Sensitivity analysis is used to illustrate the changes in projections of prices, production, consumption, and trade flows when we reduce demand projections for several regions in Asia.

The study uses the CINTRAFOR Global Trade Model (CGTM) to produce projections of consumption, production, prices, and trade flows to 2010 and reports the effects of environmental and demand growth pressures, and wood fiber resource and processing capital investment constraints on the global forest sector. The following section presents a description of the trade model. Section 4 describes results and provides a discussion of major implications drawn from the results. A final section summarizes the work and indicates future directions in trade modeling and assessment procedures.

3. Methods

The assessment uses the CINTRAFOR Global Trade Model (CGTM). The trade model is a spatial equilibrium model. A technical description can be found in Cardellichio et al. (1989). Essentially, producers of lumber and plywood meet demand projections period by period in a least-cost fashion

Low cost, achieved by trade activity, determines regional production shares and prices given that production levels are constrained by the available capacity and price of raw logs regionally. Log supply constrains the production activity through its inventory constraint and wood fiber cost. Since 1987, the CGTM has been used to provide outlooks of the lumber and plywood markets from a global perspective, concentrated mostly on the Pacific Rim and North American regions. Since the initial work with the CGTM several modifications have been introduced. More important modifications include: exogenous projections of public timber harvests in the United States West and the use of inventory projection models such as ATLAS (Mills and Kincaid 1992) and SRTS (Pacheco et al. 1996).

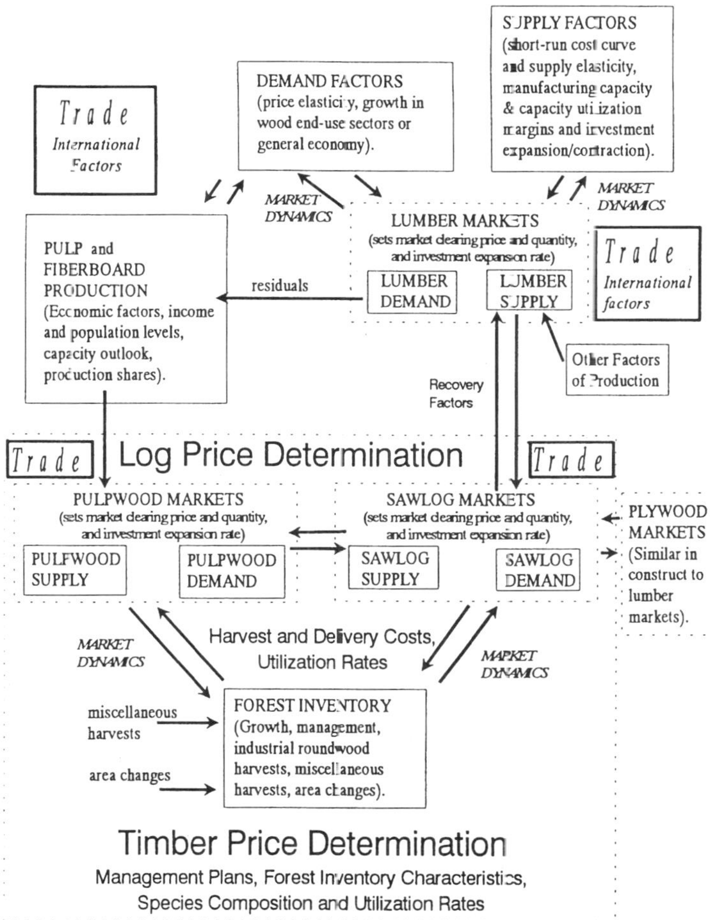

Figure 1. Factors affecting timber price determination

Figure 1 presents a regional flow diagram describing economic relations in CGTM. The economic model consists of regions with supply and demand for forest products. Figure 1 gives details of the lumber market; plywood markets are similarly modeled in CGTM. The model solves for market-clearing prices, quantities, and trade flows in the lumber, plywood, pulpwood, and sawlog markets. Factors that affect these markets are demand and supply factors, pulp and fiberboard production, other factors of production, and forest inventory. Figures 2-4 provide the geographical detail of the model.

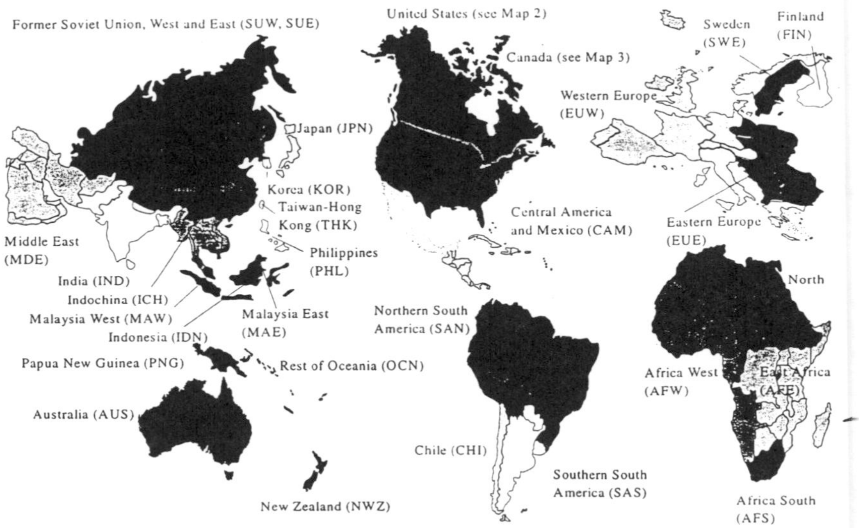

Figure 2. Regional definitions for the CGTM with three letter acronyms in parenthesis

The CGTM employs two demand specifications. One specifies consumption as a function of prices and is used to model softwood lumber demand in the United States, Canada, Japan, Korea, Chile, New Zealand, and the CGTM-defined European regions, as well as major hardwood consuming regions of the world. The second approach uses projections of consumption that are unresponsive to price changes. China and the former Soviet Union are examples of this second application. Demand end-use factors and projections of economic and population growth are used to shift demand from period to period providing a dynamic linkage between yearly calculations of quantities consumed.

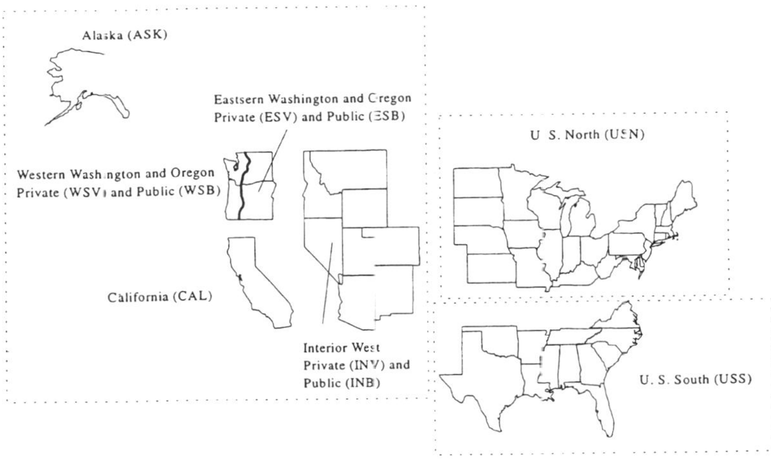

Figure 3. U.S. Regional timber supply regions with three letter identifiers in parenthesis. Public and private timber supplying regions from a single product supply region (WSV and WSB => WST; ESV and ESB => EST; INV and INB => INT (not shown on map). The three U.S. product demand regions are identified within boxes (i.e., the U.S. West demand regions combines all western regions).

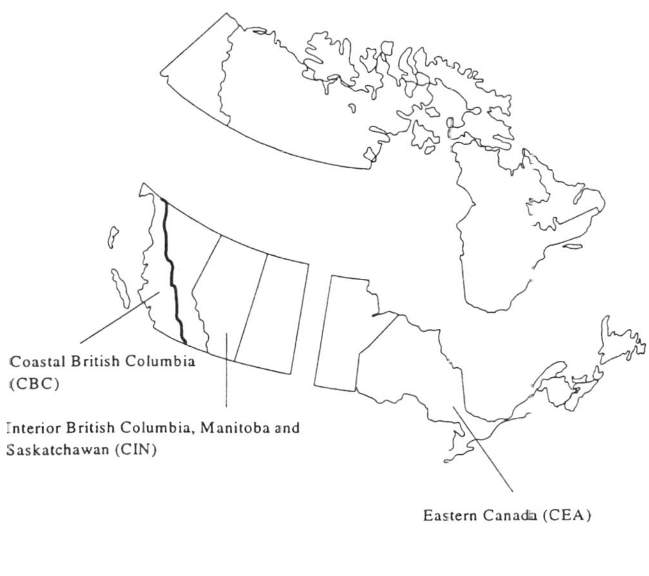

Figure 4. Canadian timber and product supplying regions (The product demand region is all of Canada)

Table 1 provides projected shifts in the demand function owing to economic activity in CGTM regions expressed in percentage growth from 1993 to 2010 (1993 = 1.000). For example 1.226 represents a 22.6 percent shift outward in demand from 1993. Adjustments are made year by year to the demand function intercept to determine the shift in demand.

Table 1. Demand growth data

		Product		
	Softwood	Hardwood	Softwood	Hardwood
Demand Region	Lumber	Lumber	Plywood	Plywood
U.S. West	1.226	1.117	1.168	1.168
U.S. South	1.226	1.117	1.168	1.168
U.S. North	1.226	1.117	1.168	1.168
Canada	1.457	1.122	0.847	0.847
Brazil	(1)	1.791	(1)	(1)
Chile	3.176	1.661	1.916	1.916
S. South America	1.000	1.325	(1)	1.088
Finland	1.080	1.000	1.314	1.314
Sweden	1.000	1.000	1.255	1.255
W. Europe	1.121	1.102	1.231	1.231
E. Europe	1.151	1.314	1.231	1.231
Japan	1.078	0.760	1.236	1.236
Korea	2.023	1.986	(1)	1.680
China	1.473	1.518	2.846	2.846
Taiwan/Hong Kong	1.281	1.575	(1)	2.227
East Malaysia	(1)	1.573	(1)	2.182
West Malaysia	(1)	1.573	(1)	2.182
Indonesia	(1)	2.039	(1)	2.613
Philippines	1.000	1.000	(1)	1.248
Papua New Guinea	(1)	1.102	(1)	1.576
Indochina	1.000	1.192	(1)	2.061
India	1.082	1.354	(1)	1.085
Middle East	1.585	1.841	(1)	2.119
East Africa	1.401	1.202	(1)	1.811
North Africa	1.716	1.716	1.972	1.972
South Africa	1.269	1.000	1.000	(1)
West Africa	(1)	1.171	(1)	1.751
Australia	1.338	1.000	1.216	1.216
New Zealand	1.037	1.000	1.236	(1)
Oceania	1.381	1.000	(1)	1.940

Notes: (1) implies no demand function or demand is exogenous.

Supply of a product is considered a function of prices, wood costs, other manufacturing costs, other wood product revenues, and milling capacity. Total processing capacity in a region acts as a constraint to total lumber and plywood manufacturing activity. Capacity is updated based on historical profitability. Table 2 reports the parameters for important processing regions for softwood lumber in the CGTM.

Log supply behavior is described by timber supply equations that specify harvests in each region given log prices, forest inventory levels, and other information. Depending on the particular region, the CGTM uses one of four log supply functions.

Table 2. Parameters for important processing regions in the CGTM: 1993

Regions	Price Elasticity	Price $US/m^3 1980	Product Quantity million m^3 1993	Installed Capacity million m^3 1992	Manuf. Costs US$/m^3 1980	Recovery Factor m^3 log per m^3 product
U.S. PNW	0.32	$188.85	11.710	16.052	$186.69	2.088
Eastside U.S. West	0.55	$290.50	3.759	6.536	$207.55	2.625
Interior U.S. West	0.45	$176.13	10.626	14.172	$163.31	2.363
U.S. South	0.19	$153.65	23.322	28.875	$122.67	2.777
U.S. North	0.34	$136.78	2.749	3.675	$123.53	2.932
Coastal B.C.	0.19	$197.85	6.432	8.100	$150.72	2.384
Interior B.C.	0.24	$173.31	20.866	23.090	$109.81	2.656
Eastern Canada	0.58	$201.50	12.287	16.461	$92.17	2.775
Chile	1.06	$180.41	2.660	3.551	$111.11	1.800
Finland	0.61	$145.73	8.300	9.762	$87.03	2.065
Sweden	0.25	$146.04	12.544	13.494	$133.37	1.996
Western Europe	0.10	$163.56	31.510	41.158	$141.19	1.650
Japan	0.11	$275.95	23.301	23.999	$289.42	1.400
Korea	0.37	$308.21	2.704	3.312	$222.46	1.400
New Zealand	0.28	$214.10	2.684	2.860	$208.14	2.147

The most common supply specification estimates the quantity supplied as a positive function of both log prices and the level of growing stock inventory. Increases or decreases in timber inventory, determined by growth and yield models and removal information, shift the supply curve in or out, period by period.

The second specification has an upward sloping curve but also places an upper harvesting limit to log supply. If a region is operating below the limits on harvests, timber supply will respond positively to increases in log prices. When harvest levels hit the limit, further increases in timber prices have no effect on harvest levels. The upper limit on harvests can be thought of as the maximum allowable annual cut (AAC) on the mature or harvestable inventory. Timber supply in Chile and New Zealand, and Canada more recently, under alternative scenarios, is modeled in this fashion.

A third specification fixes log supply at a predetermined level for each period of the simulation. In any one period, log supply is wholly unresponsive to price changes. Examples of regions using this strict AAC supply specification include China and the CGTM-defined Eastern European region planning rather than prices determines harvest levels. Public timber supply regions in the United States also specify using projected annual cut levels. The specified levels reflect a combination of constraints based on physical productivity of the forest, the infrastructure available to access the forest, and policies that restrict industrial access to timber.

A fourth alternative for log supply assumes that log supply is perfectly elastic, so log supply is determined by product supply. An example is Brazil's solidwood hardwood forest sector where log supply is no a constraint on the production of hardwood solidwood products. In the case of Brazil, log supply behavior is implicit in product supply behavior.

Table 3 reports timber supply parameters from 1993 data for important softwood producing regions around the globe. Each of the alternative supply representations identifies a special case of a generally rising cost curve in which harvesting limits are imposed to model the most appropriate supply behavior for each region and public ownerships.

Table 3. Parameters for important softwood log-supplying regions in the CGTM: 1993

Region	Log Price $US1980/m3	Price Elasticity	Log Production million m^3	Previous Period Inventory Level million m^3
U.S. PNW[a]	$79.55	0.79	32.458	811.5
Eastside U.S. West[a]	$58.74	1.22	5.581	230.1
Interior U.S. West[a]	$50.09	0.74	17.674	785.1
U.S. South[a]	$37.73	0.56	93.669	2529.0
U.S. North[a]	$36.95	2.16	8.374	1038.0
Coastal B.C.[b]	$67.52	4.04	18.184	25.1
Interior B.C.[b]	$39.54	7.35	56.355	67.1
Eastern Canada[b]	$38.92	3.12	30.535	65.0
Chile[b]	$55.74	2.84	7.923	9.4
Finland[a]	$19.04	2.88	16.232	1550.0
Sweden[a]	$41.63	0.41	25.400	2284.0
Western Europe[a]	$56.62	1.00	55.618	3995.0
Japan[a]	$162.37	0.95	16.099	1904.0
New Zealand[b]	$72.44	1.00	10.466	9.5

a: supply is a function of both prices and the level of growing stock.
b: upward sloping supply with allowable annual cut constraint.

Pulp and fiberboard production act as a fiber accounting sector in the CGTM. Projections for each product are made using economic factors, population levels, capacity outlook, and production shares. The production levels are then met individually using either residuals from solidwood product production or pulpwood harvests. The pulp and fiberboard production projections impose material balance between residual production and pulpwood harvests. If solidwood production is reduced, lowering output of wood residuals, then pulp and fiberboard production levels are met using more pulpwood roundwood.

The timber supply dynamics are captured in CGTM by changes in merchantable inventory caused by timber removals and growth in each region. Several inventory models have been linked with the CGTM to describe the development of the merchantable inventory in each region including ATLAS (Mills and Kincaid 1992) and SRTS (Pacheco et al. 1996). Both models provide detailed inventory descriptions and information on how the structure of the inventory in a region changes as each region meets its harvest requirements.

In CGTM, prices are determined period by period with no anticipation of future price levels. For each region, 1) the supply of logs, 2) processing capacity, and 3) product demand are constructed from observations of the price and quantity produced and consumed of each product. The demand for logs is derived from the product demand assuming a fixed ratio of logs used to produce a given quantity of lumber. International price levels define excess supplies and excess demands. Excess supply exists when international prices are higher than domestic prices after deducting transfer costs. Excess demand exists when international prices are lower than domestic prices including transfer costs. At the international price where excess demand and supply intersect, equilibrium is reached. The interaction among regions is defined by trade activity--a trade flow of logs, lumber, and plywood between two regions. Trade activity exists when the sum of the domestic price of a product and the transfer cost associated with shipping the product to the importing region is lower than the importer's price. Single-period solutions in the model are derived by maximizing the consumer and producers welfare (i.e., for all regions, the triangular area found under the demand curve and above the supply curve).

4. Results and Discussion

The study results indicate that timber supply restrictions and continued growth in demand have led to increased prices in North American and Asian softwood lumber markets. A relocation of processing capital has occurred with production shares redistributed from the Pacific Northwest to the U.S. South and elsewhere. Sensitivity analysis suggests that capital investments in processing capacity in Chile and New Zealand for the next several years are required to expand existing capacity by an annual amount greater than 5% per year to maintain the supply projected from their expanding plantation inventory in balance with existing processing capacity supply. Sensitivity analysis result of an Asian recovery depends on the length of time it will take for markets to come back. A quick recovery suggests that resource constraints in the Pacific Northwest region will push prices back upwards rapidly prior to any price increase in Chile and New Zealand.

Cyclical demand upswings and tight wood fiber markets are likely to increase price variability and push prices upwards above trend for short periods. High price expectations, along with an increase in price variability will affect short-term processing capacity utilization and produce fluctuating supplies with periods of oversupply and undersupply until product prices stabilize and the outlook for sawlog supply becomes secure. Sensitivity to demand analysis suggests the downturn in Asia will result in falling prices in North American markets, including the US southern region.

The price projections in real terms and before the Asian financial crisis, are more than 50% larger than the low price period observed during 1982 to 1987. Asia's recession has effectively lowered wood prices for an undetermined period of time. With high solidwood product prices, traditional forest products faced market competition from new wood products that use less wood, and from non-wood substitutes. These products are now facing stronger competition from solid wood products. Sensitivity analysis on resource availability in the US western region suggests that the continued supply constraints on sawlogs in North America are likely to limit the retreat of product prices to any large extent.

Regionally, the lack of alternative, low-cost supply sources for processors in the West indicates that tight log supply markets will likely to continue into the future. As a result, log supply constraints will contribute to upward price pressure in the national and international softwood lumber markets.

An alternative view of sawlog production where no federal timber harvests are allowed and southern hemisphere softwood plantations realize only 75% of their projected output reveals further reductions of milling capacity in western mills and a slower rate of capacity expansion in the southern hemisphere. Processing activity in western states would decline. In the Washington and Oregon coast region, 10% of the processing activity would be lost. In these states, east of the Cascade Range, 26% of the processing activity is projected to be lost. In the Interior West region, 24% of the processing activity is likely to be lost. The lost processing capacity is the result of mill dependency on federal timber and tight log inventories on private timber lands. Processing capacity growth in Chile is projected to be affected by the lower harvest potential, while New Zealand continues to expand its processing capacity under the reduced harvest scenario by at least 5% annually.

The outlook for log prices in Canada indicates a doubling of historical levels for the three major log producing regions, prior to the Asian crisis. Asia's crisis has brought low product prices. These low product prices and high resource costs indicate that a restructuring in capacity is likely to take place in western Canada.

Emerging areas of New Zealand and Chile will surpass industrial roundwood output from western Washington and Oregon by 2010 under the baseline outlook. Annual cut for the two regions is estimated to be at [illegible]1 million m^3 by 2010. The annual cut represents an 18% increase over western Washington and Oregon. However, lack of demand in Asia presents a problem to these emerging regions. Large volumes of excess supplies with no demand push Chile and New Zealand products into U.S. markets. A quick recovery shifts the trade flows back to Asia.

The growth in log prices for Chile and New Zealand over the past two and a half decades has substantially improved the return on forestry investments in these two countries. Higher prices paid to timber owners are the result of active log trade with Southeast Asian consuming regions. Continued forestry investments are attractive in the two countries irrespective of Asia's recession because sawlog prices are likely to remain high in the Pacific Rim region, particularly if additional supply restrictions on public and private lands in the Pacific Northwest regions of the United States and Canada continue.

The rise of log costs has changed purchasing patterns in Japan and Korea and reduced their level of log imports. However, projections of China's inventory suggests it will have log supply problems in the near future, and recent announcements indicate that a total ban on harvesting is likely to be implemented. The log supply shortage is assumed to be imported mainly from New Zealand and Chile. However, other problems may exist in China that would substantially reduce its level of log imports, including a lack of milling capacity particularly in the traditionally wood-producing northeast provinces.

5. Conclusions

Discounting the current recession in the Asian economies, it seems likely that supply issues will continue to dominate price formation over the medium term. There is an upward trend in prices over the medium term governed by declining inventories of mature timber. Cyclical fluctuations in demand are likely to make prices volatile around this price trend.

Much of the timber supply pressures can be associated with the growth in demand for environmental services. Environmental factors are becoming of greater importance in determining timber supplies. The growth in personal income has generated a large demand for environmental services from the forest resources in the United States, particularly public lands. This trend is likely to continue and likely to further constrain timber supplies from native forests. Greater restrictions on public lands lead to less processing capacity in the U.S. West and more investments in the Southern Hemisphere.

6. References

Cardellichio, P.A., Youn, Y., and Adams, D. 1989. A preliminary analysis of timber and timber products production, consumption, trade and prices in the Pacific Rim until 2000. CINTRAFOR Working Paper 22, Center for International Trade in Forest Products, College of Forest Resources, University of Washington, Seattle.

Mills, J.R., and Kincaid, J.C. 1992. The aggregate timberland assessment system--ATLAS: A comprehensive timber projection model. General Technical Report PNW-GTR-281. Portland, OR: U.S. Department of Agriculture, Forest Service, Pacific Northwest Research Station.

Pacheco, G., Abt, R.C., and Cubbage, F.W. 1996. Southwide timber supply projection assessment. *in* Southern Forest Economics Workshop Proceedings. Gatlinburg, TN.

Acknowledgements

This modeling work has involved many people. We would like to acknowledge their contributions and in particular the contributions of Michael Pederson. Michael is responsible for CGTM programming since its development at the University of Washington. Funding for the CGTM has come from many sources. In particular we acknowledge the funding support of the U.S.D.A. Cooperative State Research Education and Extension Service (CSREES).

The Economics of Carbon Emission and Sequestration in the Forestry Sector of Peninsular Malaysia

Ismariah Ahmad and John Perez-Garcia

Key words: Log supply, CO_2 emission, forest sector modeling

Abstract: This paper demonstrates the use of a forest sector model to evaluate the economic implications of the policy to reduce CO_2 emission by reducing timber harvest in Peninsular Malaysia. Harvest reduction causes reduced available inventory, hence shifting in of the log supply curve. The economic effect is reduced availability of logs for plywood production. Preliminary results suggest that the resulting land conservation leads to increased forest biomass, hence carbon content in the forest.

1. Introduction

1.1 Background

Most of the current increase in atmospheric carbon is caused by burning fossil fuel. Emission from this source contributes between 5.2 and 5.9 Pg of Carbon per year to annual flux of atmospheric carbon. The next most important source is land use change that releases 0.5 to 2.6 Pg of Carbon per year to the atmosphere, produced mostly by shifts in forest mass that result from land clearing or other changes in land use (Hall and Uhlig 1991). The change in forestland use is important because forests contain about 90% of the carbon in terrestrial vegetation and because forests are being cleared at a very high rate. The net land-use flux of carbon from terrestrial ecosystem to the atmosphere is primarily caused by the large-scale clearing of the tropical forests. The reasons for tropical forest clearing are complex and include expansion, rural land development, logging, shifting cultivation, mining, and cattle ranching.

In tropical Asia, approximately 1.74 million ha of Asian undisturbed closed forest are logged each year for the production of veneer and sawlogs (Brown et al. 1991). Historically, harvests in tropical timber exporting countries have followed a boom-and bust pattern (Vincent 1992). The benefits and costs of this pattern of resource exploitation remain poorly understood. To the extent that booming harvests reflect myopic behavior induced by short, uncertain concession contracts and low royalties, they are indeed overly rapid. It would be in the exporting country's best interest to reduce harvests to more sustainable levels. On the other hand, because most tropical logging occurs in old-growth forests, rapid "mining" of the resource might actually maximize the net present value of current and future harvest. In this case reducing harvest would impose economic costs on the exporting country. Such a country would understandably be reluctant to reduce its harvests unless the rest of the world compensates it. In turn, the rest of the world would be willing to do this only if the reduced harvest generated global benefits related to lower CO_2 emissions or other public goods at least as large as the compensation costs.

1.2 Study objective

The annual carbon flux from changes in forestland use depends on the rate the forests are cleared. Reducing deforestation and logging rate would reduce CO_2 emission. Consequently, a policy affecting deforestation or harvesting would affect CO_2 emission. Such a policy would have economic trade-offs throughout the economy. Reduced logging and deforestation would decrease timber supply, raise timber price, reduce timber consumption in the processing industry, decrease product output, raise product prices, decrease employment, generate lower revenue and foreign exchange, and finally reduce CO_2 emissions. In the long run, the condition may encourage substitution between wood and non-wood and between local and imported sources. The policy has its welfare implications too.

This paper demonstrates the use of a forest sector model to evaluate the economic implications of the policy to reduce CO_2 emission in the forestry sector in Peninsular Malaysia. The objective of this paper is to evaluate the overall effect of the policy of reducing harvest by 30% on CO_2 emission and supply and demand for logs in Peninsular Malaysia. How would rates of harvests, log prices, and log consumption be affected? How much does conservation in logging areas affect CO_2 emission? What is the net effect of the policy? To answer these important questions, we need a model to simulate the production and consumption of logs in the Peninsula with and without the harvest restriction.

The policy to reduce CO_2 emission by reducing harvest levels or logging rate affects the timber supply component of a forest sector model. Changes in timber supply produce changes in important economic variables: timber prices, harvest levels, and timber consumption. On the other hand, reducing the harvest increases the physical forest inventory, hence the biomass and the carbon balance.

1.3 Forest in Malaysia.

Malaysia is the world's largest producer and exporter of tropical logs and tropical hardwood sawn wood. The rapid rate of logging in Malaysia has attracted international attention comparable to the attention to deforestation in the Amazon. Reviews of forestry in Sarawak by the International Tropical Timber Organization (ITTO) in 1990 and Malaysia by the World Bank in 1991 concluded that current harvest far exceeds sustainable rates. In response, Malaysia has aggressively defended its forestry practices, and it played a leading role in defeating the international convention on tropical forestry originally proposed for UNCED. Because Malaysia is a small country, most of its timber harvest is ultimately consumed in importing countries. Changes in Malaysia's harvest levels would have a significant effect on tropical timber supplies in the world market. World demand for temperate hardwood timbers, which can substitute for tropical timbers in many end uses, would be affected too. Similarly, the policy measure to reduce carbon emissions would have significant effect on the economics of timber harvesting and timber processing in Malaysia.

2. Method

Evaluating the policy implications of reducing CO_2 emission by reducing timber harvests involves several procedures. First, estimate timber supply and demand. Second, use this supply and demand model to evaluate the change in economic variables: prices, production, and consumption, resulting from the harvest policy. Third, estimate forest biomass and the subsequent carbon account. Fourth and finally, evaluate the subsequent effect of the policy on the carbon balance.

2.1 The model

We needed to construct an economic model of the supply and demand for logs in the Peninsula. We estimated the supply function for logs, hence the price elasticity of supply. We relied on the predetermined functional forms for supply as in Cardellichio et al. (1988) but the database covers the period between 1970 and 1995. In the case of timber demand, we relied on existing estimates of demand elasticity, instead of estimating functional forms and parameters by econometric methods.

2.1.1 Demand for logs

Demand for logs is a derived demand, i.e., dependent upon the quantities of sawnwood, plywood, and other products that are produced from logs. Log consumption is related to product output by a fixed-coefficient production technology. Hence log demand is a simple statement of the log raw material required for any given level of a single product.

The domestic consumption for logs in Malaysia is assumed to be determined by the level of sawn wood and plywood consumption multiplied by the respective wood recovery rates.

[1] $D_L = S_S / \tau_S + S_P / \tau_p$

where D_L is log consumption, S_S and S_P are sawn wood and plywood consumptions, τ_S and τ_P are recovery rates for sawn wood and plywood. The recovery rate is the number of cubic meters of logs needed to produce one cubic meter of the required products (sawnwood or plywood). The domestic log consumption is modeled as the sum of the consumption of veneer logs (α_2 fixed at historical value in each year) plus consumption of sawlogs for sawnwood S_S / τ_S. The equation can be represented as

[2] $D_L = \alpha_1 P^{\beta 1} / \tau_S + \alpha_2$

where P is the domestic price of sawnwood and β is the elasticity of demand for sawnwood. When solving the demand equation, we recalculate the log equivalent price of the sawnwood price considering the marketing margin and the recovery factor. The elasticity was drawn from the University of Washington's revised version of the Global Trade Model (Cardellichio et al. 1989).

2.1.2. Supply of logs

The timber supply model has two components: harvest and delivery costs, (net delivered price) and the fees levied by the government. Net delivered price is defined as the delivered price of logs minus government charges assessed on those logs. Net delivered prices are modeled as a marginal cost curve based on the assumption that timber concessionaires maximize profits by minimizing costs at their output predetermined by the government.

One can postulate several factors that will affect the position of the supply curve, with respect to harvest and delivery costs. Because of several problems regarding the use of delivery costs, Cardellichio et al. (1988) simplified the estimation of harvest and delivery costs by modeling these costs as a function of cut, or the cut-inventory ratio. A similar model of timber supply is adopted in this paper. The same model was also used by Vincent (1986) to estimate timber supply in Malaysia.

In the short run, harvest and delivery costs are expected to increase with increasing harvests, owing to a factor of production that is essentially fixed, e.g., logging equipment. The cost curve would shift upward owing to an increase in factor prices for variable inputs—labor and energy. The short-run cost would also shift upward as logging proceeds to distant and more difficult terrain. We should note that a reduction in inventory (declining virgin forest area or depletion of old growth stands) increases the costs of stumpage because of increased timber scarcity, and increases harvest and delivery costs because of more expensive access.

The specification of the supply function is as follows:

Net price = f(harvest/inventory)

Express in a simple power function:

[3] $NP=\alpha Q^{\beta}$

The general functional form is

[4] $\ln NP_t = \alpha_3 + \beta_2 \ln Q_t + u_t$

where NP_t is the net delivered price of logs and Q_t is the cut-inventory ratio, and α_3 and β_2 are parameters. t = 1970…1995

The equation is estimated in log-linear form. The logarithmic transformation means that the price will equal zero when the quantity equals zero. Time series data often present the problem that the variance in the data increases over time. One method of dealing with heteroscedasticity problems is to transform the data via logarithms. The logarithmic transformations simplify the interpretation of regression results because the coefficient on the quantity term is the inverse of the supply elasticity.

The relevant data to estimate the function are real log price, royalty rates, premium, harvest levels, and inventory levels between 1970 and 1995. "Inventory" refers to growing stock (industrial stemwood net of defect) of all hardwood species of merchantable size (above 45-50 cm DBH). The term "harvest" and "cut" refer to log removals, that is, logs actually extracted from the forest. All prices are expressed in domestic currencies and were converted to real terms (1980 base year) using a general price deflator—the consumer price index for Peninsular Malaysia. All equations are estimated using ordinary least squares (OLS). All units of measurements are expressed in metric terms—m^3. The next section discusses the method to develop the timber account, the basis for point estimates of timber inventory for individual years.

2.1.3. Timber account

Timber inventory data is derived from the timber account that calculates the net change in timber stock. The timber physical accounts keep track of both areas of forest and stock of timber because changes in areas affect changes in stock. Areas and stocks are recorded separately for virgin (primary or unlogged) and logged forests (secondary or logged-over forests). Areas are measured in thousand hectares, and stocks in thousand cubic meters. The timber accounts cover the natural forests only. Timber plantations have been established only recently, and their area and output are still small relative to those of the natural forests. Timber stocks are defined as the volume of wood in standing trees of commercial species with a dbh (measured at 1.5 m above the ground) greater than or equal to 45cm. For the timber account, we adapt the method and accounting procedure in Economic Planning Unit Malaysia (1993).

The accounts were based on general assumptions about changes in areas and stocks. The area of virgin forests declines due to logging. Following logging, virgin forests are either converted to non-forest uses or reclassified as logged-over forests. Unlike virgin forests, the area of logged-over forests can expand, if the area added through logging of virgin forests exceeds the area of existing logged-over forests that are converted to other uses.

The accounting identities for unlogged and logged areas are:

<u>Unlogged forests</u>

CLOSING AREA = OPENING AREA - AREA LOGGED

AREA TRANSFERRED TO LOGGED FOREST = AREA LOGGED – AREA DEFORESTED

Logged forest area:

CLOSING AREA=OPENING AREA + AREA TRANSFERRED FROM UNLOGGED FOREST - AREA DEFORESTED

OPENING AREA refers to area at the beginning of a given year (= area at the end of the previous year), and CLOSING AREA refers to area at the end of the same year (= area at the beginning of the next year). The accounts assume that the forest is always logged before being deforested. Hence, unlogged forest area declines only due to logging (AREA LOGGED), which precedes conversion of unlogged forests to either logged forests (AREA TRANSFERRED TO LOGGED FORESTS) or other land uses (AREA DEFORESTED). In turn, area of logged forests can expand if the AREA TRANSFERRED FROM UNLOGGED FOREST (= AREA TRANSFERRED TO LOGGED FOREST) exceeds the area of logged forests converted to other uses (AREA DEFORESTED).

The accounting identities for timber stocks are similar to those of areas, but there are a few additional terms.

Timber stock in unlogged forests:

CLOSING STOCK=OPENING STOCK
- QUANTITY HARVESTED
- QUANTITY TRANSFERRED TO LOGGED FOREST
- QUANTITY DESTROYED BY DEFORESTATION
- OTHER REDUCTION IN STOCK

Timber stock in logged forests:

CLOSING STOCK = OPENING STOCK
+ TIMBER GROWTH
+ QUANTITY TRANSFERRED FROM UNLOGGED FOREST
- QUNATITY HARVESTED
- QUANTITY DESTROYED BY DEFORESTATION
- OTHER REDUCTION IN STOCK.

When an area of unlogged forests is logged, some portion of the timber stock on it is harvested (QUANTITY HARVESTED). The remaining stock is retained on the part of the logged area that is reclassified as logged forest (QUANTITY TRANSFERRED TO LOGGED FOREST STOCK), but it is lost on the part that is deforested (QUANTITY DESTROYED BY DEFORESTATION). The stock may also be reduced by other factors, such as logging damage or fire damage (OTHER REDUCTION IN STOCK).

In logged forest, timber stock can increase for two reasons: TIMBER GROWTH and QUANTITY TRANSFERRED FROM UNLOGGED FOREST (=QUANTITY TRANSFERRED TO LOGGED FOREST). As in unlogged forest, timber stock decreases owing to QUANTITY HARVESTED, QUANTITY DESTROYED BY DEFORESTATIOIN, and OTHER REDUCTIONS IN STOCK.

The account is generated for the period 1970 to 1995.

2.2. Policy analysis

We chose 1993 as the base year for instituting the policy of reducing harvest by 30% of the historical level. The model of supply and demand for logs was solved using the spreadsheet. The supply and demand equations were calibrated to the historical data in 1993 by adjusting their constant terms (α_i). This was done by solving each equation for α_i and inserting the formula thereby obtained into a spreadsheet in Excell 7.0. Then we imposed a 30% reduction in the volume harvested and recalculated the new equilibrium. The simulated results were compared to the historical data that infer the effects of the harvest reduction.

2.3. Carbon emission

Changes in biomass of tropical forests are one of many sources of carbon dioxide release. The carbon accounts provide information on the amount of carbon stored in Peninsular Malaysia's forests and the amount released when they are harvested or converted.

The estimate of the physical timber stocks and flows in the timber accounts provide most of the information needed to construct physical accounts for carbon. The next step is to convert the timber quantities to biomass, and biomass to carbon. For the first step we resort to the ecologist to provide the conversion factors. To convert from biomass to carbon content we simply need to divide the biomass weight by two, as 50% of the biomass is made up of carbon (Brown et al 1991).

Applying this procedure would imply that the entire amount of a negative net change in the timber stock translates directly, and immediately, into a net release of carbon. This may not be the case. Logs are processed into lumber, plywood, moldings, and furniture, which store carbon for decades or even centuries. Therefore, the procedure above provides an upper bound on the net carbon release. A lower estimate can be generated by assuming that only half of the carbon in harvested timber is released, to reflect that logs are typically converted into processed wood products at a recovery rate of 50%.

3. Data

Published data on total forest areas, area opened in Reserve and State Land forests, area converted from Permanent Forest Estate (PFE), quantity of timber harvested, government charges—premium, timber royalty, and silviculture cess (a flat-rate charge on each m^3 of timber extracted but the rates vary among states in the Peninsula and the revenues are solely for forest rehabilitation and development) for each year were available from the Forestry Department and the Ministry of Primary Industry (1993). The Malaysian Timber Industry Board (MASKAYU) provides monthly data on domestic log prices in three regions—East Coast, West Coast and Central region. Thang (1986, 1987) provided estimates of the gross growth rate for commercial timber (2.2 m^3 ha per year), the defect rate (30%), and the damage rate (20%).

Forest Inventories 1, 11, and 111 provided data on virgin and logged-over timber stocks and forest areas in the middle of 1972, and at the beginning of 1982 and 1993 (FAO 1973, Ibu Pejabat Perhutanan Semenanajung Malaysia 1987, 1995). The accounts used the revised estimates of forest areas from Brown et al. (1991)'s geographical information system (GIS) analysis of the original inventory maps that provide additional information on the total areas of virgin and logged-over forests deforested between the first two inventories (1 and 11).

Harvests were allocated between virgin and logged-over forests by assuming their shares were proportional to the timber stocks in each type of forest. The area of virgin forests logged in a given year equals harvest volume divided by the harvest per hectare. Areas logged were calculated by determining the harvest intensity that reconciled the point estimates of areas from the inventories. This means that the cumulative area logged during 1972 to 1993 equaled the actual change in area as given by the inventories.

To calculate total growth, the growth rate 2.2 m^3 ha per year was multiplied by the average of the areas of logged forest at the beginning and end of the year.

Timber stocks in virgin forests were determined by multiplying virgin forest areas by Food and Agriculture Organization of the United Nations (FAO) estimates of per-hectare growing stocks (FAO/UNEP 1981, citing earlier FAO studies). Growing stock was defined as the volume of net industrial stemwood per hectare in trees with a minimum diameter of 45 cm. Timber stock in logged-over forests was determined in the same way.

The conversion factor for timber stocks to biomass were drawn from the Economic Planning Unit (1993) report, adapting from Brown et al. (1991) cited earlier. They provided conversion factors specific to the data in Forest Inventories 1 and 11. The conversion factors are 3.39 and 3.60 metric tons per m^3 in primary forests in 1972 and 1981, and 4.17 and 5.03 metric tons per m^3 in secondary forests in the same two years. Because we are not able to generate the conversion factor from data in Forest Inventory 111 (1993), we continue to apply the conversion factor for 1981 for both forest types to calculate carbon content in 1993.

4. Results and Discussion

This section presents the results of the estimates of supply and demand for logs and the policy implication of harvest reduction.

4.1 Log supply

Figures 1 and 2, and Table 1 reveal the most important features of the analysis. Figure 1 shows the plots of delivered net price (NPL) and the cut-inventory ratio (CI)—the price and "quantity" variables in the model. The price is in real term (1980=100) and in Malaysian Ringgit per m^3.

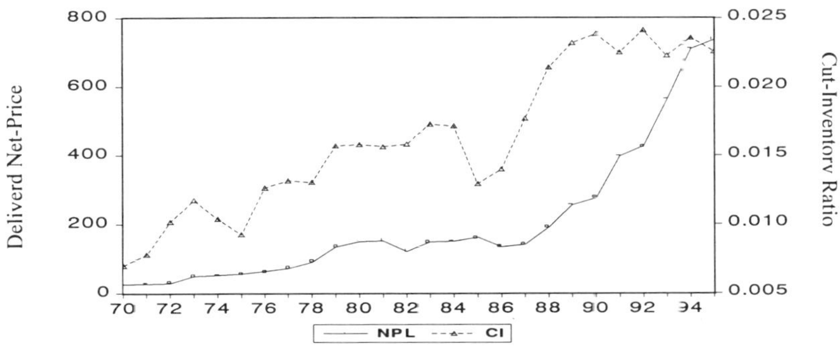

Figure 1. Delivered net price and cut-inventory ratio

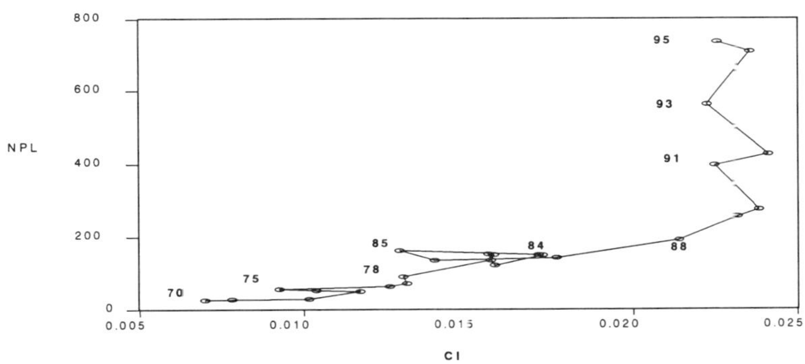

Figure 2. Scatter plot: net price and cut-inventory ratio

The historical trends show that both net delivered log price and the cut-inventory ratio were trending upward. NPL was relatively stable from 1970 to 1977, increased steadily from 1978 to 1985, and declined slightly during 1985 to 1987. The period between 1988 to 1995 saw a tremendous increase in the price. The plot of cut-inventory ratio was more volatile. It declined sharply during 1973 to 1975 and 1984 to 1985, and rose sharply during 1986 to 1989, then generally flattened out afterwards. The nature of CI relates to the trend in the inventory and log harvests. Throughout the period of analysis, the inventory declined rapidly and by 1995 the remaining inventory was slightly more than 50% of that in 1970. In contrast, log harvests were very volatile, harvests increased tremendously beginning in 1986 from 7 million m^3 to reach the all time high peak in 1989 surpassing 12 million m^3, and harvests remained at that average level up till 1995. The Peninsula's harvests come from both virgin and secondary-growth forests. Peninsular Malaysia imposed log export restrictions starting in 1972, by 1985 there was a complete export ban except for smaller-diameter logs. In 1993, the nation imposed an export levy on its lumber exports.

Table 1. Regression results
Dependent Variable: LNPL, Method: Least Squares, Sample: 1971-1995,
Included observations: 25

Variable	Coefficient	Std. Error	t-Statistic	Prob.
C	14.08798	0.736496	19.12839	0.0000
D9395	0.839427	0.172922	4.854376	0.0001
D8485	0.340968	0.187585	1.817669	0.08[illegible]4
LNCI	2.228389	0.174775	12.75007	0.0000

R-squared	0.932737	Mean dependent var	4.964345
Adjusted R-squared	0.923128	S.D. dependent var	0.912275
S.E. of regression	0.252935	Akaike info criterion	0.234278
Sum squared resid	1.343498	Schwarz criterion	0.429298
Log likelihood	1.071528	F-statistic	97.06969
Durbin-Watson stat	1.694815	Prob(F-statistic)	0.000000

The scatter plot between the two relevant variables is shown in Figure 2. The scatter plot shows a positive slope, an expected behavior between price and "quantity." It is obvious that harvests shifted in during 1983 -1985 and that 1993, 1994 and 1995 are outliers.

The next step is to rule out structural shifts during those periods mentioned earlier. We want to know if there was a supply shift or movement along the supply curve, which may be brought about by parallel shifts in demand that caused the price along the supply curve to change. We analyzed the influence of both domestic and export demands on log supply. We considered the Malaysian GDP and Japan's GDP and housing starts to proxy for domestic and export demands, respectively. We plotted the growth in log harvest with the growth in Malaysia's GDP, and Japan's GDP and housing starts. In all, the main observation shows that the period between 1984 to 1985 and 1990 to 1995 may possibly show a change in supply not really caused by the demand. A supply shift may be brought about by the condition of the available resource inventory.

We restricted the estimation sample to 1971 to 1995, and included two intercept dummies to account for the shifts that commenced in 1984 and 1993. The regression results and fit are provided in Table 1. The elasticity of log supply with respect to the net delivered price is 0.45. This differs somewhat from the previous estimate of 0.91 in Cardellichio et al. (1988).

4.2 Timber demand and simulation results

In the estimate of demand for lumber, the value of the elasticity of lumber consumption for Peninsular Malaysia drawn from the Global Trade Model's revised version is –0.5. The recovery rate for log to lumber is 0.5. Using equation 1.2 for sawnwood consumption, we solved for alpha and the quantity of logs demanded for lumber. We calibrated the supply and demand to historical data in 1993 and then solved the equations under the harvest constraint.

The effect of reducing harvest by 30% from the 1993 level is to reduce available timber inventory by the same amount. In economic terms, that produces a 30% leftward shift in the supply curve. With the assumption of constant elasticity supply curves, this will give a constant proportional change in quantity at any given price. At any price, fewer logs will be offered for consumption by the sawmilling and plywood milling industry (equivalently, the old volume can be purchased, but at a higher price).

The magnitude of change in production and consumption is presented in Table 2. The results show that a 30% harvest reduction is equivalent to conserving about 56,000 ha of forest. Harvest restriction reduces the resource base available for the primary processing industry. Apparently, sawnwood production can be sustained but plywood production will be greatly reduced. Before the reduction, about 58% of log production went to sawnwood and 42%, for plywood and veneer. After the reduction the proportion changes to 82 and 18%, respectively.

Table 2. Simulated economic effects of harvest reduction in Peninsular Malaysia: 1993

	Restrictions[a]		
	With	Without	Difference[b]
		million m^3	
Log production	7.8	11	-3.3
Log consumption for:			
Sawnwood	6.4	6.4	-
Plywood	1.4	4.6	-3.2
Forest area ($x10^3$ ha)	133	189	56
Carbon Balance			
		million m^3	
Change in stock	468.8	456.6	12.3
Change in biomass ($x10^6$ tons)	2046.7	1774.3	272.3
Change in carbon balance ($x10^6$ tons)	1023.3	887.1	136.2

a. "Without" = actual (historical) values. "With" = simulated values, based on modeling the effects of the reducing harvest volume by 30% from the 1993 level. b. Values in "Without" case minus values in "With" case.

4.3 Carbon storage

Preliminary results of the simulation show that reducing the volume of timber harvest by 30% led to a net increase in inventory of 3%, increases in biomass and carbon content of 15% while reducing CO_2 emission.

5. Conclusion

In summary, this paper illustrates the use of a forest sector model to evaluate the magnitude of economic effect of the harvest policy to reduce CO_2 emission. The analysis of log supply in Peninsular Malaysia shows that in the early 1990s the available inventory caused supply to shift inward. The economic effect of reduced harvest is a reduced consumption of logs in the plywood industry. Additionally, conservation of timber stock led to conservation of forest biomass and carbon.

6. References

Brown, S., Iverson, L., and Lugo, A. 1991. Land-use biomass changes of forests in Peninsular Malaysia during 1972-1982: Use of GIS analysis. *in* Dale, V. (ed), Effects in land-use change on atmospheric CO_2 concentrations: Southeast Asia as case study, Springer-Verlag.

Cardellichio, P.A., Youn, Y.C, Binkley, C.S., Vincent J.R., Adams D.M. 1988. An economic analysis of short run timber supply around the globe. CINTRAFOR Working Paper [illegible], University of Washington, College of Forest Resources, Seattle.

Cardellichio, P.A., Youn, Y.C., Adams, D.M., Joo, R.W., and Chmelik, J. 1989. A preliminary aanalysis of timber and timber products production, consumption, trade and prices in the Pacific Rim until 2000. CINTRAFOR Working Paper 22, University of Washington, College of Forest Resources, Seattle.

Economic Planning Unit Malaysia 1993. Malaysian national conservation strategy-towards sustainable development. Volume 4. Natural Resource Accounting. Prime Minister Department. Kuala Lumpur.

FAO/UNEP 1981.Tropical forest resources assessment project: Forest resources of tropical Asia . UN 3216.1301-78-04. Technical Report No.3.FAO, Rome.

FAO 1973. A national forest inventory of west Malaysia, 1970-72.FO:DP/MAL/72/00[illegible], Technical Report No.5.Rome.

Hall, C.A.S., and Uhlig J. 1991. Refining estimates of carbon released from tropical land-use change. Can. J. of For. Res. 21: 118-131.

Ibu Pejabat Perhutanan, Semenanjung Malaysia (Forestry Department Headquarters Peninsular Malaysia).1987. Inventori Hutan Nasional 11, Semenanjung Malaysia: 198[illegible]-1982. Kuala Lumpur.

Ibu Pejabat Perhutanan, Semenanjung Malaysia (Forestry Department Headquarters Peninsular Malaysia). 1995. Inventori Hutan Nasional 111, Semenanjung Malaysia: 1991-1993. Kuala Lumpur.

International Tropical Timber Organization. 1990. The promotion of sustainable forest management: A case study in Sarawak, Malaysia. Mimeo, ITTO Eighth Session, Denpasar, Bali, Indonesia. 16-23 May 1990.

Malaysian Timber Industry Board. Maskayu, Monthly Timber Bulletin. Various Issues.

Ministry of Primary Industries, Malaysia. 1993. Statistics on primary commodities: Forestry. Kuala Lumpur.

Thang, H.C. 1986. Can the existing wood resources meet future domestic requirements? Unpublished Paper presented at FRIM Symposium, 13 September 1986.

Thang, H.C. 1987. Forest management system for tropical high forest, with special reference to Peninsular Malaysia. Forest Ecology and Management. 21:3-20.

Vincent, J.R. 1992. The tropical timber trade and sustainable development. Science. 256: 1651-1655.

Vincent, J.R 1986. Growth of the Forest Product Industry in Malaysia, 1961-1985. Malaysian Forester Vol. 49. No 3: 223-240.

World Bank. 1991. Malaysia: forestry subsector study. Report 9775. MA. Washington, DC.

Biofuels and the Forest Sector

- An econometric model of the Swedish forest sector -

M. Ankarhem, R. Brännlund and M. Sjöström

Key words: Forest economics, demand and supply, biofuels

Abstract: The main objective of this paper is to analyze potential effects on the Swedish forest sector of a continuing rise in the use of forest resources as a fuel in energy generation. The background to the problem can be found in the commitments Sweden has made concerning energy policy. Two such commitments are the phase-out of nuclear power, and a decision that the Swedish energy system should be sustainable, i.e., it should be based on renewable resources. However, an increasing use of forest resources as an energy input may have effects outside the energy sector. In this paper we attempt to consider this by estimating a system of demand and supply equations for the four main actors on the Swedish roundwood market; forestry, sawmills, pulpmills, and the energy sector. The purpose of the paper is thus to analyze the effects on each of these markets as a result of a switch in energy policy, towards an increase in the use of biofuels.

1. INTRODUCTION

One of the main objectives in current Swedish energy and environmental policy is that the energy system be sustainable. From a Swedish perspective this means that energy should be supplied by renewable resources as opposed to non-renewable resources. A first step in this direction has been taken in the decision to successively abandon nuclear power as an energy source. The main policy to achieve the goal of a renewable-based energy system is to promote the use of bioenergy. Since Sweden is well endowed with forest land, an increase in the use of bioenergy is more or less equivalent to an increase in the use of the forest resources as energy. Using biofuels as an energy source is nothing new in Sweden. In 1970 less than 10% of the total energy supply was derived from bio-fuels. The corresponding figure in 1996 is almost 20%. However, so far the main source of biofuels has been wood residues and sawmill residues that have little or no alternative use. In order to go ahead with a massive substitution from, say, nuclear power towards biofuels, forest resources that may have an alternative value will have to be considered for energy purposes.

The objective of this paper is a first attempt to analyze the possible effects of such a policy on the Swedish forest sector. To do this we specify and estimate an econometric forest sector model in the spirit of Brännlund and Kriström (1996). The model consists of four main components; the forest owners who supply sawtimber, pulpwood and biofuels; the sawmill industry, which demands sawtimber; the pulp and paper industry, which demands pulpwood; the energy sector, which demands biofuels.

As has already been pointed out, the Swedish government has decided to phase out the nuclear power, which currently supplies 16 % of the total energy supply and approximately 50% of the electricity supply.

This in combination with the fact that the Swedish government also has a commitment to stabilize carbon dioxide emissions at the 1990 level before the year 2000, and the limited possibilities to extend the use of hydropower indicates that the use of biofuels may increase in the future. The utilization of Swedish forest resources in terms of wood supply is displayed in Figure 1 below.

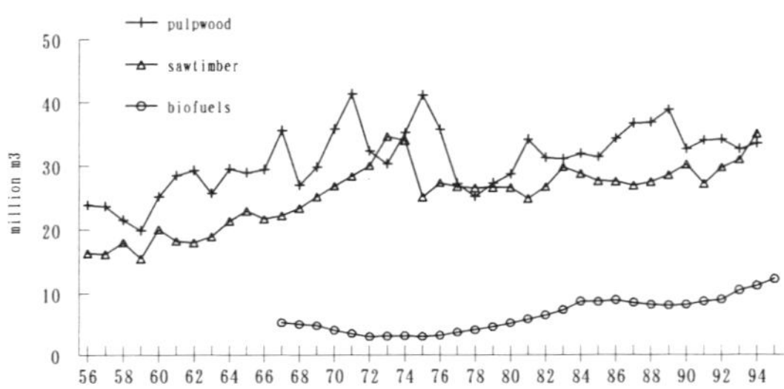

Figure 1. Supply of pulpwood, sawtimber, and biofuels 1956-1995, million m^3

In Figure 1 it can be seen that pulpwood and sawtimber have been and are the major products coming out from the Swedish forests. However, it can also be seen that the supply of biofuels has had two major periods of increase. The first period was in connection to the oil crisis (OPEC I and OPEC II), which started in the beginning of the seventies. The major reason for the increase in biofuels during this period was concern about energy shortage. The second upswing started in the beginning of the nineties and is still going on. The major reason for the increase this time is a combination of energy shortage and environmental concerns. In addition, energy taxation has to a large extent favored biofuels during the nineties. The question posed in this paper is thus whether this rising trend, if it continues, affects other parts of the forest sector.

2. The Model

We assume that a forest owner is supplying three different products: pulpwood, that is used as an input in the pulp industry, sawtimber, that is used as an input in the sawmill industry, and biofuels, that are used as an input in energy generation. This division is not a too poor description of reality, since almost all wood consumption is for those purposes. To produce these three products we assume that labor and the standing inventory of timber are used as inputs. Labor is viewed as a flexible input while forest land is regarded as fixed, at least in the short run.

Assuming perfect competition, the representative forest owners profit function can be written as:[1]

[1] $\Pi^f = \Pi^f(w_{pv}, w_{sv}, w_{bv}, w_{fl}; \overline{K}_f)$

where w_{pv} is the price of pulpwood, w_{sv} is the price of sawtimber, w_{bv} is the price of biofuels, w_{fl} is the wage rate in the forest industry and $\overline{K}_f$ is the fixed capital stock in the forest industry, which means that we regard the model as a short run model. By applying Hotelling's lemma to the profit function [1], we obtain the supply of pulpwood, sawtimber, and biofuel, as well as the demand function for labor in forestry.

Similarly, the profit function for the energy firm is expressed as:

[2] $\Pi^b = \Pi^b(P_b, w_{bl}, w_{bv}, w_{boil}; \overline{K}_b)$

where P_b is the output price on generated energy, w_{bl} the wage rate in the energy sector, w_{bv} the price of biofuels, or fuelwood, w_{boil} the price of oil for heating, and $\overline{K}_b$ is the fixed capital stock. Again, applying Hotelling's lemma gives us the supply of energy as well as demand for labor and biofuel as a function of the energy price, wage rate, price of biofuel, and the fixed capital stock. Correspondingly we define the profit function for the pulp firm as:

[3] $\Pi^p = \Pi^p(P_p, w_{pl}, w_{pe}, w_{pv}; \overline{K}_p)$

Equation [3] gives us the supply of pulp, and the demand for the various production factors as a function of the price of pulp, P_p, the wage rate in the pulp industry, w_{pl}, the electricity price, w_{pe}, and the fixed capital stock, $\overline{K}_p$. Finally we define the profit function for the sawmill as:

[4] $\Pi^s = \Pi^s(P_s, w_{sl}, w_{se}, w_{sv}; \overline{K}_s)$

where P_s is the price of sawn products, w_{sl} is the wage rate in the sawmill industry, w_{se} is the price of electricity, and $\overline{K}_s$ is the fixed capital stock.

[1] The profit function is obtained as the solution to the following problem: Max $\{pf(q) - wx\}$, where q is output, p is the price of the output, x is the input quantity, and w the corresponding price, and f the production function (see for example Varian 1992). The profit function is assumed to possess the usual properties, such as nondecreasing in p, nonincreasing in w, convex and continuous in prices, and homogeneous of degree 1 in prices.

In this framework it is assumed that each firm acts as a price taker, i.e., no firm in any of the industries can affect output or input prices by its own actions.[2] In addition we assume that labor and electricity demand in these sectors is small in relation to total supply of labor and electricity, which means that the wage rates and the electricity price is exogenous to these industries. On the other hand, we reasonably assume that forest owners supply of pulpwood, sawtimber, and biofuels are used entirely by the pulp industry, sawmills, and energy producers. This can be expressed as:

$$[5] \quad \sum_{i=1}^{n_p} x_{pv}^i(P_p; w_{pl}, w_{pe}, w_{pv}, \overline{K}_p) = \sum_{i=1}^{n_f} y_{pv}^i(w_{pv}, w_{sv}, w_{bv}; w_{fl}, \overline{K}_f)$$

$$[6] \quad \sum_{i=1}^{n_s} x_{sv}^i(P_s; w_{sl}, w_{se}, w_{sv}, \overline{K}_s) = \sum_{i=1}^{n_f} y_{sv}^i(w_{pv}, w_{sv}, w_{bv}; w_{fl}, \overline{K}_f)$$

$$[7] \quad \sum_{i=1}^{n_b} x_{bv}^i(P_b; w_{bl}, w_{bv}, w_{boil}, \overline{K}_b) = \sum_{i=1}^{n_f} y_{bv}^i(w_{pv}, w_{sv}, w_{bv}; w_{fl}, \overline{K}_f)$$

where x_j^i is the i-th firm's demand for good j, and y_j^i is the i-th forest owner's supply of good j, and where n_p, n_f, n_s and n_b are the number of pulp producers, forest owners, sawmills and energy producers. Solving for w_{pv}, w_{sv} and w_{bv} gives us the equilibrium price on pulpwood, sawtimber, and biofuels as:

$$[8] \quad \begin{aligned} w_{pv}^* &= w_{pv}^*(P_p, P_s, P_b, \tilde{w}_p, \tilde{w}_s, \tilde{w}_b, \tilde{w}_{fl}, \overline{K}) \\ w_{sv}^* &= w_{sv}^*(P_p, P_s, P_b, \tilde{w}_p, \tilde{w}_s, \tilde{w}_b, \tilde{w}_{fl}, \overline{K}) \\ w_{bv}^* &= w_{bv}^*(P_p, P_s, P_b, \tilde{w}_p, \tilde{w}_s, \tilde{w}_b, \tilde{w}_{fl}, \overline{K}) \end{aligned}$$

where $\tilde{w}_p, \tilde{w}_s, \tilde{w}_b$ are vectors of input prices, other than pulpwood, sawtimber, and biofuel, referring to the pulp industry, sawmill industry, and energy industry respectively, and $\overline{K}$ is a vector including the capital stock in each industry.

We can now define the industry, or aggregate, profit functions as:

$$[9] \quad \Pi\Pi^f = \Pi\Pi^f(w_{pv}^*, w_{sv}^*, w_{bv}^*, \tilde{w}_f; \overline{K}_f)$$

$$[10] \quad \Pi\Pi^b = \Pi\Pi^b(P_b, \tilde{w}_b, w_{bv}^*; \overline{K}_b)$$

$$[11] \quad \Pi\Pi^p = \Pi\Pi^p(P_p, \tilde{w}_p, w_{pv}^*; \overline{K}_p)$$

[2] This is a questionable assumption at least for the pulp industry, which consists of a small number of companies. For a discussion concerning the oligopoly power of the Swedish pulp industry we refer to Wiberg (1987). Bergman & Brännlund (1995) found that the pulp industry seems to have, or at least utilize, market power on the Swedish roundwood market.

$$[12]\quad \Pi^s = \Pi^s(P_s, \tilde{w}_s, w_{sv}^*; K_s)$$

By differentiating equations [9] – [12] with respect to output and input prices we obtain the industry supply and demand functions.

For our purposes it is not necessary to estimate all parameters in all profit functions. Endogenous prices in our model are the prices of pulpwood, sawtimber, and biofuels, which means that the minimum requirement is estimates of the supply and demand functions for pulpwood, sawtimber, and biofuels. By estimating these equations as a system we can calculate the elasticities that are of interest in this study.

3. Empirical Specification

In order to calculate elasticities we need to parameterize equations [9] to [12]. Our objective is to choose a parameterization that imposes as few technological restrictions as possible, i.e., we want a flexible parametric representation of the technology inherent in equations [9] to [12]. One such flexible form is the Generalized Leontief (GL) profit function (see Diewert 1973). In this case the GL profit function can be expressed as:[3]

$$[13]\quad \Pi^f = \sum_i \sum_j \alpha_{ij} w_i^{1/2} w_j^{1/2} + \sum_i \alpha_{ik} w_i \bar{K}_f$$

i=pv, sv, bv, fl, j=pv, sv, bv, fl

$$[14]\quad \Pi^b = \sum_i \sum_j \lambda_{ij} w_i^{1/2} w_j^{1/2} + \lambda_{ik} w_i \bar{K}_b + \lambda_{bb} P_b + \sum_i \lambda_{ib} w_i^{1/2} P_b^{1/2} + \lambda_{bk} P_b \bar{K}_b$$

i=bl, bv, boil, j=bl, bv, boil

$$[15]\quad \Pi^p = \sum_i \sum_j \beta_{ij} w_i^{1/2} w_i^{1/2} + \sum_i \beta_{ik} w_i \bar{K}_p + \beta_{pp} P_p + \sum_i \beta_{ip} w_i^{1/2} p_p^{1/2} + \beta_{pk} P_p \bar{K}_p$$

i=pl, pv, pe, j=pl, pv, pe

$$[16]\quad \Pi^s = \sum_i \sum_j \gamma_{ij} w_i^{1/2} w_j^{1/2} + \sum_i \gamma_{ik} w_i \bar{K}_s + \gamma_{ss} P_s + \sum \gamma_{is} w_i^{1/2} P_s^{1/2} + \gamma_{sk} P_s \bar{K}_s$$

i=sl, sv, se, j=sl, sv, se

By applying Hotelling´s lemma on equation [13] we obtain the supply of pulpwood, sawtimber, and biofuel, as well as demand for labor in forestry as a function of all prices:

$$[17]\quad y_i = \frac{\partial \Pi^s}{\partial w_i} = \sum_j \alpha_{ij} \left(\frac{w_j}{w_i} \right)^{1/2} + \alpha_{ik} \bar{K}_f, \; i = pv, st, bv, \quad j = pv, st, bv, fl$$

[3] Note that the fixed capital stock for the sawmill industry and the energy sector are omitted in the empirical specification. The reason is simply that data for these variables are lacking.

$$[18] \quad -x_i = \frac{\partial \Pi^s}{\partial w_i} = \sum_j \alpha_{ij} \left(\frac{w_j}{w_i} \right)^{1/2} + \alpha_{ik} \overline{K}_f, i = fl, j = pv, st, bv, fl$$

Similarly from equation [13] we obtain supply of energy, and demand for fuelwood and labor in the energy sector as:

$$[19] \quad y_b = \frac{\partial \Pi^b}{\partial P_b} = \lambda_{bb} + \sum_j \lambda_{ib} \left(\frac{w_i}{P_b} \right)^{1/2} + \lambda_{bk} \overline{K}_b$$

$$[20] -x_i = \frac{\partial \Pi^b}{\partial w_i} = \sum_j \lambda_{ij} \left(\frac{w_j}{w_i} \right)^{1/2} + \lambda_{ib} \left(\frac{P_b}{w_i} \right)^{1/2} + \lambda_{ik} \overline{K}_b$$

$i = bv, bl, boil, \quad j = bv, bl, boil$

Similarly we obtain supply of pulp and demand for pulpwood, electricity, and labor in the pulp industry as:

$$[21] \quad y_b = \frac{\partial \Pi^p}{\partial P_p} = \beta_{pp} + \sum_j \beta_{ip} \left(\frac{w_i}{P_p} \right)^{1/2} + \beta_{pk} \overline{K}_p$$

$$[22] \quad -x_i = \frac{\partial \Pi^p}{\partial w_i} = \sum_j \beta_{ij} \left(\frac{w_j}{w_i} \right)^{1/2} + \beta_{ip} \left(\frac{P_p}{w_i} \right)^{1/2} + \beta_{ik} \overline{K}_p$$

$i = pv, pl, pe, j = pv, pl, pe$

Finally, we obtain the supply of sawnwood and the demand for sawtimber, labor, and energy in the sawmill industry as:

$$[23] \quad y_s = \frac{\partial \Pi^s}{\partial P_s} = \gamma_{ss} + \sum_j \gamma_{is} \left(\frac{w_i}{P_s} \right)^{1/2} + \gamma_{sk} \overline{K}_s$$

$$[24] \quad -x_i = \frac{\partial \Pi^s}{\partial w_i} = \sum_j \gamma_{ij} \left(\frac{w_j}{w_i} \right)^{1/2} + \gamma_{is} \left(\frac{P_s}{w_i} \right)^{1/2} + \gamma_{ik} \overline{K}_s$$

$i = sv, sl, se, \quad j = sv, sl, se$

The data we use are annual time series data covering the period between 1967 and 1994 and are collected from Swedish official statistics (mainly SOS Industry but also from the Yearbook of Forest Statistics). Since equations [17] to [24] are linear in the parameters ordinary least squares (OLS) could in principle be used. There are, however, two problems with OLS in this case. The first is that it is very likely that the errors within each system are correlated. This implies that a simultaneous method would be more efficient. We estimate the demand and supply functions as a system using 3SLS (Three Stage Least Squares) and GMM (General Method of Moments). That is, we do not need to estimate all the parameters in the profit functions, for our particular purpose.

The price of energy produced by the energy sector (P_b) is measured by the output price from Swedish district heating plants. The choice of the price is motivated by the fact that most biofuels are used in district heating plants. For this variable the supply of relevant data is limited to the period 1972-1995. By regressing the price of district heating on the price of heating oil and electricity we 'backcast' the price of district heating for the missing periods.

5. Results

Since there really isn't much one can tell by observing the parameter estimates of our supply and demand equations, we omit them and instead focus on the t-ratios, partially given in Table 1 below. Out of a total of 31 parameter estimates, 19 are significantly different from zero at the 10 percent level. This is the case for both 3SLS and GMM estimates. According to the Durbin Watson (DW) estimates in Table 1, we have some problem with autocorrelation, which means that the variance of the parameters may be underestimated.

Table 1. GMM estimates of t-ratios for the restricted profit functions

Forestry								
	Sawtimber	Pulpwood	Biofuel	Labor	Capital	T	DW	R^2
Sawtimber	0.30	-1.30	-2.48	-2.05	2.37	-1.63	1.55	0.28
Pulpwood		3.11	2.40	-2.15	-0.04	0.07	1.47	0.52
Biofuel			-1.65	1.21	0.23	3.15	0.58	0.82
Pulp Industry								
	Pulp	Pulpwood	Energy	Labor	Capital	T	DW	R^2
Pulpwood	3.08	1.62	-1.07	-0.96	-2.87	3.22	1.39	0.46
Sawmill Industry								
	Lumber	Sawtimber	Energy	Labor	Capital	T	DW	R^2
Sawtimber	6.85	4.54	-9.41	8.24	2.05	-3.89	2.01	0.78
Energy Industry								
	Energy	Biofuel	Labor	Oil			DW	R^2
Bio fuel	7.28	-8.09	-4.17	-4.96			0.82	0.84

The elasticities are evaluated at the quantities and prices of the last period, 1994. All price elasticities corresponding to our demand and supply functions are given in Table 2. Starting with wood supply, we notice that the own-price elasticities have the expected positive sign. According to the results, the own-price supply elasticity for biofuels is close to zero, implying that forest owners' supply of biofuel is not very sensitive to price changes. The cross-price elasticities of pulpwood and biofuel have an unexpected sign. For example, a price increase of pulpwood would, according to the results, lead to an increase of the supply of biofuels, i.e., biofuel and pulpwood seem to be complementary products. One possible explanation of the latter is that there is joint production between pulpwood and biofuel, i.e., a higher price of pulpwood increases pulpwood supply, but also increases the supply of wood residues, as a side effect. This explanation is supported by the fact that most of the biofuels supplied from forestry consists, so far, of wood residue, which to a large extent is a side product of industrial wood supply. Another possible explanation may be related to the use of biofuel. Until the mid seventies the use of biofuels (i.e., wood for heating) has increased as a consequence of changing heating systems and urbanization. After the mid seventies there has been a structural change in the use of biofuels as a result of the oil price shocks in the seventies, but also due to environmental concerns in the late eighties. It may very well be the case that the simple model used in this study is incapable of taking this structural change into account.

Furthermore, it can be noted that labor and pulpwood, and labor and sawtimber seem to be substitutes in the production of pulp and lumber respectively, while fuelwood and oil seem to be complements in the production of energy.

Table 2. Output supply and input demand price elasticities (GMM-estimates), evaluated for 1994

Forestry				
	Sawtimber	Pulpwood	Biofuel	Labor
Sawtimber	0.50	-0.07	-0.12	-0.30
Pulpwood	-0.11	0.23	0.12	-0.25
Biofuel	-0.55	0.33	0.001	0.21
Pulp industry				
	Pulp	Pulpwood	Electricity	Labor
Pulpwood	0.13	0.11	-0.11	-0.12
Sawmill industry				
Sawtimber	Lumber	Sawtimber	Electricity	Labor
	0.51	-0.58	-0.75	0.81
Energy industry				
	Energy	Biofuel	Oil	Labour
Biofuel	0.50	0.05	-0.23	-0.32

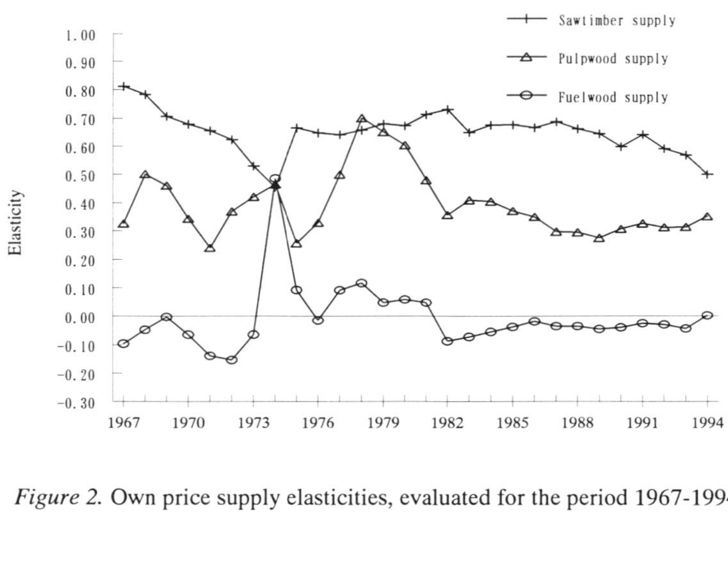

Figure 2. Own price supply elasticities, evaluated for the period 1967-1994

The own-price elasticities of wood supply for each year is displayed in Figure 2. Here we can see that the price elasticity of industrial wood has been fairly stable, and positive, over the whole period. The own-price elasticity of fuelwood, on the other hand, fluctuates more between positive and negative values. Specifically interesting is the period 1974-1982 where the elasticity is positive.

6. Conclusion

The objective of this study is to analyze the interrelations between the supply of pulpwood, sawtimber, and fuelwood from Swedish forestry. To do this we specify an econometric model for the Swedish forest sector where it is assumed that the forest owners supply pulpwood, sawtimber, and biofuel. Furthermore it is assumed that pulpwood is used by the pulp industry, sawtimber by the sawmill industry, and fuelwood by the "energy" industry. The data used in the econometric analysis are time series covering the period 1967-1994.

According to the results, the own-price elasticities are positive for sawtimber and pulpwood, but close to zero for biofuels. Thus, biofuel supply seems to be less price sensitive than supply of industrial wood. According to the results concerning demand for pulpwood, sawtimber, and fuelwood, we conclude that the own-price elasticities have the expected negative sign. The main conclusion concerning the interrelation between supply of industrial wood and fuelwood is that we can reject the hypothesis that fuelwood and pulpwood are substitutes in Swedish forestry during the period 1967-1994. This conclusion follows from the positive cross-price elasticity between fuelwood and pulpwood.

The results in this paper should, however, be interpreted with some care since there are a number of problems which have to be considered in future research. One of them is the absence of capital stock data for the energy industry. This may cause inefficiency in the parameter estimates and even worse, bias, if the capital stock is correlated with some of the included explanatory variables. With capital stock data we would not only obtain better estimates, but we would also be able to formulate the model in a dynamic framework and incorporate costs of adjustment[4]. Another problem is that the model adopted here is too simple in its structure in order to be able to describe reality. For example, we have not considered the fact that the

[4] Naturally there could also be costs of adjustment in changing the labor stock, but Lundgren and Sjöström (1999) finds no evidence supporting that theory in there study of the Swedish pulp industry. For a nice overview of adjustment cost see Hamermesh and Pfann (1996).

sawmill industry "produces" woodchips that is sold both to the pulp & paper industry and to the energy sector.

7. References

Brännlund, R., and Kriström, B. 1996. Welfare measurement in single and multimarket models: Theory and application. American Journal of Agriculture Economics. 78:157-165.

Bergman, M.A., and Brännlund, R. 1995. Measuring oligopsony power: An application to the Swedish pulp and paper industry. Review of Industrial Organization. 10: 307-321.

Diewert, W. E. 1973. Functional forms for profit and transformation functions. Journal of Economic Theory. 6: 284-316.

Hamermesh, D.S., and Pfann G.A. 1996. Adjustment costs in factor demand. Journal of Economic Literature. 34:1264-1292.

Lundgren, T., and Sjöström, M. 1999. A Dynamic factor demand model for the Swedish pulp industry. Journal of Forest Economics. 5: 45-68.

Varian, H. R. 1992. Microeconomic analysis, Norton.

Wiberg, A. 1987. Svensk massa- och pappersindustris marknadsställning. Department of Forest Economics, Swedish University of Agricultural Sciences, Report No. 71.

Have Forest Sector Models Changed Forest Policy in the United States

Richard W. Haynes and Darius M. Adams

Key words: Models, policy analysis, price effects

Abstract: The United States has a 100-year history of broad scale assessments of forest conditions and timber trend types of studies (see Haynes 1993 for a more complete discussion). What has changed over that period has been both new types and amounts of data and our ability to develop more explicit information from the available data. How have the forest sector models developed in the last 20 years changed forest policy in the United States? The answer is threefold: they shifted the discussions from quantities to prices. They have allowed the use of scenario planning to discuss different views of the future, and they have changed the perceptions of decision makers through collaborative learning in the sense that we have been able to demonstrate how various projections are based on the underlying assumptions. The role that forest sector models have played in United States forest policy serves as a reminder that they are a means to inform policy debates, industrial decisions or landowner decisions. The models are not an end in themselves.

1. Introduction

The United States has a 100-year history of broad scale assessments of forest conditions and timber rend types of studies (see Haynes 1993 for a more complete discussion). What has changed over that period has been both new types and amounts of data and our ability to develop more explicit information from the available data. The early efforts involved estimating trends in timber availability based on resource statistics and trends in consumption in final products and established a strong science tradition as the basis for discussions of forest policies. This policy basis expanded again in the 1970s to include price trends.

How have the forest sector models developed in the last 20 years changed forest policy in the United States? The answer is threefold: they shifted the discussions from quantities to prices. They have allowed the use of scenario planning to discuss different views of the future, and they have changed the perceptions of decision makers through collaborative learning in the sense that we have been able to demonstrate how various projections are based on the underlying assumptions.

These models have been used to organize various views of the forest sector and to derive useful information that can improve and expand the perceptions held by land managers, the public, and government and nongovernment agencies. In the United States, forest sector models have been used to assess present and future resource conditions, demands for forest products and, in the last several decades, prospective price trends. These assessments have played several roles. First, they have helped guide land stewardship decisions. Second, they have provided a basis for including science in policy making in the sense that improved information will improve decisions. These evolving roles have been evident in the United States where 20 years of continued support for the maintenance and further development of forest sector models demonstrates their value.

Before going further we need to define what we mean by the terms forest sector and forest sector models. By forest sector, we mean the consideration of all aspects of forestry from the basic land tenure issues, to harvesting, production and consumption decisions. Such a view is no different than similar views of various agricultural and industrial sectors in an economy. It is important to remember that informal forest sector analysis methods have long been part of the tools available to foresters. Forest sector models are characterized by a systems view of forestry's biological and social processes. The importance of this systems view is that it can be used to explore the biological development of forest resources and the attendant dynamics of the forest industry over time, and under alternative futures and management strategies. A basic premise is that we can specify various relationships that describe (or mimic) biological and socioeconomic processes.

2. The Emergence of Price as a Policy Variable

The first area where forest sector models have changed forest policy involves the need to estimate price effects associated with specific policies or lack of policies. Changes in policy debates from describing policy needs in terms of quantity shortfalls (that were never observable with real world data) to trying to identify policy needs in terms of price impacts stimulated the evolution of forest sector models. This shift in the United States occurred in the mid 1970s during a period of rapid changes in commodity prices (see Figure 1) and concerns about the effectiveness of various policies to alter selected price trajectories. Prior to the 1970s, there had been relative stability in prices for protracted periods of time (for example 1910 to 1945 and 1953 to 1968). Up until this time the prevalent forest sector model was the gap model where policies and emerging trends were discussed in the context of the gap between the trajectories of the demands for forest products (expressed as roundwood equivalents) and the prospective availability of timber.

The implication of this interest in prices for model builders was that models needed to be built around supply and demand representations that directly included price as an independent variable and that model solution processes needed to seek the price and quantity equilibrium. In the US, this led to forest sector models based on market equilibrium concepts. These models evolved from the earlier Gap models. The Gap model still has much intuitive appeal because policy makers can gauge the amount of timber that is needed if the goal of a policy action is "to fill the gap". The Equilibrium model appeals to policy makers interested in price changes and estimates of consumer or producer impacts. Within the United States Forest Service, all policy discussions in the last decade have been based on the Equilibrium models but decision makers still find it useful, at times, to frame a policy discussion around the notion of some quantity (for example, the level of the timber sale program) associated with a particular issue.

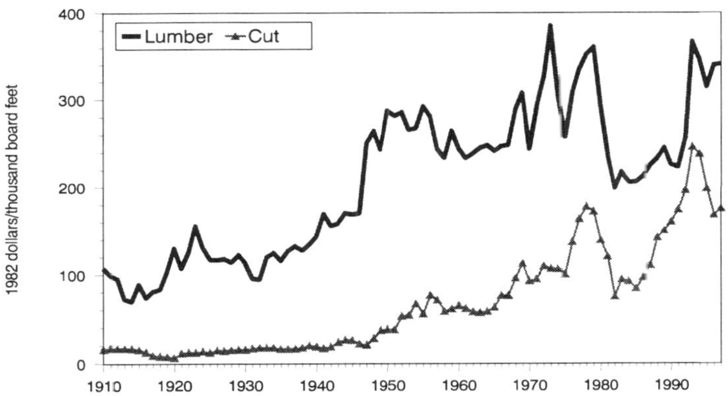

Figure 1. Deflated price of coast Douglas-fir lumber and Pacific Northwest national forest cut stumpage

With the evolution of the equilibrium model, price and quantity variables are available for use as policy variables. Often the projected path of prices is used to identify potential problems or to measure the effectiveness of specific policy actions including various forms of market interventions. In addition, there are potential policy implications associated with interpreting the solution of equilibrium models. First, while these solutions do not represent intertemporal production or consumption strategies that are in some sense optimal, they do represent how the markets might evolve from considerations of economic efficiency. Second, economic models allow the estimation of various welfare measures that can be used to address the economic equity questions of who gains and who loses.

The additional power of looking at prices as a policy variable can be illustrated in Figure 2 using data from a United States Forest Service

Assessment (USDA Forest Service 1982). Figure 2a illustrates the results of a gap analysis for projected supply-demand balances for U.S. softwoods. Figure 2b illustrates a stumpage price projection for softwoods in the southern states of the U.S. associated with balancing supply and demand in the stumpage market.

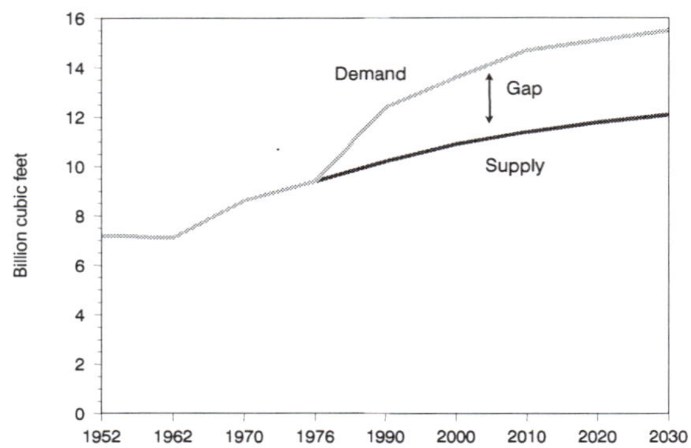

Figure 2a. Determining policy needs based on quantity differences

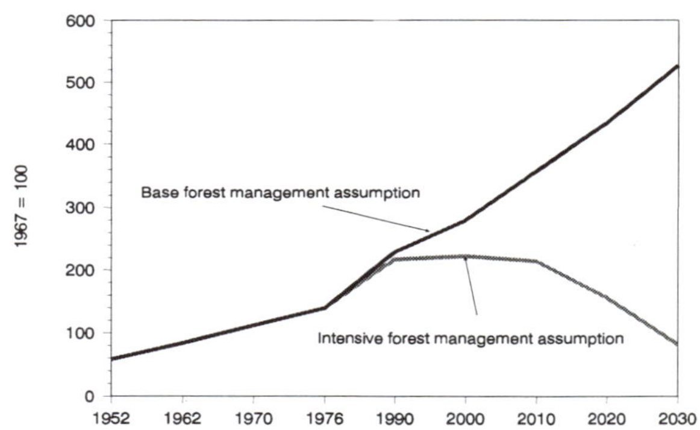

Figure 2b. Determining policy needs based on prices

To a policy maker, the inference in Figure 2a is that projected demands will exceed available supplies and that policy actions will be needed to reduce the gap. Often the policy discussion is framed around reducing some shortfall of, say 2.7 billion cubic feet in the year 2000. Implied although seldom addressed is the associated price changes expected with failures to fill the gap. Figure 2b illustrates a case where stumpage prices are used as the policy variable of interest. For example, if we assume continuation of recent timberland management practices, stumpage prices are expected to increase (net of inflation) at 2.1 percent per year between 1990 and 2030. The policy issue is framed around the societal acceptance of that rate of price increase. If for example, it is not acceptable we could institute programs to increase softwood tree planting and conversion of low value hardwood stands on privately owned timberlands that would stabilize stumpage prices in the near decades (1990-2010) and as the planted stands mature, eventually lower stumpage prices (see Figure 2b).

Both cases illustrate similar policy inferences but for different reasons. In the case of the gap analysis, perceived quantity shortfalls drive policy actions. In most countries, these shortfalls are not observable as price changes cause market adjustments that offset supply shortfalls. In the equilibrium analysis, perceived price changes drive policy actions. Usually, the lack of policy actions is observable in the sense of undesirable price changes.

3. Scenario Planning

There has been an evolution in the use of forest sector models in strategic planning by both government and non-government organizations. This is particularly evident in the use of simulation approaches to analyze alternative futures and management strategies. One goal of these efforts is to derive useful information that can improve and expand the perceptions held by decision-makers and the public. As such, this differs from traditional modeling (and particularly forecasting) approaches by both treating uncertainty directly in the analysis and stressing information needs of the public.

In the modeling literature this has been called scenario planning (Wack 1985). Scenario planning, as it usually applied, does not attempt to predict the future. Instead it postulates a set of plausible futures, each dependent on the assumptions underlying that future. By doing so, the technique focuses on what might happen (either positive or negative) and how to deal with it. This can be contrasted with planning for a predetermined future. Scenario planning has been around both in the general literature and forestry literature since the mid 1980s. In the United States, scenario planning has been part of the RPA efforts since the early 1980s (see USDA Forest Service 1982, Haynes 1990, Haynes et al. 1995). Most forestry applications take a classical sensitivity analysis approach, a limited number of key exogenous and endogenous elements are varied and key projection results are examined for differences. These differences allow the identification of emerging problems and provide a way to measure the effectiveness of possible solutions to various problems.

Figure 3 illustrates five alternative futures for the United States forest sector that were developed as part of the 1993 RPA Timber Assessment Update (see Haynes et al. 1995). These futures illustrate the differences in one selected variable (the softwood lumber price index) that is indicative of overall market behavior under different views of the future. These views include a base run that generally represents a consensus view, the effects of lower harvests (and exports) in Canada, the effects if alternative macroeconomic forecasts (lower Gross National Product [GNP] but higher housing activity--CEA demand projections), the effects of increased tree planting on 2 million acres (Molton and Richards, and the effects of climate change which acts to increase net productivity for most United States forest types.

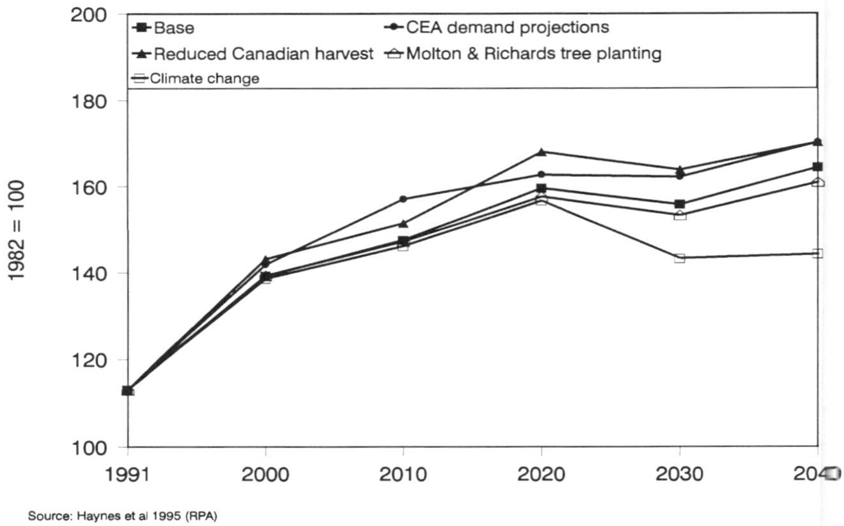

Figure 3. Alternative futures and the softwood lumber price index

The results shown in Figure 3 have illuminated U.S. policy discussions. For example, the often cited prospective of "catastrophic ecological declines" associated with statements about climate change have been exposed as over stated. In the near-term reductions in Canadian harvest and alternative demand assumptions lead to the greatest differences in the outlook. In the longer-term (after 2030) increases in net productivity due to climate change lead to some of the largest differences. This illustrates the delayed effects of changes in growth rates of timber inventories where these changes eventually lead to increases in available inventories and to increases in timber harvest.

4. Shifting Perceptions

The third area was a change that we did not initially foresee. Although we did design a forest sector model to be extremely flexible for the range of questions for which it could be employed we did not anticipate its usefulness for educating the interested publics. Given its robust economic framework, we found ourselves soon considering various shifts in supply and demand functions both in the product and stumpage supply markets. These numerous alternative futures or "what ifs" soon lead to extensive debates within the policy community.

Former researchers who were now decision makers became our advocates for the use of forest sector models to help explain various market or resource impacts. Over a relatively short period of time (about 10 years) we changed the perceptions of top decision-makers and various stakeholder groups about the relative impacts associated with most common forest policy issues. We also reduced the barrier between model builders and decision makers in these discussions.

Much of this success depended on our ability to mix the art and craft of policy analysis. Much of this depends on the robustness of both the empirical and the underlying conceptual models. Our use of a spatial equilibrium model that simultaneously solves both the product and stumpage markets allows us to consider questions in three spatial and explicit temporal scales. For example, we can consider national issues that set context for what happens in different regions, we can consider regional differences, and we can reflect differences among the regions in various supply and demand relationships and other determinants. The explicit treatment of time leads to a robust examination of temporal trajectories of policy effects.

5. Closing

In many ways we were fortunate that we adopted a relatively fully specified economic model to serve as the underpinning of the forest sector model. This had several advantages. First, economics places an inherent importance on systems types of thinking. Second, by the mid 1970s there were a number of alternative approaches already developed to consider complete industrial sectors. We had the advantage of being able to select the features that seem to fit our situation most closely. Third, we had a specific set of questions guiding model development that were mostly economic in nature. For example, could we develop stumpage price projections for five decades for regions of traditional interest that bore the correct relationship to product prices and could we develop these stumpage projections to be sensitive to changes in National Forest harvest flows (which initially was the main policy variable)?

Finally, the role that forest sector models have played in United States forest policy serves as a reminder that they are a means to inform policy debates, industrial decisions or landowner decisions. The models are not an end in themselves.

6. References

Haynes, R.W. 1990. An analysis of the timber situation in the United States: 1989-2040. A Technical Document Supporting the 1989 USDA Forest Service RPA Assessment. Gen. Tech. Rep. RM-199. Fort Collins, CO: U.S. Department of Agriculture, Forest Service, Rocky Mountain Forest and Range Experiment Station. 249p. [plus appendices].

Haynes, R.W. 1993. Forestry sector analysis for developing countries: Issues and methods. Gen. Tech. Rep. PNW-GTR-314. Portland, OR: U.S. Department of Agriculture, Forest Service, Pacific Northwest Research Station. 46p. [plus appendix].

Haynes, R.W., Adams, D.M., and Mills, J.R. 1995. The 1993 RPA timber assessment. Gen. Tech. Rep. RM-GTR-259. Fort Collins, CO: U.S. Department of Agriculture, Forest Service, Rocky Mountain Forest and Range Experiment Station. 66p.

U.S. Department of Agriculture, Forest Service. 1982. An analysis of the timber situation in the United States 1952-203. For. Res. Rep. 23. Washington, DC. 499p.

Wack, P. 1985. Scenarios: Uncharted Waters Ahead. Harvard Business Review. 63(5): 73-89.

Impacts of South-East Asian Log Export Ban Policies on Japanese Plywood Market

Satoshi Tachibana and Shin Nagata

Key words: Southeast Asian countries, log export ban policies, Japanese plywood market, econometric model, and economic surplus

Abstract: This study aims to assess the effects of the log export ban policies of Southeast Asian countries on the Japanese plywood market, whose supply mainly comes from Southeast Asian and domestic plywood industries. Both plywood industries process tropical raw logs. Our econometric analysis reveals the following: The log export ban policy of Indonesia, along with its plywood industry promotion policy, has increased plywood exports from Southeast Asia to Japan, leaving the log supply rather untouched. Sabah's log export ban policy has decreased log exports from Southeast Asia to Japan. The producers' surplus of Southeast Asian plywood increased after the log export ban of Indonesia. Consumers' surplus in the Japanese plywood market increased accordingly. On the other hand, the producers' surplus in the Japanese plywood industry decreased significantly from both export bans.

1. Introduction

Timber export restriction policies of major exporting countries have had great influences on the world-wide timber trade. Some of these timber-exporting countries have reinforced their timber export restriction policies since the 1970s, for example Canada, United States, and Southeast Asian countries. These restriction policies are introduced because of forest degradation, deforestation, and environmental movements in some cases, and for economic development and resource nationalism in others.

We use softwood from temperate forest and subarctic forest mainly as lumber, paper, and boards. Hardwood from tropical forest is processed for plywood and lumber. In Japan, the softwood demand for lumber is bigger than that for hardwood in value. Considering this situation, we analyzed the effects on Japanese softwood timber markets of the restrictions imposed both on timber harvest and log exports in North America (Tachibana and Nagata 1996, Yukutake et al. 1996). It revealed that Japanese consumers paid much more for these restrictions than North American producers did.

Southeast Asian countries have played an important role as suppliers in Japanese timber markets since the 1960s. We have used their hardwood mainly as plywood. These countries started log export restriction in the late 1970s. Especially the log export ban policies of Indonesia in 1985 and of Sabah in 1993 had serious influences both on these exporting countries and on timber importing countries, including Japan. Togu (1995) studied economic effects of the log export ban on the development of forest product industries of Indonesia. His study showed that the effects on net employment and foreign earnings from the policy changes were not large.

This study aims to assess the effects of log export ban policies of Southeast Asian countries on the Japanese plywood market. Because the Japanese plywood market is a major outlet for Southeast Asian logs, this analysis partly reveals the effects on the Southeast Asian countries as well.

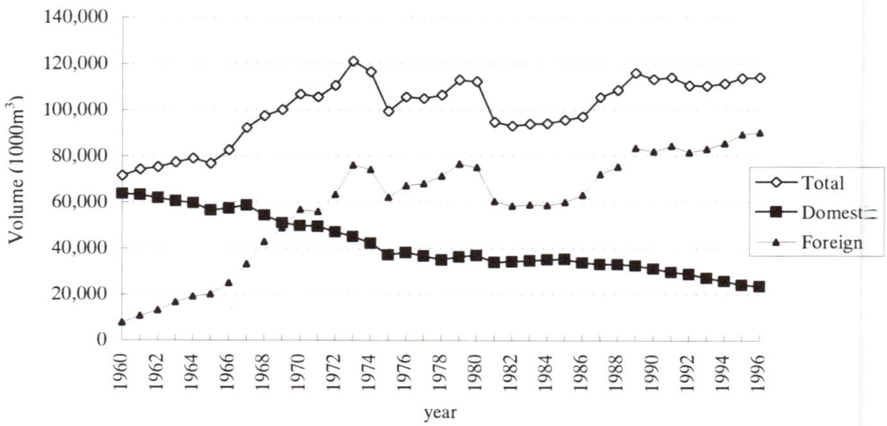

Figure 1. The trend of Japanese timber market
(Source: the Forestry Agency, Mokuzai Jyukyu Hyo)

2. Trends of Japanese Timber Trade

2.1 Japanese timber trade

Figure 1 shows the trend of Japanese timber trade after World War II. Because of the high economic growth before the first oil crisis, log imports had been increasing rapidly in Japan in that period.

Especially, housing construction starts in Japan contributed strongly to the increase. Foreign supply exceeded that of domestic forest owners in the Japanese timber market in 1969, and domestic supply has been decreasing since then. Japanese log imports have been slowly diminishing since the oil crisis, with noticeable increases of processed timber imports. These changes are partly a result of log export restrictions, for example the regulations on timber harvest and log exports from the end of the 1980s in the United States and a log export ban from 1985 in Indonesia. In 1997, Japanese domestic timber resources supplied less than 20% of timber consumption in Japan.

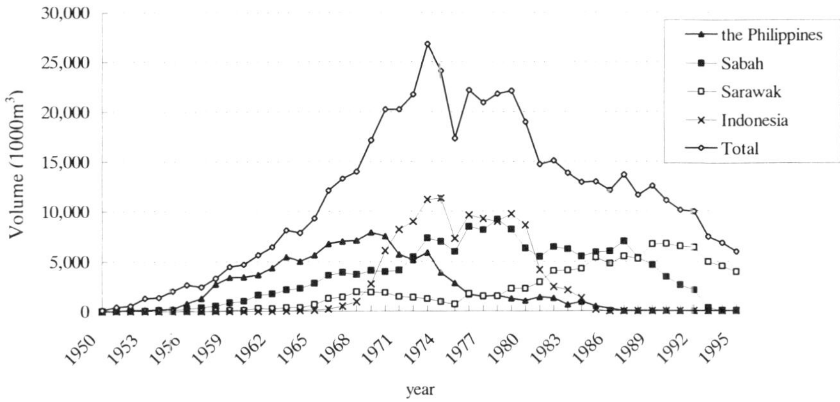

Figure 2. Japanese imports of south sea timber (Source:the Ministry of Finance, Boeki Tokei)

2.2 Timber export and its restriction policies of Southeast Asian countries

Among Southeast Asian countries, forest exploitation started in the Philippines, because of their comparatively high-quality logs and the proximity to Japan, the largest timber importer. The Philippines had been the main supplier of tropical logs to Japan under the free timber trade scheme (Figure 2).

Both Indonesia and Sabah have taken over the role of main suppliers to Japan since the middle of the 1970s. Their forest exploitation started in the late 1960s, when forest exhaustion in the Philippines became apparent. Indonesia intensified log export regulations in 1979 to 1985, to promote exports of plywood and lumber. They reinforced these regulations and banned log exports in 1985, giving subsidy for plywood exports to "new markets"; especially Japan and Korea. These policies made Indonesia one of the biggest plywood suppliers in the world.

On the other hand, Sabah had played an important role as a major tropical log supplier to Japan till the beginning of the 1990s. The first restriction on log exports from Sabah was planned in 1976, after the establishment of SEALPA (South-East Asia Lumber Producers' Association) in 1974. Sabah, however, had not enacted any log export restrictions in spite of many plans to do so. The main reason was the financial dependence of the State government on timber-related revenues, especially from log royalties. In Sabah, they did not want to cut off the log royalty, which has provided almost half of State revenue. Sabah had continuously harvested in excess of forest growth resulting in forest exhaustion (Tachibana et al. 1995), and Sabah had to ban log exports in 1993.

Sarawak has become a main tropical log supplier in Southeast Asia since the late 1980s. The extent of forest exploitation in Sarawak fell far behind the above countries because of the lower quality of its logs. Sarawak, however, must play a role of major log exporter in the region in place of Indonesia. Papua New Guinea has increased log exports at the same time as Sarawak.

Recently Japanese tropical log imports have been decreasing remarkably under the two log export ban policies and because of the exhaustion of forest resources in Southeast Asia. To mitigate this situation, compound plywood that is made of both hardwood and softwood has increased in the 1990s, including that made from Siberian logs and radiata pine logs.

2.3 Japanese plywood market

The structure of Japanese plywood markets is given in Figure 3. There are two main suppliers to the plywood markets, Southeast Asia and the domestic industry. Both process tropical raw log to plywood. Domestic log supply is very small; 228m^3 in 1995. The majority of plywood logs come from the tropics.

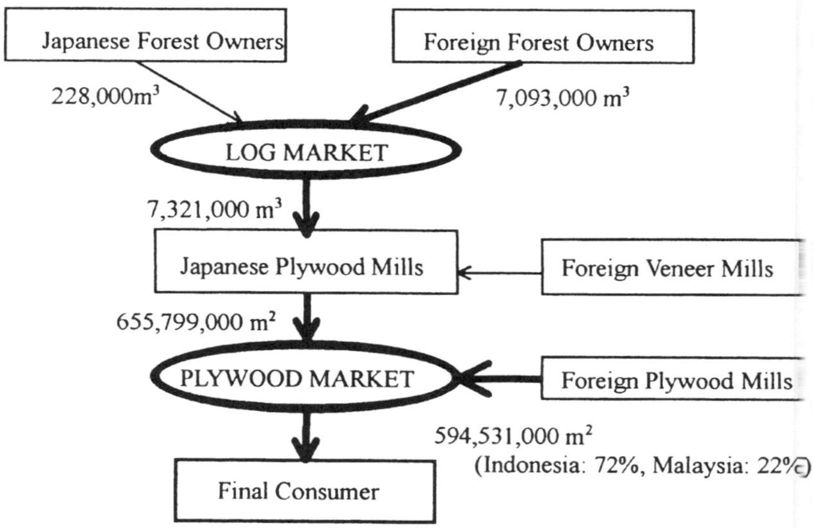

Figure 3. Plywood market structure in Japan - timber flow (Source: the Ministry of Agriculture, Forestry and Fisheries, Mokuzai Jukyu Hokukusho 1995)

The plywood is demanded by builders and final consumers. Supply from Southeast Asia region is mainly from Indonesia and Sabah: 72% and 22%, respectively, in 1995.

3. Method and Data

3.1 Model framework

We postulate an econometric model framework for the Japanese plywood market, including both trades of raw logs and plywood. Considering the above plywood market structure, we will hypothesize that the model consists of the raw log sector and the plywood sector. Especially, we want to analyze structural changes caused by the two log export ban policies of Indonesia and Sabah by using two dummy variables.

3.1.1 Raw log sector

Hypothetical supply function of foreign logs is given in equation [1]. The suppliers are forest owners, concessionaires, contractors and log traders. The forest owners in Southeast Asia are usually the State governments. Log production is carried out under basically the same royalty systems. In the royalty system, concessionaires or contractors get timber licenses under the log production plan of their government, to harvest logs.

The contractors decide whether they export logs to Japan or to other countries or sell them to domestic buyers, based on export log prices and domestic log prices. But in our model it turned out to be enough to use the export log price to Japan (P_{RL}) only, because logs exported to Japan are of higher quality than any other export logs. Other log prices seem irrelevant compared to the log exports to Japan. Besides, it is difficult to get domestic log prices.

We use an oil price (C_{OIL}) as a cost element in logging operation and transportation. Generally an exchange rate has a significant influence on trade behavior. We pick up the exchange rate between the U.S. dollar and the Japanese yen (R_{JA}), because the dollar is usually used in trade settlements. The suppliers will promote log exports, with increases in standing volume or timber harvest (SV_{SE} / TH_{SE}). Two dummy variables (DM1, DM2) (see Table 1) are included to capture effects of trade policy changes. The signs under the variables in equation [1] and the following equations, show their direction of effects on the independent variables.

[1] $SQ_{FRL} = f(C, P_{RL}, P_{OTHER}, C_{OIL}, R_{JA}, SV_{SE} / TH_{SE}, DM1, DM2, TT)$

$\qquad\qquad + \quad + \quad - \quad - \quad - \quad + \quad - \quad - \quad +/-$

Plywood mills in Japan demand logs (DQ_{RL}) as raw material. We assume that the demand function is made up like equation [2]. The mill owners are confronted with two prices (P_{RL}, P_{PLW}). P_{RL} is a price of logs as raw material and P_{PLW} is a price of plywood as product. It is very difficult to get wage data of mill workers, so we use oil price (C_{OIL}) instead as a cost variable on the demand function. The demand may be a function of Gross Domestic Products (GDP_J) and housing starts (HS_J) of Japan.

[2] $DQ_{RL} = f\,(C, P_{RL}, P_{PLW}, C_{OIL}, GDP_J / HS_J, TT)$

+ - + - + +/-

The log demand (DQ_{RL}) is equal to the sum of foreign log supply (SQ_{FRL}) and domestic log supply (SQ_{JRL}). SQ_{JRL} is taken as exogenous here.

[3] $DQ_{RL} = SQ_{FRL} + SQ_{JRL}$

3.1.2 Plywood sector

Equation [4] is a supply function of domestically processed plywood. Plywood mills in Japan must decide the levels of plywood production and supply of the products, seeing the plywood price and cost variables. Main components of cost variables are raw log price (P_{RL}) and oil price (C_{OIL}). Two dummy variables (DM1, DM2) are included to capture structural changes of the plywood supply behavior.

[4] $SQ_{PLWJ} = f\,(C, P_{PLWJ}, P_{RL}, C_{OIL}, DM1, DM2, TT)$

+ + - - - - +/-

We assume that the exporters of plywood consist of forest owners, concessionaires, contractors, plywood mill owners, and plywood traders in Southeast Asia. The supply function is given in equation [5].

Exporters decide the amount of exports to Japan, based on a price vector of some export plywood prices and domestic plywood prices. In this case, it is sufficient to use a plywood export price to Japan (P_{PLWJ}) because the plywood quality for exports to Japan is much better than that for export to any other country, including the Middle Eastern countries, and better than that for domestic use. So we can regard the Japanese plywood export market as independent from other plywood markets. Besides, plywood exports to Japan are rather controlled.

We take the oil price (C_{OIL}) as an explanatory variable reflecting costs. Wage rate of workers is excluded here because of absence of data. We take exchange rate (R_{JA}), standing volume (SV_{SE}), and timber harvest level (TH_{SE}) as exogenous variables. Two dummy variables (DM1, DM2) are included to measure the effect of the policy changes.

[5] $SQ_{PLWF} = f\,(C, P_{PLWJ}, C_{OIL}, R_{JA}, SV_{SE} / TH_{SE}, DM1, DM2, TT)$

+ + - - + - - +/-

In Japan, plywood is mainly used in concrete shuttering, housing structure, and furniture by builders and our final consumers. They decide plywood demands, considering the plywood price (P_{PLWJ}), the price of gypsum board (C_{CSCP}) substitutional goods, and a wage rate of construction workers (W_{CNST}). Gross Domestic Products (GDP_J) and housing starts (HS_J) of Japan have an important effect on the demand. We can think of GDP_J as Japanese income, and HS_J as an immediate scale factor.

[6] $DQ_{PLW} = f\,(C, P_{PLWJ}, C_{CSCP}, W_{CNST}, GDP_J / HS_J, TT)$

+ - - + + +/-

The log demand for plywood (DQ_{PLW}) is equal to the sum of domestic plywood supplied (SQ_{PLWJ}) and foreign plywood supplied (SQ_{PLWF}).

[7] $DQ_{PLW}=SQ_{PLWJ}+SQ_{PLWF}$

3.2 Method and data

Firstly, we estimate the coefficients of models described in the previous section, to reveal the quantified structure of the Japanese plywood market, including the magnitudes of structural changes resulting from Southeast Asian log export bans. The model framework constructs a system of simultaneous equations, which has seven equations with seven endogenous variables. These equations are expressed in the logarithmic linear in both sides. To estimate the coefficients of these equations, the two-stage least-squares (2SLS) on TSP 4.4 for Windows 95 is used to mitigate the simultaneous equation bias (Johnston 1991) with annual data during 1975-95 (Table 1).

Table 1. Variables and data

Endogenous variable	D_{RL}	: Demand for raw logs (1,000 m^3)
	D_{PLW}	: Demand for plywood (1,000 m^2)
	P_{FRL}	: Real price of imported raw logs on the basis of 1990 (Japanese yen / m^3)
	P_{PLWJ}	: Real price of plywood on the basis of 1990 (Japanese yen / sheet)
	S_{RLF}	: Supply of raw logs from foreign countries (1,000 m^3)
	S_{PLWJ}	: Supply of domestically processed plywood (1,000 m^2)
	S_{PLWF}	: Supply of plywood from foreign countries (1,000 m^2)
Exogenous variable	C_{GSCP}	: Real price index of gypsum board on the basis of 1990
	CV_{FRL}	: Industrial log production in South-sea countries (1,000 m^3)
	C_{OIL}	: Real price index of oil on the basis of 1990
	DM1	: Dummy variable, putting 1 during 1985-95 and 0 otherwise
	DM2	: Dummy variable, putting 1 during 1993-95 and 0 otherwise
	GDP_J	: Japanese Gross Domestic Products (billion yen)
	R_{JA}	: Exchange rate between US dollar and Japanese yen (¥/$)
	S_{RLJ}	: Domestic supply of raw logs
	TCS_J	: Construction starts in Japan (1,000 m^2)
	TT	: Time Trend
	W_{CNST}	: Real wage rate of construction workers on the basis of 1990
Source	Bank of Japan, Comparative Economic and Financial Statistics : Japan and Other Major Countries Bank of Japan, Price Indexes Annual The Economic Planning Agency, Annual Report on National Accounts The Forestry Agency, Ringyo Tokei Yoran (in Japanese) The Management and Coordination Agency, Japanese Statistical Yearbook The Ministry of Agriculture, Forestry and Fisheries, Mokuzai Jukyu Hokokusyo (in Japanese) The Ministry of Labor, Year Book of Labour Statistics Food and Agricultural Organization, Yearbook of Forest Products	

Secondly, we calculate changes in economic surplus of the two markets after the log export ban policies. Producers' surplus is nothing but profit to the producers, and consumers' surplus is a measure of welfare of the consumers. If producers' surplus increases after a log export ban, it indicates that profits to the producers increase. Table 1 is the list of our variables.

4. Estimation Results of the Structural Model

4.1 Raw log market

The estimation results shown in Table 2 indicate that we have successfully modeled the structure of the Japanese plywood market. The estimation points out the structural changes induced by the two export bans.

Estimation result [1] shows that the coefficients of P_{FRL}, R_{JA}, DM2 and TT are significant at 5% level with expected signs. The export ban of Sabah (DM2) had a significant negative effect on Southeast Asian log exports to Japan, probably because Southeast Asian countries did not have enough forest resources to offset the ban in the 1990s. The coefficient of DM1 turned out to be insignificant and thus dropped out, indicating that there was little impact from the Indonesian log export ban on the Japanese raw log market. The reason is that the other exporting countries, such as Sabah, Sarawak, and Papua New Guinea, could increase raw log supply to Japan, counterbalancing Indonesian supply reduction. The coefficients of SV_{SE} and TH_{SE} turned out to be insignificant probably because forest resources and timber harvest in Southeast Asian countries were not properly reflected in SV_{SE} and TH_{SE} data because of technical problems.

Table 2. Estimation results of Japanese plywood market

= Raw log sector =

Export supply from foreign countries

[1] $\log S_{RLF} = 18.76 + 0.259*\log P_{FRL} - 0.366*\log R_{JA} - 0.266*DM2 - 0.039*TT$
$(17.38^{**})\ (2.53^{**})\quad (-4.13^{**})\quad (-5.22^{**})\quad (-7.89^{**})$
R-squared = 0.947 D-W = 2.106

Demand of Japanese plywood mills

[2] $\log D_{RL} = 1.585 - 0.429*\log P_{FRL} + 0.642*\log P_{PLWJ} + 1.794*\log GDP_J - 0.093*TT$
$(0.329)\ (-2.690^{**})\quad (2.872^{**})\quad (3.794^{**})\quad (-5.272^{**})$
R-squared = 0.930 D-W = 1.166

Equilibrium condition

[3] $D_{RL} = S_{RLF} + S_{RLJ}$

= Plywood sector =

Supply from domestic plywood mills

[4] $\log S_{PLWJ} = 19.58 + 0.488*\log P_{PLWJ} - 0.245*DM2 - 0.019*TT$
$(20.321^{**})\ (3.567^{**})\quad (-5.118^{**})\quad (-8.509^{**})$
R-squared = 0.940 D-W = 1.934

Supply from foreign countries

[5] $\log S_{PLWF} = -74.38 + 2.395*\log P_{PLWJ} - 0.762*\log C_{OIL} + 3.995*\log TH_{FRL} + 2.178*DM1$
$(-3.175^{**})\ (2.575^{**})\quad (-2.320^{**})\quad (3.314^{**})\quad (7.911^{**})$
R-squared = 0.967 D-W = 1.667

Demand of Japanese consumers

[6] $\log D_{PLW} = 11.01 - 0.202*\log P_{PLWJ} + 0.917*\log TCS_J + 2.145*\log C_{CSCP} - 1.157*\log W_{CNST}$
$(5.065^{**})\ (-1.730^{*})\quad (7.027^{**})\quad (3.605^{**})\quad (-6.354^{**})$
R-squared = 0.904 D-W = 1.709

Equilibrium condition

[7] $D_{PLW} = S_{PLWJ} + S_{PLWF}$

Note: The numbers in the parenthesis under the estimated coefficient indicate t-statistics of the coefficient. Superscripts ** indicates significance at 5% level, * at 10% level.

In result [2], the coefficient of C_{OIL} was not significant and thus dropped out. We should have included some cost variables, but we have failed to find appropriate ones yet.

4.2 Plywood market

Result [4] shows that the log export ban policy of Sabah (DM2) had a significant negative effect on the domestic plywood supply. We can say that it was a reflection of result [1]. The coefficients of P_{RL} and C_{OIL}, were not significant and thus dropped out. Combining with the result of [2], might show that the mill owners decide both raw log demand and plywood supply at the same time.

In result [5], the coefficients of P_{PLWJ}, C_{OIL}, TH_{SE} and DM1 are significant with expected signs. The result of DM1 indicates that the log export ban of Indonesia had a significant effect to promote plywood exports to Japan from Southeast Asia, for the most part exports from Indonesia. The coefficients of P_{PLWJ} and TH_{SE} turned out to be extraordinarily high. These probably reflect the fact that plywood promotion policy was successful in that plywood exports expanded without increasing either price or harvest level.

In estimation result [6], most coefficients we postulated are significant with expected signs. The coefficient of P_{PLWJ} which is the price elasticity of plywood demand, is very low with only 10% level significance. Combining the coefficients of TCS_J and W_{CNST}, we see that plywood demand is strongly affected by construction activities. The coefficient of C_{CSCP} as substitution goods is significant as well.

Our modeling of the Japanese plywood market could reveal that the log export ban of Indonesia had a positive effect on plywood exports from Southeast Asia, and that of Sabah a negative effect on log exports from Southeast Asia to Japan, and similar effects on the Japanese domestic supply of plywood.

5 Calculation Results of Economic Surplus

5.1 Surplus changes on raw log market sector

Considering the estimation results of the Japanese plywood market as stated above, the log supply curve must shift to the left after the log export ban of Sabah (Figure 4). It means that both producers' and consumers' surpluses decreased. The decrease of producers' surplus is larger than that of consumers' surplus, because the price elasticity of log supply is smaller than that of demand.

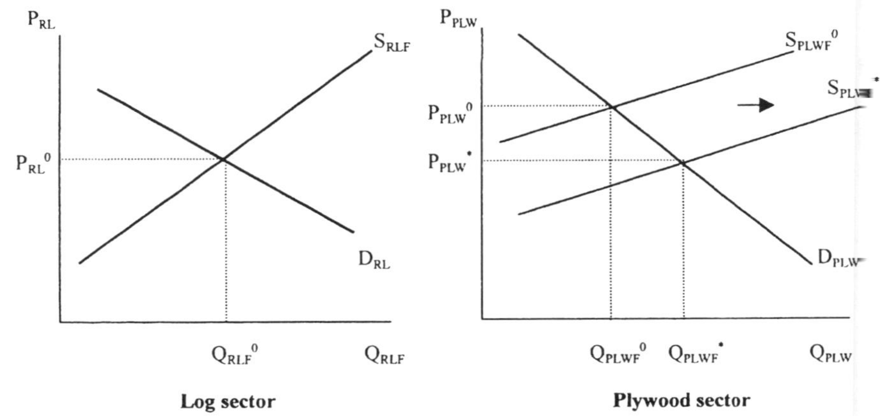

Figure 4. Shift of supply curve after Sabah's log export ban -DM2-

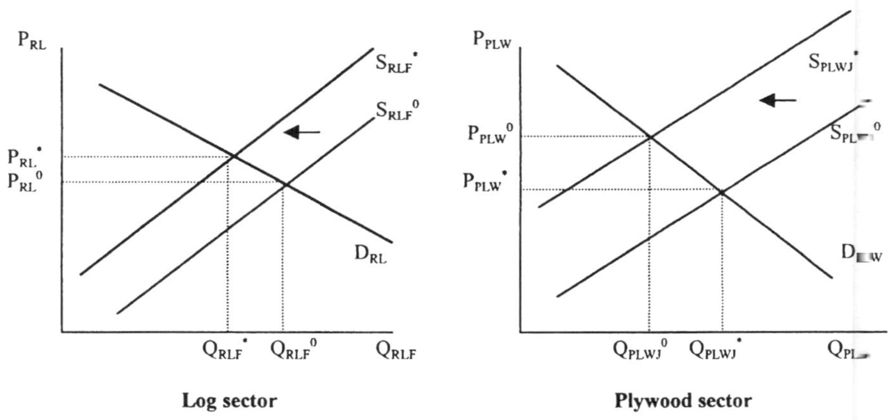

Figure 5. Shift of supply curve after Indonesian log export ban -DM1-

5.2 Surplus changes on plywood market sector

Using the estimation results, the supply curve from Southeast Asia shifted to the right after the Indonesian log export ban (Figure 5). It means that producers' surplus of Southeast Asian plywood industries and consumers' surplus increased in the Japanese plywood market.

On the other hand, the supply curve moved to the left after Sabah's log export ban (Figure 4). So, the producers' surplus and consumers' surplus in the Japanese plywood industry decreased. Total consumers' surplus increased because of plywood price decline.

6 Conclusion

The following conclusions are derived from our econometric analysis. The log export ban of Indonesia from 1985, along with a plywood-industry promotion policy, increased plywood exports from Southeast Asia to Japan. Sabah's log export ban from 1993 decreased log exports from the region to Japan. We think that the depletion levels of forest resources in the countries cause the difference of effects between these bans. In the late 1980s, Sarawak and Papua New Guinea could increase log exports to counterbalance Indonesian exports decrease. On the other hand, no country could take Sabah's place as a main log exporter to Japan in the 1990s. Furthermore, Southeast Asian countries have been promoting strongly their timber processing industries including powerful encouragement of the plywood industry in Indonesia.

On the raw log market, both surpluses of producers and consumers might decrease as a result of the log export ban of Sabah. It indicates that both Southeast Asian producers and Japanese plywood mills as consumers of raw logs paid costs of the log export ban of Sabah. On the plywood market, the producers' surplus of Southeast Asian countries increased considerably after the log export ban of Indonesia probably because of Indonesia's plywood promotion policy. The consumers' surplus in the Japanese market increased accordingly, reflecting reduction in the equilibrium price. Besides these surplus changes, the producers' surplus of the Japanese plywood industry decreased significantly and drastically as a result of both export bans.

In summary, log export bans led to large profits by plywood industries of Southeast Asia, but, enforced costs, were levied on the Japanese plywood industry.

7 References

Johnston, J. 1991. Econometric methods, 3rd ed., McGraw-Hill Book Company. 568p.

Tachibana, S., and Nagata, S. 1996. Impacts of the timber harvest restrictions in North America on the Japanese Timber Market (in Japanese). P.102-104 *in* Proceedings of EEPS Annual Conference 1996.

Tachibana, S., Kato, T., Furuido, H., and Yamamoto, N. 1995. Impacts of royalty system on timber production: A case study of Sabah (in Japanese). AJIA KEIZAI. 37(1): 21-39.

Togu M., E.G. 1995. Economic impacts of the log export ban policy on the development of forest products industries of Indonesia. Dissertation at University of Wisconsin-Madison: 147p.

Yukutake, K., Yoshimoto, A., Nagata, S., and Tachibana, S. 1996. Forest sector modeling in Japan. P. 111-134 *in* Proceedings of Project Group P.6.11 Foresea Meeting at the 20th IUFRO World Congress (Tampere, Finland, 1995).

Acknowledgements

This research supported by a Grant-in-Aid for International Scientific Research (No. 08041048) from the Ministry of Education, Science, and Culture of Japan.

Modelling Japanese Import Demand for European Construction Lumber

Hans Nilsagård

Key words: Timber trade, import demand

Abstract: European construction lumber was introduced to the Japanese market in the early 1990s as a reaction to decreasing supply from North America and a negative trend of domestic supply of construction lumber. The market share for European construction lumber has since then increased rapidly from close to zero in 1992 to more than 15% of the imported volume in 1997. This paper develops a simple model for the Japanese demand for European construction lumber during the last 6 years. Traditionally used variables such as relative import price and Gross National Product (GNP) are combined with an information-based model, where time is the independent variable. To model the demand for a new product, diffusion of information has to be considered. Presumptive buyers need information about the product, such as available volumes and characteristics. This decides whether or not the product will be accepted by the market actors. During the introduction period an increasing number of actors are gaining information about the product resulting in an increasing acceptance of the product. The model explains the growing market share of European lumber for construction during recent years.

1. Introduction

After World War II, Japan was more or less self-supporting in timber materials, owing to a timber control system. During the fifties the economy went through a period of extreme growth, which led to increased demand of timber and, as a consequence, an increase of timber prices. In the end of the fifties, the policy changed and imports of timber were promoted. Tariffs on timber were lowered in the late fifties and infrastructure was built to handle the import of forest products (Akai 1988). Since then, the imports of timber products increased from 7.5 million m^3 in 1961 to peak in the beginning of the 1990s. In 1991, Japan imported 84 million m^3 of timber material.

Conifer lumber has increased in a similar way. Japan imported almost nothing in 1960, but more than 10 million m^3 of sawnwood in 1995 (FAO 1998). The largest exporting countries are Canada, the United States, and Russia. During the 1970s, import activities from New Zealand and Chile began as well. The increase of lumber imports is most often explained by the economic growth in Japan during this period, resulting in a rising exchange rate. The economic growth has also resulted in rising cost of labor, which has affected the Japanese forest sector. The forest sector has not been able to increase its productivity as much as the manufacturing sector, leading to a substitution of domestic saw timber and lumber for imported lumber and saw timber. Today the Japanese import of lumber is 60% of the import of saw timber and the trend is that the lumber share is growing and that the saw timber share is decreasing (Anon 1998).

During the beginning of the 1990's, the supply from North America decreased sharply owing to environmental restrictions in the Pacific Northwest. The fall of the Soviet Union also decreased Russian imports. This was the starting point for the trade from the European countries to Japan, both for coniferous saw timber and for coniferous lumber (Pesonen 1993). The main part of the volumes is used for construction purposes. From a market share of zero in 1990 the trade has grown to more than 15% of the market for coniferous lumber (Anon 1998). Finnish and Swedish exports of coniferous lumber compose the major part of the northern European export.

McKillop (1973) analyzed the trade of forest products from North America to Japan. Quarterly data for 1950 to 1970 was used to estimate demand for U.S. and Canadian lumber. The demand for U.S. lumber was estimated using residential construction in Japan, the index of manufacturing production in Japan, and the prices of Canadian and Philippine lumber imported into Japan, while the demand for Canadian lumber was a function of residential construction, price of domestic lumber, and stocks of lumber. Owing to the quarterly data, lagged variables were to some extent used. Pesonen (1993) pointed out the possibilities of increased future imports from Sweden and Finland. The lack of information was also mentioned to be one of the main barriers to increasing imports from Sweden and Finland. Blandon (1994) modelled the Japanese market for lumber. The joint demand for imported lumber and the domestic species sugi was estimated using residential housing and a joint price for sugi and imported lumber.

This paper provides an attempt to explain the growing exportation of lumber from Finland and Sweden to Japan during the last decade. The main goal is to test whether a logistic function can be used to model the entrance of Swedish and Finnish lumber. To be successful in introducing an unknown species from a new exporting region, you have to invest in marketing activities to disseminate information about the product. If the development of the lumber imports from northern Europe can be modelled in this way, it will be possible to detect whether or not the market can be said to have received full information about this new product. The paper develops a simple model for the Japanese demand for European construction lumber during the last 8 years. Typical variables such as relative import price and Gross National Product (GNP), are combined with an information-based model where time is the independent variable.

Section 2 explains the import demand model, and the data is described 3. Section 4 presents the results followed by a short discussion in section 5 and conclusions in section 6.

2. Model Specification

Traditionally, import demand models include the real price of the good, prices of other products, substitutes and complements, and income (see for example Leamer and Stern 1970). Normally, the product can be substituted for a domestically produced product, which price therefore should be included. In this paper the price of imported lumber and the price of domestically produced lumber is combined in a relative price variable, and the GNP represents income.

The European entrance on the Japanese market is assumed to be an introduction of a completely new product for the Japanese users. To model the demand of a new product, diffusion of information has to be considered. This is done by using a function, where time is the independent variable and the market share of European lumber is the dependent variable.

Thus, a traditional import-demand model is combined with a time-based function where the function for total import is multiplied with the time function to result in an import-demand function for European lumber on the Japanese market.

Total import of conifer lumber is assumed to be a function of the import price, the price of domestic lumber, and real GNP,

$$[1] \quad D_{TOT,t} = f\left(P_{IMP,t}, P_{DOM,t}, GNP_t, d_{2,t}, d_{4,t}\right)$$

where:

$D_{TOT,t}$ Total import demand of conifer lumber in period t

$P_{IMP,t}$ Price of imported conifer lumber

$P_{DOM,t}$ Price of domestic conifer lumber

GNP_t Real GNP (1990), adjusted for seasonal effects by 12-month moving average

t Quarterly index, ranging from 1 to 25

d_2 Dummy variable for seasonal changes in the second quarter

d_4 Dummy variable for seasonal changes in the fourth quarter

It should be noted that to use a unit import price for all countries, we have to assume that import lumber is a homogenous good.

The import price and the domestic price is combined in a relative price,

$$[2] \quad P_t = {P_{IMP,t}}\Big/{P_{DOM,t}}$$

thus making P_t inflation independent.

By insertion of [2] into [1] we obtain,

$$[3] \quad D_{TOT,t} = f\left(P_t, GNP_t, d_{2,t}, d_{4,t}\right)$$

When a new product is introduced on a market, it is assumed that no one knows about it. As more and more potential buyers receive information about the new product, sales will increase accordingly. At some point, the rate of information diffusion starts to decrease. This type of information diffusion model was suggested by McKillop and Wibe (1987) (Figure 1).

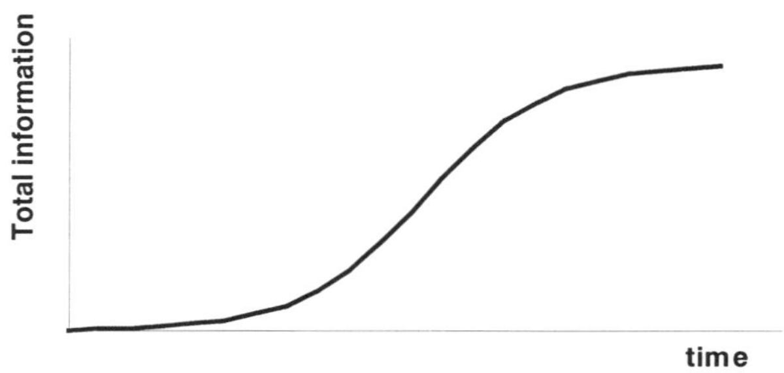

Figure 1. A logistic curve describing diffusion of information

The growth of information, dI/dt, is assumed to be a function of

1. Total information already spread, *I*(*t*). The higher amount of potential users with information, the more users will receive information from the ones that already know.
2. The size of information not spread, i.e., the amount of potential buyers without information, *B*-*I*(*t*) where *B* represents the total amount of potential users.

Assume a multiplicative model,

$$[4] \quad dI/dt = \gamma_0 I(t)(B - I(t))$$

Solving [4] for *I*(*t*) and normalizing B to one, results in the formula for a logistic function. $I(t)$ can then be interpreted as the market share at time t.

$$[5] \quad I(t) = \frac{B}{1+\gamma_1 e^{\gamma_2 t}} = \frac{1}{1+\gamma_1 e^{\gamma_2 t}}$$

where $\gamma_1 > 0$ and $\gamma_2 < 0$

By multiplication of equations [3] and [5], the import demand ($D_{SF,t}$), for lumber from Sweden and Finland is given by:

$$[6] \quad D_{SF,t} = f\left(P_t, GNP_t, d_{2,t}, d_{4,t}\right)\frac{1}{1+\gamma_1 e^{\gamma_{2t}}}$$

Two versions of [6] have been estimated using non-linear ordinary least squares regression estimation, using the LIMDEP (Greene 1995) software package. In the first one, model 1, import demand is assumed to be linear in parameters of price and GNP:

$$[7] \quad D_{SF,t} = \left(\beta_0 + \beta_1 P_T + \beta_2 GNP_T + \beta_3 d_2 + \beta_4 d_4\right)\frac{1}{1+\gamma_1 e^{\gamma_{2t}}} + \varepsilon$$

while, in the second, model 2, the logarithm of import demand is assumed to be linear in parameters of the logarithm of price and GNP:

$$[8] \quad \ln D_{SF,t} = \beta_1 \ln P_t + \beta_2 GNP_t + \beta_3 d_2 + \beta_3 d_4 - \ln\left(1 + \gamma_1 1 e^{\gamma_2 t}\right) + \varepsilon$$

3. Data

Data on prices were collected from Japan wood-products information and research center (Anon 1994, 1996, 1998). The prices are wholesale price indexes for imported lumber and for domestically produced lumber. Gross National Product is official Japanese data, and the import data was collected from the Swedish Wood Exporters Association (Stef 1998).

4. Results

The result is not good enough to tell whether this model is a possible method of describing the recent increase in imports of lumber from northern Europe. Table 1 lists the parameter estimates and the standard errors. Neither the GNP nor the relative price has the expected sign in either of the tested models.

Table 1. Parameter estimates of Model 1 and Model 2. In both models, n=25

	Model 1 Coefficient	Standard Error	Model 2 Coefficient	Standard Error
CONSTANT	-421.56	(886.85)		
GNP	-1.4658	(8.5061)	-1.0255	(1.9000)
PRICE	647.20	(187.53)	2.0835	(1.9003)
DUMMY2	63.401	(22.565)	0.32612	(0.23349)
DUMMY4	90.603	(22.868)	0.22555	(0.23474)
γ_1	527.99	(1084.73)	3843866.6	(4864469.1)
γ_2	-0.3601	(0.15168)	-0.93761	(0.10409)

However, running the model without relative price as independent variable results in a significant parameter estimation and the expected sign for the GNP. Figure 2 shows the time-series for actual imports and the resulting time-series for model 1 and 2.

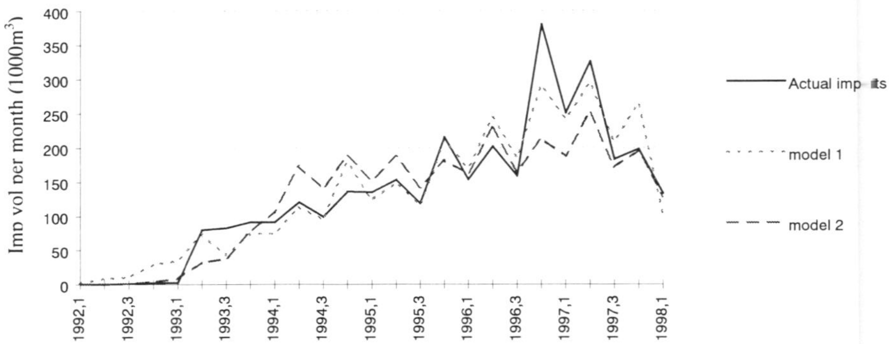

Figure 2. Actural imports and the results from Model 1 and 2

5. Discussion

We do not know whether we truly have found the demand function or not, i.e., the identification problem (Green 1993). It seems like a supply function will be necessary to add into the system in order to get the expected signs on GNP and relative price.

Another problem with studies like this is that there is insufficient data to satisfy our needs. A longer time series should have improved our result, but using monthly data is not feasible owing to large variations in imports between single months.

The assumption of a unit price for all conifer lumber imports may not hold. There are differences in the properties between the different species imported from Europe and from, for example, North America.

In 1997 the Japanese value added tax (VAT) increased from 3% to 5%. This increased consumption prior to the tax raise and decreased consumption after the tax raise. This effect can be seen in the data for housing starts, with a decline in monthly housing starts and in the demand for import lumber after the tax raise. However, the time-series is too short to allow for a structural dummy variable.

6. Conclusions

This model could not explain the recent increase of imports from Europe satisfactorily. A further attempt should include a supply function, solving the identification problem. This may improve the results. Also, using data that includes the total imports from Europe may improve the model further.

7. References

Akai, H. 1988. Policy for the stabilisation of the supply, demand and price of timber, in forest policy in Japan (Handa, R.). Association for Research & Publishing on Japanese Forestry. Tokyo.

Anon. 1994. Wood and supply demand information service, June 1994. JAWIC. Tokyo.

Anon. 1996. Wood and supply demand information service, June 1996. JAWIC. Tokyo.

Anon. 1998. Wood and supply demand information service, June 1998. JAWIC. Tokyo.

Blandon, P. 1994. A simple econometric model of Japanese sawlog market. *in* Proceedings of Project Group P.6.11 Foresea Meetings at the 20th IUFRO World Congress Tampere, Finland. University of Washington. Seattle.

FAO. 1998. FAOSTAT statistic database. (http://apps.fao.org).

Greene, W.H. 1993. Econometric analysis. Prentice-Hall. New Jersey.

Greene, W.H. 1995. Limdep Version 7.0 user's manual. Bellport: Econometric Software, Inc.

Leamer, E., and Stern, R. M. 1970. Quantative international economics. Allyn and Bacon, Boston.

McKillop, W. 1973. Structural analysis of Japanese-North America trade in forest products. Forest Science. 19 (1): 63-74.

McKillop, W., and Wibe, S. 1987. Demand for sawnwood and panels. P. 306-327 *in* The global forest sector: An analytical perspective. Kallio, M., et al. (eds.). John Whiley & Sons.

Pesonen, M. 1993. Japanese market for Scandinavian wood products. University of Helsinki Department of Forest Economics Reports. Helsinki University Press. Helsinki.

STEF. 1998. Data from the Swedish Wood Exporters Association's database. Stockholm.

The Joint Production of Timber and Environmental Values

- Demonstration for the west side of Washington State -

Bruce Lippke, B. Bruce Bare, Weihuan Xu, Jeffrey Moffett and Bill Turner

Key words: Timber supply, forest economics, habitat protection, biodiversity, non-market valuation, linear programming

Abstract: Regulatory requirements to protect habitat have increased, requiring forest managers to include measures of biodiversity, habitat, and streamside protection in their plans. A demonstration of environmental and economic effects for a range of timber management alternatives was developed for western Washington. Economic and environmental effects were characterized for current and prospective regulations as well as for several management alternatives designed to achieve environmental objectives similar to those produced by the regulations. Tradeoffs between environmental and economic effects were developed for the landowners' decision process. As a basis for policy analysis, tradeoffs between the benefits and costs to rural and urban communities were identified, demonstrating who gains and who loses under various management alternatives. Integration of timber growth and yield models with habitat models in a mathematical programming framework was used to produce the biological and economic measures associated with different management strategies. Experimental choice analysis among management alternatives was then used to develop public preference values for measures of forest biodiversity, aesthetics, rural job losses, and costs as surrogates for the economic-environmental tradeoffs. Results showed that regulations and proposals that focus on preservation without considering long-term motivation to produce non-market forest amenities result in significantly less value to the public and less equitable treatment across urban and timber rural communities than do alternatives that seek to jointly produce timber and non-market amenities.

1. Introduction

Regulations affecting U.S. forest practices on state and private forestland (WFBD 1996) have changed several times since 1990, driven largely by the Endangered Species Act (1973) and other efforts to conserve environmental amenities valued by the general public. Different owners have been required to meet different standards, and the economic effects have been substantial. However, until now no comprehensive assessment of the cumulative effectiveness of policies to meet environmental goals has been completed.

This study examines the cumulative economic and environmental effects of forest practice and policy changes on the forested lands of the west side of Washington State (referred to as western Washington). Government regulations on forested land in western Washington generally have required the preservation of existing habitat. This study evaluates the effects of these regulations as well as the effectiveness of active management strategies in achieving environmental objectives. Results show that while environmental amenities are increasing with regulatory changes to forest practices, it is at a significant cost. Recent research on the public's value preferences for ecosystem management is used to determine to what extent the public has experienced a net increase in their utility value from the economic and environmental changes.

Assessing how much the public values environmental benefits relative to economic costs provides a new method in determining the best strategies for environmental restoration while also being responsive to community equity issues.

Measures of economic and biological progress are crucial to such an analysis. Therefore, economic comparisons are benchmarked to the market potential of a "base" case to determine the cumulative economic loss. A biodiversity index of partial restoration to pre-European conditions is used to benchmark environmental progress. These economic and environmental measures provide the setting for assessing public preferences for economic and environmental tradeoffs.

2. Simulation of Management Alternatives

Forest inventory data were updated to 1996 as a starting point for analysis. Acreage categories were defined as either upland or riparian management zone (RMZ). RMZs are based on distance from streams and for this study are defined as areas within 150 feet of fish-bearing streams and 50 feet of non-fish-bearing streams.

Average site classes were selected for each owner forest type and land category to reflect the average site index noted in the west side timber supply analysis (Adams et al. 1992). With these starting inventories: by land category, age class, and by owner for three sub-regions of the west side, each of several possible management alternatives was simulated with growth and yield simulators. Stands over 40 years of age were assumed to be essentially unmanaged with growth projections taken from empirical yield tables. Stands under 30 were assumed to be under an active management regime with restocking. In this study, they were simulated by DFSIM (Curtis et al. 1981) adjusted for empirically observed stocking levels. The 30-40 age class was assumed to be half in one category and half in the other. Biological measures were linked to the growth and yield simulations by including management requirements for leave trees, snags, downed logs, and debris. The functional stage of the forest over time is then determined as a consequence of growth and yield and these management requirements.

The management treatment alternatives used in this study included a range of practices from commercial to habitat management strategies to no-management. They are defined as follows: (1) no-management with aging of stands without disturbances; (2) natural growth with final harvest; (3) commercial rotations with pre-commercial thinning (PCT) and commercial thinning (CT) at age 30 or as appropriate, and final harvest at age 50 or more; (4) biodiversity management pathways (Lippke et al. 1996) with 3 periodic thinnings that leave debris, downed logs, and snags culminating in forests with the functional-equivalent features of old forest in about 100 years, and with rotations of 100 or more years; (5) a partial cut in 60- to 70-year-old stands, followed after 20 or more years by either commercial or biodiversity pathways to accelerate old-forest conditions more quickly; and (6) partial cut sequences like the biodiversity pathway for the riparian zone (with no clearcut of the overstory), including retention of large trees for stream recruitment.

Revenue, economic activity, jobs, and state and local tax receipts were then determined for each treatment alternative in every decade in addition to stand structure classifications, habitat and biodiversity indices, and aesthetic measures.

A harvest-treatment schedule was then determined over a 200-year time horizon with the objective of maximizing the landowner's net present value subject to no-management zones required by regulations or habitat goals (minimum standards over time) and operational harvest flow constraints (restricted decade to decade change in harvest level). Thus, for each policy and management strategy scenario, a time profile of economic attributes and a time profile of environmental attributes were determined for a 200-year future.

Measuring changes between different management scenarios provided an analysis of the effectiveness of each strategic scenario. The output for each scenario included a rich array of economic effects and environmental attributes over time as well as cumulative effects. Hence, quantitative measures of progress driven by regulatory or alternative policy goals are provided. Since the treatments were limited to statistical characterizations within a region, management zone, and ownership, spatially sensitive characteristics within these categories are not available. However, for studies of this scale (almost 10 million acres), it is the description across the total region that is important. Operational scale implementation of spatial features should not deviate significantly from these simulated results at this larger scale.

3. Management Scenarios and Resulting Physical Attributes

Scenarios-Alternatives include a base case representative of commodity management with low environmental constraints (Case 1: Base); minimum 1996 Washington State Forest Practices regulations (Case 2: Minimum Regulations); habitat conservation plans and state proposed riparian buffers (Case 3: Proposed Regulations); and finally two active management alternatives: one on state and private land (Case 4: Alts. Non-Federal) and one that includes Case 4 lands but also adds unreserved, federal timber land hereafter referred to as federal timber land (Case 5: Alts. All Owners).

For each case, Table 1 summarizes the effects for five different economic measures (landowner net present value [NPV]), average harvest, rural employment in the first two decades, the second century sustainable employment rate, and average state and local tax receipts for the first two decades. Effects for four biological measures are also shown for current conditions and longer-term averages (late-seral % of acres, late-seral % of RMZ by 5th decade, biodiversity index and mature and over-mature stands as a % of acres).

Table 1. Economic and environmental impacts of alternative management strategies

CASES:	1 Base	2 Minimum Regulations	3 Proposed Regulations	4 Alts. Mgmt-Non-Federal	5 Alts. Mgmt-All Owners
Economic Measures:		Change from Base (Case 1)			
Timber Rev. NPV (US$ billions)	48	-10.5	-20.6	-9.8	-5.1
Harvest (average mmbf)	6,380	-1,040	-1,991	-1,090	-540
Rural jobs, first 20 years (000's)	134	-31	-63	-31	-18
Rural jobs sustained 100+ years (000's)	130	-19	-37	-5	14
State/Local taxes, first 20 years (US$millions/year)	1,490	-320	-680	-310	-150

CASES:	Current	1	2	3	4	5
Environmental Measures:						
Late-seral % (average)	11	3	17	25	29	21
Late-seral RMZ % by 5th decade	11	3	12	16	56	48
Biodiversity Index (average)	56	56	67	70	73	66
Mature & Over-mature % (ave.)	15	22	34	42	46	39

3.1 Base--Low constraints commodity management (Case 1)

Green tree retention and green-up requirements (5% harvest loss) as well as 85-foot, no-management buffers on fish-bearing streams (2.5% of all acres) are included in the Base case. To avoid rapid liquidation of the excess mature inventory, harvest constraints were imposed on federal timberland (no change in harvest level from decade to decade) and other public, and private (plus or minus 5% per decade change), resulting in economic losses of 52% federal timber land, 11.5% other public, and 4% private from a less-restrictive 25% flow constraint. These different losses reflect the potential of each group to convert excess mature inventory into product in the first several decades, potentials they have not exercised. Any excess mature private inventory shown is largely non-industrial.

Under the Base case, the NPV of future income of west-side lands is 48 billion US$, using a consistent 5% real discount rate assumption to discount future revenues to the present. This harvest schedule supports 134,000 in rural employment over the first two decades with 130,000 sustained over the long term. In addition, the forest sector Gross Domestic Product (GDP) of 11 billion supports 1.5 billion US$ in state and local taxes.

The Base case harvest level of 6.4 billion board feet (with commercial management over the 200-year period) does not retain forest structures of importance to some species owing to the late-seral structures declining from 11% currently to only 3% (also 3% in the RMZ) in 200 years. This, of course, was the motivation for regulations that reserve more of these critical habitat structures, which are in decreasing supply. The Biodiversity Restoration Index—a measure of the diversity across the mix of stand structures benchmarked to 100 for pre-European forest conditions—falls to 56, a substantial loss in the diversity of forest types.

3.2 Minimum Regulations (Case 2)

The Minimum Regulations (case 2) characterizes the effect of placing federal timberland under FEMAT (the management option that was selected by the U.S. Government Administration and developed by the Forest Ecosystem Management Assessment Team 1993), with additional reserves included on state and private lands as required by the 1996 Washington State Forest Practices Board for the protection of endangered species. Private and other public harvest flow constraints were relaxed (to 25% change per decade) so as to not overly constrain meeting the regulatory minimums, thus being conservative in understating the economic effect of the regulatory change from the base.

The most significant change under current regulatory minimums is the removal of federal timberland from the available harvest. Under FEMAT, only 21% of federal timberland appears to be available for harvest activities, and the requirements for volume left as debris were estimated to reduce any harvest volume an additional 10% over the other management alternatives. The unmanaged federal acres outside of the defined RMZ are 66% of the total acres. Because the simulation using these assumptions supports a 233 million board foot harvest, which appears to be well above USFS projections, even these assumptions may overstate the volume of future Federal harvests. For state and private lands, prior studies estimated the effect of owl circles under different regulations. For state managed lands, 105,000 acres of mostly mature forestland were estimated to be in owl circles (6.3%) (Bare et al. 1997) and these reserves were adopted for other public lands. The comparable private acreage was 82,000 acres (1.5%). These estimates may understate the full effect of minimum owl and murrelett regulations, as they assume a high degree of overlap between owl and murrelett habitat and may not consider the long-term affects of other acres being affected as forestland matures. Nevertheless, this estimate should be adequate to demonstrate the effect of different policy-driven management alternatives in providing comparable habitat protection.

In this Minimum Regulations case, the NPV to timberland is reduced by 10.5 billion US$ from the base (-22%), with 80% of the loss on federal timber lands. The 1 billion board foot reduction in harvest reduces the support for rural employment by 19,000 sustained jobs and by 31,000 in the first 20 years. The early losses are greater as the set-asides remove a larger share of currently mature and harvestable timber. Even so, these early losses may be understated as the harvest flow constraint was relaxed from the Base case to avoid excessive constraints in substituting available mature timber for mature reserves. The state and local tax receipts are reduced by [illegible]20 million US$.

The environmental benefits are quite notable, with the share of late-seral acres rising from 3% in the Base case to 17%, 1 ½ times the current 11% level. Improvement in the RMZ by the fifth decade is not as notable, increasing from the current 11% to only 12% because most all forestland except the federal reserves, are uplands. The biodiversity index increases from 56 to 67, a 20% improvement.

In comparison to the Base case, this Minimum Regulations case characterizes the tradeoff between biological gains and economic losses with most of the landowner losses on federal timberlands but with substantially reduced employment in rural communities. The 14% improvement in late-seral (1.3 million acres) resulted in a 7,950 US$ NPV loss per acre of late-seral improvement as a gauge of effectiveness.

3.3 Proposed Regulations (Case 3)

Increasing the reserves on other public timberlands under the Washington State Department of Natural Resources Habitat Conservation Plan (DNR HCP [WDNR 1993]) along with no-management within the RMZ on private timberlands as one proposal to improve salmon habitat, increases the unmanaged reserve acres substantially. Reserve acres are increased by 23% on other public timberland and 13% on private timberlands.

The additional 0.95 billion board foot reduction in harvest reduces the NPV another 10.1 billion US$, all on non-federal lands. Support for rural employment is reduced by an additional 18,000 sustained jobs and 32,000 jobs in the first 20 years. Again, the early losses are greater as the set-asides remove a larger share of currently mature and harvestable timber. The state and local tax receipts are reduced an additional 360 million US$.

The environmental benefits are greater, with the late-seral percentage rising to 25%. The RMZ late seral by the fifth decade improves to 16% of the RMZ, substantially less than across all acres. The biodiversity index increases another 3 points, to 70.

In comparison to the Base case, the Proposed Regulations case characterizes a more severe tradeoff between biological gains and economic losses. The 22% improvement in late seral (2.1 million acres) resulted in a 9,930 US$ NPV loss per acre of late-seral improvement, a substantially higher cost per acre than the Minimum Regulations.

3.4 Management Alternative (Biodiversity management) - Non-Federal (Case 4)

In this scenario, instead of placing the primary emphasis on no-management reserves, minimum habitat goals are enforced to produce an equivalent quantity of habitat as produced under Minimum Regulations; however, management strategies are used to accelerate the creation of that habitat. Nesting-Roosting-Foraging habitat, as defined under forest practice rules, was used for the minimum goal to insure that the mix of old and young habitat was biased toward the more fully functional older habitat. In the RMZ, biodiversity pathways were used to accelerate the restoration of old-forest functionality up to a 20-foot, no-management stream bank buffer. No clearcuts were allowed in this RMZ. Treatments designed to produce large trees for stream recruitment and other old-forest functionality were used.

The environmental benefits were greater than for either the Minimum Regulations or the Proposed Regulations cases and perhaps reflect targets higher than necessary. Late-seral rises to 29%, nearly three times current levels and more than double Minimum Regulations. Improvement in the RMZ late seral by the fifth decade is even more notable, rising to 56%, essentially full restoration of old-forest functionality (within the RMZ). The biodiversity index increases to 73, 6 points above that of the Minimum Regulations scenario.

Even more notably, in case 4 the economic losses were generally found to be less than those under Minimum Regulations and substantially less than those under Proposed Regulations. The harvest loss was essentially the same as under Minimum Regulations, with the sustained job losses much lower (-5,000 vs. -19,000). This is a result of the biodiversity management treatments (case 4) producing larger trees and higher growth per acre over the long term and supporting increased employment in forest management, and primary and secondary processing as well as indirect employment. The employment support (-31,000) for the first two decades, however, is no better than that found for the Minimum Regulations, but far better than for Proposed Regulations at -63,000.

In comparison to the Base case, this Alternative Management case demonstrates more cost-effective achievement of the biological goals implicit in the regulations. The 26% improvement in late seral (2.5 million acres) results in a 4,000 US$ NPV loss per acre of late seral improvement, about half the cost per acre as from minimum regulations.

3.5 All-Owner Management Alternative (Case 5)

The benefits noted in case 4 are extended to all owners in case 5 by opening one-third of the federal timberlands to the alternative management regimes, thereby accelerating the rate at which these stands take on late-seral functionality. In this demonstration, a mixture of reserve and active management strategies are applied to these federal timberlands.

The environmental benefits provided by this scenario are comparable to those of the Minimum Regulations scenario and substantially better in the RMZ. Late-seral increases to 21%, or almost double current levels.

The economic losses are substantially less than those found under Minimum Regulations, representing an improvement from current conditions. The harvest loss is only half of that found under the Minimum Regulations scenario, with the sustained job support showing a net gain over the Base case assumptions, and a first-two-decades loss of 18,000 jobs is 13,000 better than the 31,000 job loss under Minimum Regulations. State and local tax losses are reduced to 150 million US$, a gain of 170 million US$ over current Minimum Regulations and 530 million US$ over Proposed Regulations.

The addition of adaptive management on one-third of the federal timberland further increases the cost effectiveness in reaching biological goals. The 18% improvement in late seral (1.7 million acres) results in a 3,000 US$ NPV loss per acre of late-seral improvement, about 30% of that from Proposed Regulations or 38% of the loss with Minimum Regulations.

Even though this scenario can be considered more cost effective in reaching critical habitat goals, it does not establish the net benefits to the public: who benefits and who pays. By using these environmental and economic attribute tradeoffs, estimates can be made of who the beneficiaries are and who bears the burden by estimating public preferences across these management alternatives.

4. Estimation of Public Utility Values

A random sample of residents from urban, timber-dependent rural, and other rural communities were administered a survey that characterized a wide array of forest management alternatives across the west side of Washington, resulting in a 65% response rate (Xu 1997). Respondents were first familiarized with the nature of tradeoffs among managing for biodiversity, aesthetics, rural employment, and out-of-pocket costs such as taxes and product costs. They were then given sets of four alternative management strategies and asked to select their preferred management alternative, knowing the predicted economic and environmental outcomes for each. Since the alternatives they were exposed to included a full range of economic and environmental tradeoffs, their preferences across economic and environmental alternatives were estimated. Their willingness to pay (WTP) for environmental gains in terms of their out-of-pocket costs and willingness to accept (WTA) job losses defines economic utility among the tradeoffs as though environmental values could be purchased in the market. The model estimated non-market utility value for both the environmental attributes and the rural employment and cost attributes so they can be compared directly and so that best tradeoffs could be evaluated. The test statistics were remarkable in that significant differences were observed between timber-dependent rural, and urban populations, and every economic and environmental attribute was deemed important (Xu 1997).

Tests of reasonableness were also noted, because the public placed a high value on restoring environmental measures but with declining returns in relation to the amount of restoration. Using out-of-pocket costs as the common denominator allows for characterizing the public's utility value of increased biodiversity or aesthetics in terms of their willingness to pay for a given amount of each attribute. The two environmental factors, biodiversity and aesthetics, provide two surrogates for the public to relate to in considering their own uses of the forest. The two economic variables, cost and rural job losses, characterize the competing economic effects that are also valued by the public.

Figure 1 shows the willingness of urban and timber-dependent rural communities to pay for biodiversity gains. Not surprisingly, it shows urbanites willing to pay more, but both groups see it as wasteful to go much beyond 70% to 80% restoration, where 100% would be full restoration to pre-European conditions. The study shows that each urban household is willing to pay as much as 600 US$ annually for improved biodiversity, meaning that if they had to pay that amount, they would see no net benefit between their cost and their non-market benefit. Since the Base case forest management strategy results in a biodiversity value of 56, rising to 67 under Minimum Regulations, the urban household biodiversity value increases from 278 to 600 US$ for a 322 US$ increase.

Similar results for willingness to pay for aesthetic improvements are shown in Figure 2, indexed to the percentage of mature and over-mature stands as a good proxy for aesthetics.

These gains in environmental and losses in economic utility values can be accumulated across the key contributing factors, which are biodiversity, aesthetics, rural employment, and cost. The willingness to accept job losses (Figure 3) demonstrates a declining willingness to pay more than a certain amount as the number of rural job losses increases. The lack of any significant differences between rural timber-dependent and urban communities is consistent with their reaction that it is other people's jobs, not their own, that are at risk.

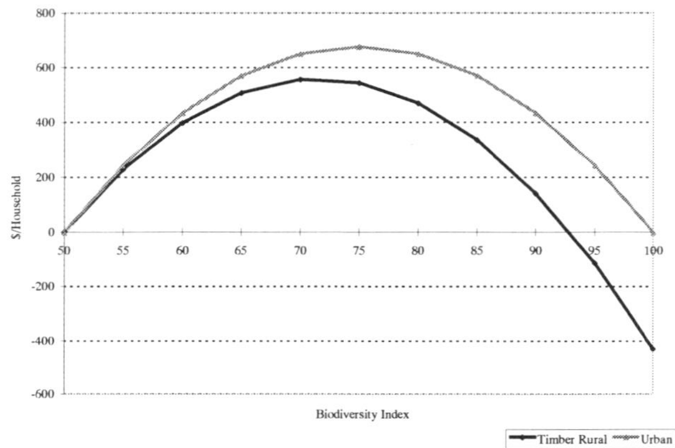

Figure 1. Willingness to pay for biodiversity

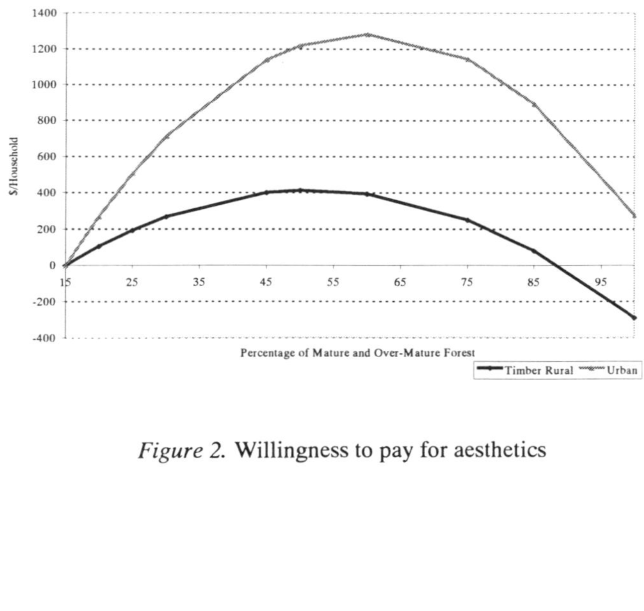

Figure 2. Willingness to pay for aesthetics

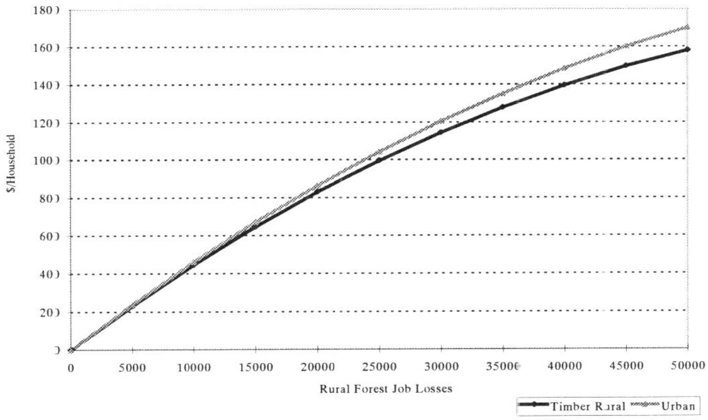

Figure 3. Willingness to accept rural job losses

5. Household Utility Values from Management Alternatives

Table 2 displays the utility values for each competing factor across the five management alternative scenarios as well as the total utility value for timber-dependent rural and urban households.

Table 2. Economic utility by household: willingness to pay for gains or accept losses (US$ per household (Hh))

CASES:	1 Base	2 Minimum Regulations		3 Proposed Regulations		4 Alts. Non-Federal		5 Alts. All Owners	
WTP Biodiversity									
Timber Rural	0	266		287		287		254	
Urban	0	322		364		387		304	
WTP Aesthetics									
Timber Rural	0	175		242		261		214	
Urban	0	486		710		791		608	
WTA Job Losses									
Timber Rural	0	-964		-1,437		-427		+73	
Urban	0	-1,008		-1,532		-441		+75	
WTA Cost (with & w/out compensation)		with	without	with	without	with	without	with	without
Timber Rural	0	-413	-1,414	-778	-3,443	-309	-127	-123	-621
Urban	0	-413	-321	-778	-532	-309	-326	-123	-192
Total Value									
Timber Rural	0	-936	-1,937	-1,686	-4,351	-138	-6	+417	+1,161
Urban	0	-612	-520	-1,236	-990	+428	+411	+864	+795

Because cases 2 through 5 all show improvement on environmental attributes compared to the Base Case, they all show utility value increases over the Base case but not much differentiation as each of the alternatives fall in a range that approaches peak biodiversity benefits. The value gains for aesthetics show a larger difference between urban and rural households as well as a larger range across the 5 cases.

No significant difference appears between urban and timber-dependent rural communities with regards to their willingness to accept employment losses. Both groups would be willing to pay a considerable amount to avoid employment losses, about 42 US$ per household for each thousand rural jobs for the first 10,000 jobs, with the amount per job declining as the job losses increase. The negative effect of these job losses on total utility exceeds the sum of the environmental values for both Minimum (case 2) and Proposed Regulations (case 3) but not for Management Alternatives (case 4).

Finally, there is the actual cost per household, which could be different by community. The study treated the change in state and local taxes as a cost and assumed an insignificant change in product costs given the potential for global trade. The remaining issues are who pays and whether rural landowner losses are compensated or not. Table 2 shows the results "with compensation," which means that the cost to the household is not different between timber-dependent rural and urban communities. It also shows the result "without compensation," which recognizes that the out-of-pocket losses to the timber-dependent rural communities are different than for urban communities. Revenue losses to timber-dependent communities include the 25% of federal timber revenues paid to counties in lieu of taxes, 50% of other public revenues that remain with the counties, and 95% of private landowner revenues (after a 5% yield tax). Those losses, when not compensated, become offsetting gains by the non-timber communities since they do not contribute their share thus leaving the total cost unchanged.

The net utility value across all factors is generally less positive for the timber-dependent rural households then it is for the urban households and is positive for both of these groups only under the All Owner Management Alternative (case 5). For the regulatory cases for which no compensation has been provided, the rural household loss is 4 times as large as that of the urban households, with the loss ranging from 2,000 to 4,000 US$ per household going from Minimum Regulations to Proposed Regulations. These losses are substantially reduced under the Management Alternative with compensation that shows only -188 US$ for timber-dependent rural, which becomes a 428 US$ gain to the urban and non-timber-dependent households. For these alternative management cases, compensation would almost certainly be required so that landowners would alter their management to achieve habitat goals. With compensation, both timber-dependent rural as well as urban and non-timber-dependent households receive net positive utility values under case 5. Even under case 4, which does not involve federal timberlands and is essentially under state control, the utility values are much improved over current Minimum Regulations, (from -1,937 US$ per timber-dependent rural household under Minimum Regulations to -188 US$ under Management Alternative-Non-federal, and from -520 to +428 US$ for urban households).

6. Community Utility Values Accumulated Across Households

Multiplying household values by the number of households in each community produces statewide utility values for west side timber management. Table 3 shows that the environmental values (WTP Biodiversity and Aesthetics) approach a 2 billion US$ increase over the Base case in case 4; however, the economic value losses (WTA Job Losses and Increased Costs) are even greater than the environmental gains under the Minimum Regulations alternatives (cases 2 and 3). These economic losses are less than the environmental gains under Alternative Management (cases 4 and 5). The net utility value ranges from -2.5 billion (case3) to +1.5 billion US$ (case 5).

Table 3. Regional economic utility by community

	1 Base	2 Minimum Regulations	3 Proposed Regulations	4 Alts. Mgmt-Non-Federal	5 Alts. Mgmt-All Owners
All Community (US$ millions)					
WTP Biodiversity & Aesthetics	0	1,366	1,802	1,969	1,535
WTA Job Losses and Increased Costs	0	-2,686	-4,349	-1,419	-88
Total Value	0	-1,321	-2,547	+550	+1,447
By Community (US$ millions)		No-Compensation		Compensation	
Timber Rural	0	-306	-688	-30	+66
Urban & Other Rural	0	-1,015	-1,859	+580	+1,382
Equity Differences by Community Households (US$/household)					
Timber Rural Value per household	0	-1,937	-4,351	-188	+417
Urban Value per household	0	-520	-990	+428	+864

Compared to Minimum Regulations or Proposed Regulations, the Alternative Management cases (4 and 5) with compensation offer substantial gains to both these communities (timber-dependent and urban/other rural), while reducing the inequitable differences on who gains and who pays.

The environmental gains of almost 2 billion US$ in utility value per year are nearly as large as the market value of the harvest. However, without compensation, there is no payment made from the largely urban beneficiaries to the rural producers of these values. Under the regulatory cases (cases 2 and 3), NPV losses to landowners of 10 to 20 billion US$ correspond to unrealized tax losses of 3 to 6 US$ billion. These funds could support annualized payments of 150 to 300 million US$ in annual incentives to motivate forest land managers to change their management practices, enough so as to provide the incentive needed to manage under the alternatives

(Lippke et al. 1996). While the utility value for environmental gains suggests that the land value should be increased over its timber market value when it also produces increased non-market amenities, the regulatory emphasis on set-asides (forestland taken out of production and put in reserves) substantially lowers the asset values, working contrary to market efficiencies. For all communities to benefit, landowners need to be provided incentives to compensate for the non-market values they produce at no cost to the largely urban beneficiaries.

7. Conclusions

The illustrated cases may not accurately reflect the effect of prior and proposed regulations, as it is sometimes difficult to determine the full effect of regulations. Nevertheless, the examples are sufficient to demonstrate the direction and range of specific management effects. The tradeoffs between economic and environmental attributes are large and affect communities and groups differently. Strict commodity management of the forests that produces economic benefits also reduces old-forest functionality to the degree that some species become endangered. Regulatory approaches that rely only on no-management set-asides to retain the forest structures that are in declining supply are more costly and take much longer to produce equivalent levels of late-seral stand structures important to endangered species. Active management approaches can restore forest structures that are in decline in a shorter period of time. As a consequence, management alternatives are substantially less costly than regulatory set-asides in restoring habitat. Estimates of the public's utility values show that there do exist alternatives with increased net utility value over the low constraints commodity market baseline. However, the regulatory alternatives fail to produce enough environmental gains to offset the negative utility from reduced timber-dependent rural employment and increased cost to each household.

To achieve utility value gains, landowners would need to be motivated to manage their forest lands to produce both environmental and economic benefits simultaneously. This could be in the form of compensation for producing structures that meet habitat goals. Urban communities could provide that compensation and still be substantial net winners, while reducing the disparity in equitable treatment under current and proposed regulations. Regulatory approaches reduce the value of the land asset in order to achieve environmental objectives instead of increasing the asset value to reflect the increased environmental attributes, making markets less efficient in producing environmental values. Even recapturing the tax losses imposed by regulations through conservation easements and in lieu of regulations would likely be sufficient to provide the incentives for alternative management. Opportunities to improve the effectiveness of achieving non-market, social benefits from forested lands are great and win/win solutions do exist.

8. References

Adams, D., Alig, R., Anderson, D., and Stevens, J. 1992. Future prospects for western Washington's timber supply. Institute of Forest Resources Contribution #74. College of Forest Resources, University of Washington. Seattle, WA USA. 200p.

Bare, B.B., Lippke, B.R., Xu, W., and Oliver, C.D. 1997. Demonstration of trust impacts from management alternatives to achieve habitat conservation goals on Washington Department of Natural Resources Managed Lands. (Available as CINTRAFOR RE-42.) College of Forest Resources, University of Washington. Seattle, WA USA. 236p.

Curtis, R.O., Clendenen, G.W., and DeMars, D.J. 1981. A new stand simulator for coast Douglas-fir: DFSIM users's guide. Gen. Tech. Rep. PNW-GTR-128. Portland, OR: USDA, Forest Service, Pacific Northwest Research Station. 157p.

(FEMAT) Forest Ecosystem Management Assessment Team. 1993. Forest ecosystem management: An ecological, economic, and social assessment. Portland, OR: U.S. Department of Agriculture; U.S. Department of the Intern [and others]. [irregular pagination].

Lippke, B.R., Sessions, J., and Andrew B.C. 1996. Economic analysis of forest landscape management alternatives: A final report of the Working Group on the Economic Analysis of Forest Landscape Management Alternatives for the Washington Forest Landscape Management Project. Sponsored by USDA Forest Service, PNW Research Station, and Washington State Department of Natural Resources. (Available as CINTRAFOR SP-21, College of Forest Resources, University of Washington, Seattle, WA USA.) 157p.

U.S. Laws, Statutes, etc. Public Law 93-205. The Endangered Species Act of 1973. As of December 28, 1973. 87 Stat. 884, as amended; 16 U.S.C. 1531-1536, 1538-1540.

(WDNR) Washington Department of Natural Resources. 1993. Draft Habitat Conservation Plan (for certain lands managed by the Washington State Department of Natural Resources). DNR. Olympia, WA.

(WFPB) Washington Forest Practices Board. 1996. Forest Practice Board Rules-WAC 222. Department of Natural Resources, Forest Practices Division. Olympia, WA.

Xu, W. 1997. Experimental choice analysis of non-market values for ecosystem management with preference heterogeneity. Dissertation. College of Forest Resources, University of Washington. Seattle, WA USA. 155p.

Paper Recycling

- A discussion of methodological approaches -

Stig Byström and Lars Lönnstedt

Key words: Systems analysis, life cycle analysis, allocation methods, waste paper, fiber recovery, bioenergy production

Abstract: When analyzing the waste paper cycle, we use a systems analytical approach. In contrast to the most common methods used in life cycle analyses, no allocation methods are needed for allocating different effects. The use of such methods always introduces great uncertainty about how to interpret the result. Our conclusion is that a systems analytical approach is to be preferred. Examples will be presented of differences between the two approaches.

1. The Waste Paper Cycle

Since the 1980s, recycling and reuse are key issues around the globe. Consumers demand environmental considerations from producers. Producers demand the same thing from their suppliers. This is also true for paper and board products. However, considering environmental and economic aspects, it is not obvious how to reuse waste paper. The main question to answer is whether it should be used for fiber recovery or bioenergy production.

A key issue when trying to answer this question is the energy use when producing pulp, paper, and board (compare Anon. 1997, Byström and Lönnstedt 1997, Leach et al. 1997, Ruth and Harrington 1997). Processing waste paper for paper and board production requires energy that is usually derived from fossil fuel, such as oil and coal. Let us compare pulp and paper production based on virgin fibers with that based on waste paper. In contrast to the production of virgin fiber-based pulp, waste paper processing does not yield a thermal surplus. Thus, thermal energy must be supplied to dry the paper web. If, the waste paper was used instead for energy purposes, the need for fossil fuel would be reduced. This reduction would have a favorable effect on the carbon dioxide balance and the greenhouse effect. Moreover, pulp production based on virgin fibers requires consumption of roundwood and electricity and causes emissions of air-polluting compounds as does the use of waste paper.

Figure 1 presents two alternatives for producing office paper from chemical pulp. If fiber recycling is used, the waste paper must first be deinked, i.e., pass a deinking process (DIP-process). When the paper production is based on virgin fibers, 2 metric tons of pulpwood is used and 5 GJ of electricity (Byström and Lönnstedt 1997). After collection of used office paper for bioenergy production, 9 GJ of heat can be produced. If instead the waste paper is recycled, the use of virgin fiber is 0.3 metric tons, 4.375 GJ of electricity and 7.95 GJ of heat. Thus, this alternative requires more external energy, produces no energy surplus, but saves trees.

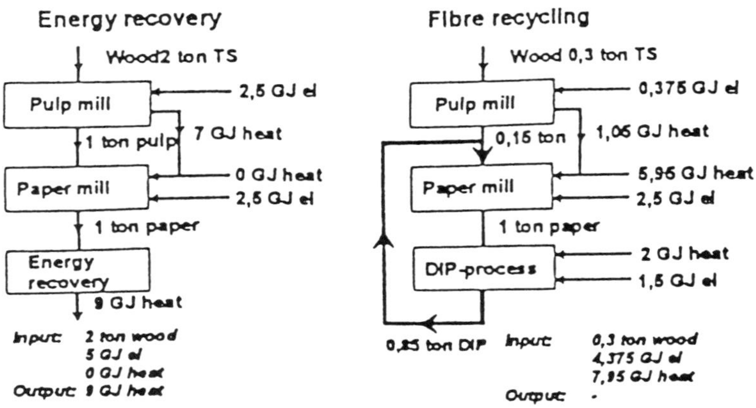

Figure 1. Use and production of energy in office paper

The waste paper cycle has been modelled by Byström and Lönnstedt 1997, Ruth and Harrington 1997, Virtannen and Nilsson 1992. In our model we compare twelve paper qualities: newsprint, Super Calandered (SC) paper, Light Weight Coated (LWC) paper, office paper (wood-free), coated paper (wood-free), tissue, white lined chipboard, "return fiber chipboard", wrapping paper, white liner, kraft-liner and fluting. Each product uses its own recipe for how to mix fibers, filler, and energy. Fibers could be virgin or recovered. Electric energy could be produced from fossil fuels or from hydroelectric power stations. Furthermore, different types of pulp exist. The two major qualities are chemical and mechanical pulp, respectively. Chemical pulp is classified depending on type of fiber, i.e., short or long fibers. Still, another pulp quality is semichemical. If the pulp mill and paper mill are on the same site, pulp is delivered as flush pulp. However, many paper mills are not integrated with a pulp mill. They have to buy market pulp delivered in sheets. For each pulp quality, the need for pulpwood (short and long fibers) and energy must be specified. At the site of the pulp mill, electricity can be produced from back-pressure power.

Pulp and paper are produced in, exported from, and/or imported to different countries. In Europe sometimes the Nordic countries (Finland, Norway, and Sweden) are described as the European lumberyard. The products produced can be delivered to the domestic market and/or to the export markets. After end-use, paper is recycled for production of paper and board, recovered for energy, or deposited as land-fill. If recycled, the paper is recovered, sorted, baled, and transported to paper mills in the country where the consumption took place or exported to countries with waste paper deficits, for example the Nordic countries. If recovered for energy use or transported to a land-fill dump, the waste paper normally follows the waste-handling system. If it is used for energy production it replaces oil or coal.

2. Purpose

This short description shows the complexity of the fiber flow system. Thus, the most appropriate approach is the system analytical and not life cycle analysis where allocation methods are used. In the latter case one often looks at just one production process and uses allocation methods for in- and outflow from or to other processes. Is it not recommended that approximations should reflect changes in input data as in reality? Has it ever been shown that the allocation methods used in traditional life cycle analyses give a good approximation? Thus, it is generally recommended that allocation should be avoided by, for example, expanding the system. If allocation cannot be avoided the allocation should be based on the way in which the inputs and outputs are changed by quantitative changes in the products or functions delivered by the system.

The purpose of this paper is to compare a systems analytical approach with a life cycle analysis approach where allocation methods are used. The different approaches will be compared by measuring the environmental load from producing the product.

We will assume that an identical product, for example newsprint, is produced in two different ways. One production line is based on virgin fibers and the other on recycled paper. After use, the virgin-based newsprint can be reused in the other process. Thus, the two production lines are linked. In a systems analytical approach both lines will be included in the same model, while the life cycle analysis approach will use allocation methods. Different methods for allocations exist and we will present two of them.

3. Model

The two productions lines, L1 and L2, and their connection are shown in Figure 2. The products produced are assumed to be identical but production line L1 uses virgin material while production line L2 uses material recycled from L1. The following notations are used:

- V The environmental load from the first production step of production line L1, for example forestry and the pulp process
- R The environmental load when using recycled material from production line L1 in production line L2, for example deinking

P The environmental load of the second production step of both lines L1 and L2, for example newsprint production

W The environmental load for material not recycled but, for example, recovered for energy production

X The amount produced (tons) by production line L1

Y The amount produced (tons) by production line L2

C The amount consumed (tons)

No inventories are considered and the following relation exists between consumption and production:

[1] C=X+Y

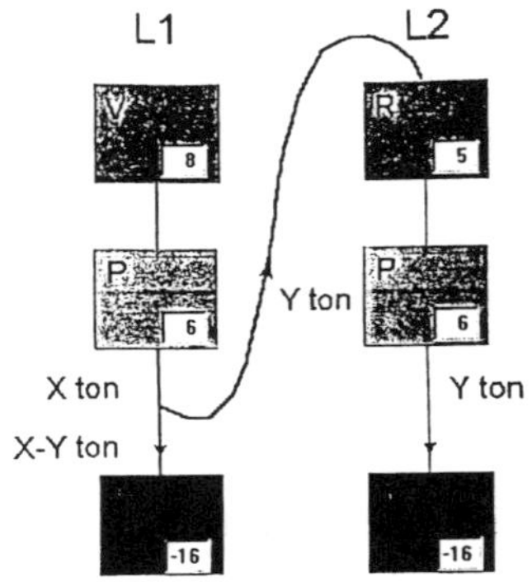

Figure 2. Two production lines and their connection

In general the systems analytical approach means a simplification of reality. For example, the processes and the transports are simplified and the number of production steps are aggregated. However, if a real structure was constructed the correspondence between the real world and the model would be perfect. As in reality, all measures used are physical. A life cycle analysis usually gives a detailed and accurate description of the reality. However, if allocation methods are used, a non-physical link is introduced. As a consequence, the answers given when changing the input data don't necessarily correspond to reality.

4. Data

To calculate the environmental effect of pulp and paper production environmental indexes can be used. The various indexes assign an environmental load value to emissions and to the use of renewable and non-renewable resources such as forest, oil, and coal. In this way, use of non-renewable resources and emissions can be added together.

An example of one such index is the Environment Load Unit-index (ELU) based on the Environmental Priority Strategies in Product Design (EPS) method (Steen and Ryding 1994). This evaluation method was developed in 1991 and revised in 1994. It is used in the realm of life cycle analyses to assess the environmental burden of many products or processes. According to this criterion, each effect is evaluated as cost and quantified in ELU (1 ELU corresponds to 1 Ecu. For example, the "value" of the use of 1 kg of fresh water in areas with a water deficiency is 0.003 ELU. The "punishment" for destroying 1kg of oil is 0.4 ELU. The ELU-index for renewable resources reflects the market value of the resources, i.e., the demand and supply conditions. For non-renewable resources, it is a theoretical calculation of the cost for "recreating" the resource or the cost for finding a substitute. This explains why oil has a higher value than, for example, coal. Depending on use and emissions, a value may be added. If oil is used as fuel, emissions from incineration are added. An overview of different environmental load indexes can be found in Bengtsson (1998).

For the numerical calculations we have used the following values:
V=8 load units per metric ton
P=6 load units per metric ton
R=5 load units per metric ton
W=-16 load units per metric ton
C=1 000 metric tons
The question to be answered is whether the total load will increase or decrease if the production in line L2 increases. When answering this question we will use both a systems analytical approach and a life cycle analyses approach where allocation methods are used.

5. Result

5.1 The systems analytical approach

When both production lines are included within the system boundaries, the environmental load can easily be calculated. This is done by adding the loads from production lines L1 and L2, respectively, and using the relationship X=C-Y.

The environmental load from production line L1 is

[2] $X(V+P)+(X-Y)W$

The environmental load from production line L2 is

[3] $Y(R+P+W)$

After simplification the total environmental load from both production lines can be expressed as:

[4] $Y(R-V-W)+C(V+P+W)$

It is obvious that if Y increases one unit, the load for the whole system will increase by (R-V-W) units. If the numerical values given above are used this means 13 load units.

5.2 The allocation method "Cut Off"

In life cycle analyses, the production lines are sometimes treated separately, and the link between them is handled by allocation methods. The environmental load is measured as units per ton. One such allocation method is the "Cut Off." However, as the name indicates, in this method the link between the processes are not considered at all, i.e., no allocation is made. In our example, the environmental load from production line L1 is calculated as

[5] V+P+W(X-Y)/X

and from production line L2 as

[6] R+P+W

According to the "Cut Off allocation method" the change in environmental load when increasing production from line L2 and decreasing production from line L1 is calculated by subtracting equation [5] from equation [6]. After simplification the difference can be expressed as:

[7] (R-V-W)+W(X+Y)/X

In Figure 3, three different graphs are shown. Each of them shows the environmental load (load units/metric ton) related to the amount of recycled fibers (metric tons), i.e., L2-production. Calculated from equation [4], one graph shows the result when the two production lines are looked at as a system. Calculated from equation [5] for L1and equation [6] for L2, the two other graphs show the result when the two production lines are treated separately.

However, it is more interesting to look at the change in environmental load when production from line L2 increases by one ton and production from line L1 decreases by one ton. In Figure 4 the answers given by the two approaches are shown. Once again, the environmental load per ton is plotted against amount of recycled fibers. Change in environmental load according to the systems analytical approach is given by equation [4] as the coefficient of Y. The answer given when using "the Cut Off allocation method" is calculated from equation [7]. It is worth noting that this graph is non-linear, which is remarkable, as the system is linear. As can be seen from Figure 4, the answers are very different! The difference between the two graphs shows the "error" introduced by "the Cut Off allocation method."

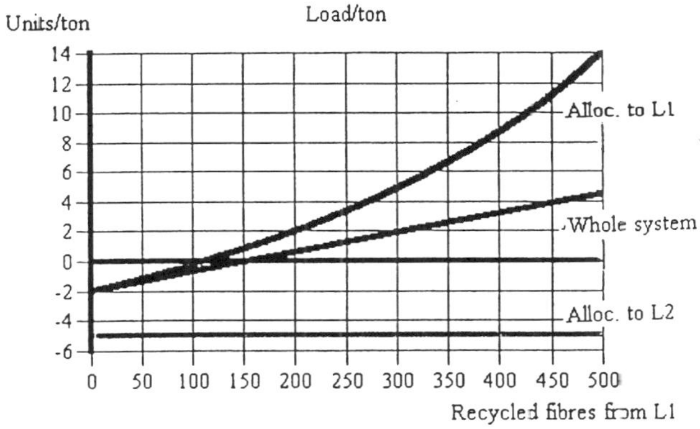

Figure 3. Environmental Load Per Ton at Different Recycling Levels according to the Systems Analytical Approach and a Life Cycle Analysis Using the Cut Off Allocation Method

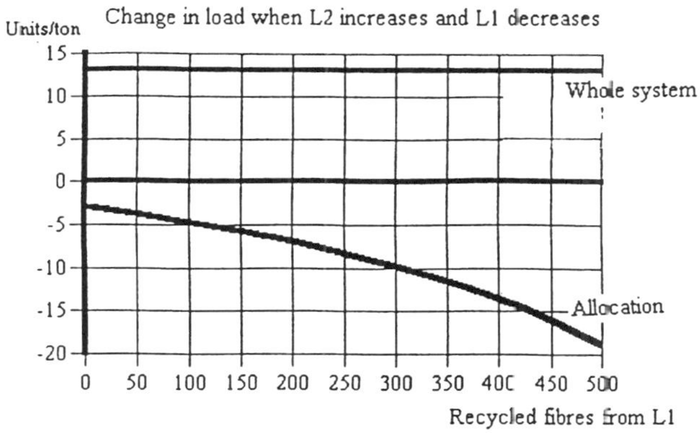

Figure 4. Change in environmental load when production from line 2 increases one ton and production from line 1 decreases one ton at different recycling levels according to the systems analytical approach and a life cycle analyses using the Cut Off allocation method

5.3 The allocation method CIT 50/50

Another allocation method sometimes used in life cycle analyses is "The CIT 50/50-method" developed by Chalmers Industrial Technique (CIT) (Ekvall and Tillman 1997). In this method the environmental load for production line L1 is reduced and the environmental load for production line L2 is increased by the term (V-W-R). This term is based on a judgement of environmental loads up- and downstream, respectively. The factor used for this correction term has in this method been set to 0.5. This explains the name "CIT 50/50".

Using this allocation method the environmental load per metric ton for production line L1 is calculated as:

[8] V+P+W(X-Y)/X-0.5(V-W-R)Y/X

and for production line L2 as:

[9] R+P+W+0.5(V-W-R)

As for "the Cut Off allocation method" the difference in environmental load between production line L2 and line L1 represents the effect of increasing production in line L2 and decreasing production in line L1. After simplification the difference can be expressed as:

[10] (R-V-W)+(V-R+W)(X+Y)/2X

It can be noted that if the factor used is set to zero instead of 0.5 equation [10] will be the same as equation [7].

As for the "Cut Off-method," in Figure 5 three different graphs are shown. Each of them shows the environmental load (load units/metric ton) related to the amount of recycled fibers (metric tons), i.e., L2-production. Calculated from equation [4], one graph shows the result when the two production lines are looked at as a system. Calculated from equation [8] for L1 and equation [9] for L2, the two other graphs show the result when the two production lines are treated separately and the link in between them is taken care of through "the CIT 50/50-allocation method."

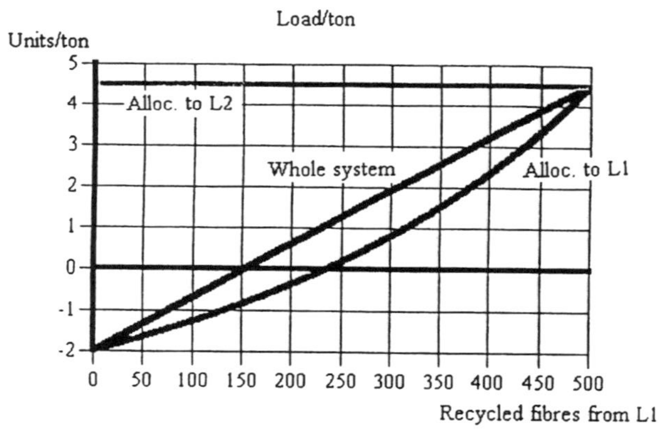

Figure 5. Environmental load per ton at different recycling levels according to the systems analytical approach and a life cycle analysis using the CIT 50/50 allocation method

However as above, it is more interesting to look at the change of environmental load when production from line L2 increases by one metric ton and production from line L1 decreases by one metric ton. In Figure 6 the answers given by the two approaches are shown. Once again, the environmental load per metric ton is plotted against amount of recycled fibers. Change in environmental load according to the systems analytical approach is given by equation [4] as the coefficient of Y. The answer given when using "the CIT 50/50 allocation method" is calculated from equation [10]. As for "the Cut Off allocation method," it is worth noting that this graph is non-linear. The difference between the two graphs shows the error introduced by the allocation method. The answers differ by 50 to 100%!

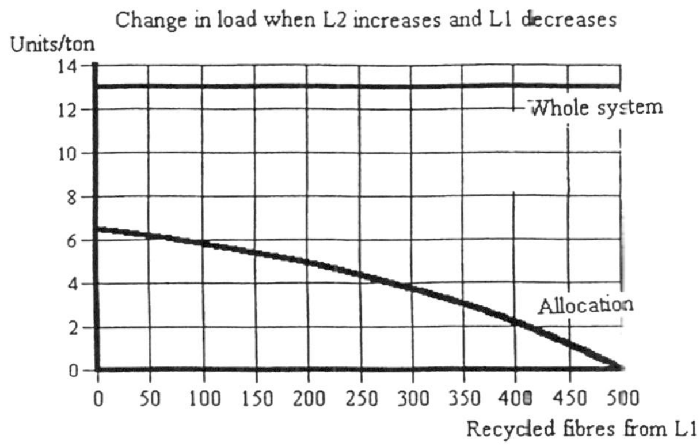

Figure 6. Change in environmental load when production from line 2 increases one ton and production from line 1 decreases one ton at different recycling levels according to the systems analytical approach and a life cycle analyses using the CIT 50/50 allocation method

6. Conclusions

Compared with a systems analytical approach, life cycle analysis with allocation uses correction factors and measures effects on the environmental load when increasing production of one line and decreasing production of another. If the purpose is to approximate a technical system, the allocation methods tested above cant in general be regarded as useful methods. At least in the case of open loop recycling the interaction between the products involved is incorrectly described by the allocation methods. Decisions concerning the use of recycled materials should therefore not be based on such methods.

7. References

Anon. 1997. Burn Me. New Scientist 22 November: 31-34.

Bengtsson, M. 1998. Värderingsmetoder i LCA. Metoder för viktning av olika slags miljöpåverkan - en översikt. (Valuation methods in LCA). Teknisk Miljöplanering/Centrum för produktrelaterad miljöanalys, CPM. Chalmers tekniska högskola, 412 96 Göteborg, Sweden.

Byström, S and Lönnstedt, L. 1997. Paper recycling: Environmental and economic impact. Resource, Conservation and Recycling. 21: 109-127.

Ekvall, T., and Tillman, A-M. 1997. Open-loop recycling: Criteria for allocation procedures. International Journal of Life Cycle Analyses. 2: 155-162.

Leach, M.A., Bauen, A., and Lucas, N.J.D. 1997. A systems approach to material flow in sustainable cities: A case study of paper. Journal of Environmental Planning and Management. 40: 705-723.

Ruth, M., and Harrington Jr., T. 1997. Wastepaper repulping versus incineration in the US pulp and paper industry: A dynamic systems analysis. International Journal of Energy Environment Economics. 5:151-182.

Steen, B., and Ryding, S-O. 1994. Valuation of environmental impact within the EPS-System. Published in Integrating Impact Assessment into LCA. Proceedings of the LCA symposium held at the Fourth SETAC-European Congress, 11-14 April 1994, The Free University, Brussels, Belgium. 6p.

Virtannen, Y and Nilsson, S. 1992. Some environmental policy implications of recycling paper products in Western Europe. IIASA, Laxenburg, Austria.

PART 3

ROLE OF JAPANESE FOREST POLICY IN A GLOBAL CONTEXT

Current States of Japanese Forestry and Timber Trade

Kiyoshi Yukutake and Atsushi Yoshimoto

Key words: Japanese forest resources, timber trade, silvicultural cost, forest economics

Abstract: Although the forest stock of Japanese forests has increased in recent years, production of domestic timber has not yet increased. By contrast, imported timber has reached approximately 80% of the total timber consumption in Japan. In this paper, we overview current states of Japanese forest resources, then analyze the relation between Japanese and foreign timber in Japanese markets in order to investigate why domestic timber production has not increased. Our main findings are that the price of domestic timber may not be high enough to cover the total production costs. Silvicultural costs in Japan are found to be almost five to ten times greater than those in the main timber exporting countries. Considering only physical forest resources, it would be possible for Japan to be self-sufficient in supplying lumber to its domestic demand.

1. Introduction

Since the "Earth Summit" held in 1992 as the United Nations Conference on Environment and Development (UNCED) in Rio de Janeiro, achieving "Sustainable Forest Management" has become an emerging global issue. At the third conference of the parties to the United Nations Framework Convention on Climate Change (COP3) held in Kyoto, Japan, in 1997, the role of forests as carbon "sinks" by absorbing CO_2 from the atmosphere, was re-emphasized and recognized internationally (United Nations 1998). This brought international attention to one common environmental problem, i.e., global warming, and provided an opportunity to work together for a common environmental goal.

In Japan, sustainable forest management ought to be carried out for the purposes of land conservation and economic development in rural areas. Forests in Japan are mainly managed in mountainous areas with steep slopes. Maintaining forests would prevent devastating landslides and at the same time provide economic benefit to the locals. As for the timber market, Japanese lumber has become competitive price-wise against imported lumber, mainly U.S. lumber. However, the price seems not high enough to retain forest management. This could make sustainable forest management difficult for Japanese forests.

During the last decade, imported timber has occupied approximately 80% of total Japanese timber consumption. Among timber exporting countries, the United States and Canada have been exporting a vast amount of timber to

Japan. Difficulty in exporting timber from these two nations arose when the wildlife protection movement in those countries began. This movement took most old-growth forests away from forests for timber production.

Because a large amount of Japanese artificial forest planted after World War II has reached maturity for harvesting resulting in an increase in forest stocks and merchantable forest resources, a careful management plan should be set up for forest resource utilization to seek sustainable use of the resources. In addition, because Japan is one of the largest timber importing countries in the world, our policies for forest management and timber trade greatly influence forest resource utilization in other countries. Searching for a balanced situation considering environmental protection and economic development, Japan has been in a position to take responsibility for leading timber importing and exporting nations toward an ideal situation.

The purpose of this paper is to overview the current states of domestic and foreign timber in Japanese markets, and to seek reasons why domestic timber production does not increase. In the following sections, forest resources and harvest level in Japan are examined. Then the characteristics of timber demand-supply and price, and change in price elasticity of domestic and foreign timber are illustrated, followed by comparison of silvicultural and transportation costs among Japan, Canada, and New Zealand. In the last section, the possibility of increase in timber harvest in Japan is discussed.

2. Japanese Forest Resources

Almost 67% of the land in Japan (25.2 million ha estimated in 1995) is covered by forests, and forested area per capita is therefore about 0.2 ha, which is less than half of the world average. There are three types of forest ownership, private, public (prefectural and municipal), and national. Private forests cover 58% of the total forested area, while public and national forests cover 11% and 31%, respectively (Rinyacho 1995). In terms of the stocking level, the proportion is almost the same.

Natural forests occupy 13.4 million ha, and the rest is artificial forest covering 10.4 million ha. In 1995, the forest stocking level of natural forests was estimated at 1,590 million m^3, and 1,892 million m^3 for artificial forests. Artificial forests mainly consist of sugi (*Cryptomeria japonica*) 58% on the stock basis and hinoki (*Chamaecyparis obtusa*) 20%. These are the main species for softwood lumber in Japan.

Japanese forest resources have a non-uniform age distribution. Forest stands younger than 35 years old cover 68% of the total forested area. Most are from 25 to 35 years old. Merchantable forest area older than 25 years in 1995 covered 78% of the total forested area, an increase of 10% since 1990

(Rinyacho 1990, 1995). Merchantable forest resources have become available gradually.

3. Japanese Timber Markets

Forests were harvested regularly before 1960, but forest harvest declined subsequently after that (Figure 1). In particular, reduction in the harvest level was remarkable after 1973 owing to a decrease in house construction affected by the oil crisis, structural economic change, and the high exchange value of the Japanese yen after 1986.

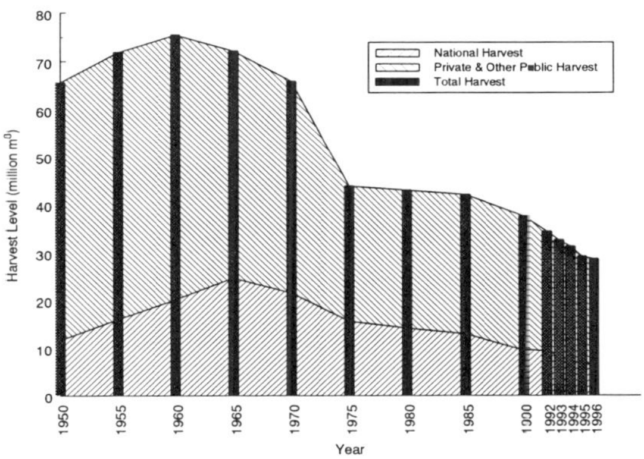

Figure 1. Dynamics of harvesting in Japan

The total demand for timber in Japan remained at 90 million m^3 from 1981 to 1986. Since then, it has recovered to a level of 100 million m^3 influenced by growing house construction activity since 1987. Although the timber demand in Japan increased by a large amount, the domestic timber supply could not make a quick response to the demand. This was because production costs were high in Japan and domestic products were of uneven quality as merchantable forest resources were limited owing to large-scale harvesting during World War II and the period of economic revival immediately after the war. Because of these high costs, domestic timber producers could not increase their production, providing an opportunity for foreign timber producers to invade the market with a large amount of good-quality timber at a low price. As a result, the share of foreign timber in Japanese markets reached about 50% in 1969, 60% in 1973, and 70% in 1987. The share was approximately 80% in 1996.

The two largest foreign timber suppliers are the United States and Canada. Their main species are hemlock (*Tsuga heterophylla*) and Douglas-fir (*Pseudotsuga menziesii)* for use of construction material, competing with sugi. Around 40% of the total timber supply (1.1 billion m^3) in 1996 was from the United States and Canada (Figure 2). Log, lumber, and pulp-chips from the United States occupied approximately 40%, 12%, and 48% of the market share, respectively, while from Canada they accounted for less than 1%, 67%, and 38%, respectively.

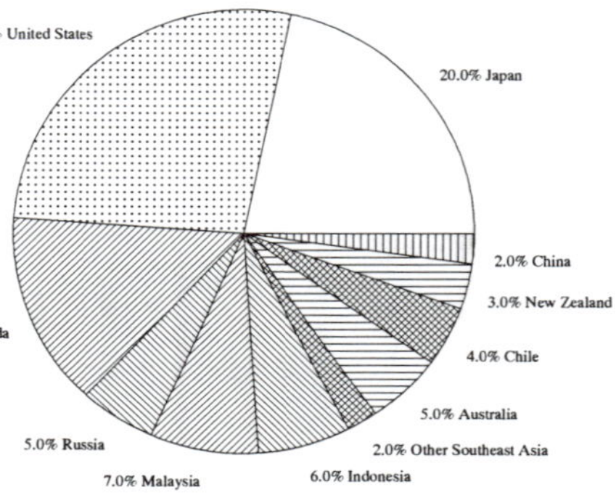

Figure 2. Market share by nations in 1996

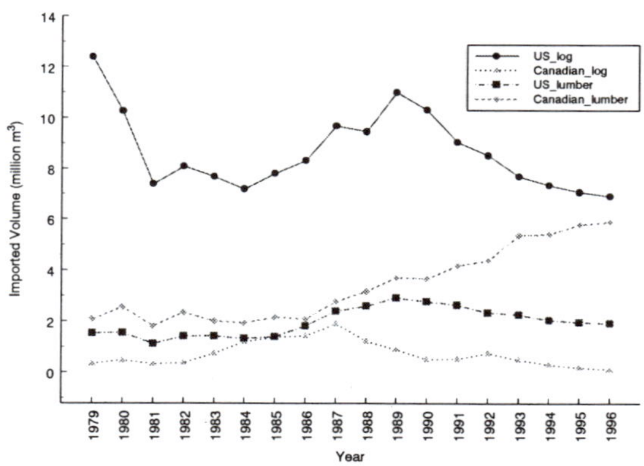

Figure 3. Imported volume of North American logs and lumber

The rest of foreign timber shares is as follows: Malaysia, Indonesia and other Southeast Asian countries occupied 15%; Russia and Australia occupied 5%; and New Zealand and Chile, where radiata pine (*Pinus radiata*) is the main export species, occupied 4% and 3%, respectively.

Since 1989, the import of U.S. logs decreased, while Canadian lumber increased in Japanese markets (Figure 3). Total U.S. logs in 1996 were estimated at 6.9 million m^3 with 37% reduction since 1989, while Canadian lumber increased rapidly up to 5.9 million m^3 in 1996. This amount is 2.8 times more than that in 1986. Since 1995, the total volume of lumber imported from the United States and Canada exceeded that of logs imported from the United States and Canada. In 1996, total imported lumber became 7.8 million m^3, and total imported logs was 7.0 million m^3. If lumber volume is converted into log volume, the total imported lumber becomes twice the quantity of imported logs.

All imported timber in the Japanese markets had almost the same trend as U.S. and Canadian timber. Since 1989, whole imported logs have been decreasing except logs imported from New Zealand and Papua New Guinea. Imported lumber has been increasing since 1985. In 1996, all imported logs amounted to 21.3 million m^3 and imported lumber was 11.5 million m^3, respectively. Considering the recovery rate of logs into lumber, the amount of all imported lumber exceeded that of all imported logs.

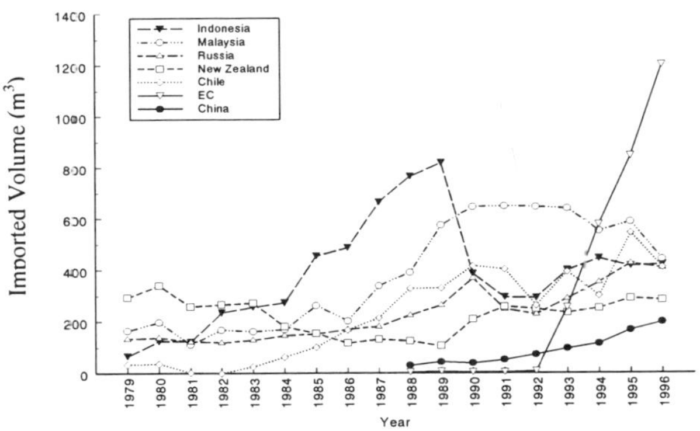

Figure 4. Imported volume from other nations

Recently, there has been a large amount of timber imported from European nations, especially from Finland, Sweden, and Austria (Figure 4). Logs imported from Europe totaled only 9,000 m^3 in 1992 but they went up to 101,000 m^3 in 1993, 517,000 m^3 in 1995, and then down to 203,000 m^3 in

1996. Lumber imported from Europe was 8,000 m^3 in 1992. It rose to 254,000 m^3 in 1993, 846,000 in 1995, and further up to 1,203,000 m^3 in 1996. Lumber imported from Chile, Russia, and China has been increasing remarkably as well.

Importing of engineered wood products, such as laminated wood, particle board, medium density fiberboard (MDF), oriented strand board (OSB), and value-added products improved through a second and third process, has been accelerated during the last decade (Nishimura 1998). Advantages of these products are preciseness, durability, and strength as construction materials. After the Hanshin earthquake disaster in Japan in 1995, Japanese consumers started paying more attention to these characteristics.

4. Demand, Supply and Price of Domestic and Foreign Timber

The domestic timber supply, demand, and price can be roughly characterized by the following three periods: 1953 to 1960, 1961 to 1973 and afterwards. From 1953 to 1960, the domestic timber market was dominated by domestic timber. This was before a large quantity of foreign timber began to be imported. During this period, the domestic demand for timber markedly increased because of the economic revival and rapid economic growth after the war. This resulted in a high price of timber in domestic markets.

During the period of high economic growth from 1961 to 1973, the domestic demand for timber increased. At the same time, because of a high price of timber in the domestic market, a large amount of imported timber penetrated into the market, pushing up the timber supply. Because increase in the timber supply was more than that of the timber demand, the timber price became lower than expected.

The oil crisis in 1973 had a profound influence on the world economy. In the third period, a leveling-off of the high economic growth in Japan took place. From 1974 to 1995, the domestic timber price decreased mainly because a large amount of foreign timber penetrated into the market, and, as a result, the domestic log and lumber supply decreased. During this period, logs from thinning activities were introduced into the markets with increase in merchantable forest resources and reduction in production costs. The timber supply from foreign countries, especially the United States and Canada, had not increased as much as expected during this period. After the oil crisis in 1973, the supply of timber products affected the timber price. Further changes in the timber supply can be predicted by considering the following: (1) fluctuation in exchange rate, (2) shortage of available forest resources and harvesting regulations owing to environmental conservation and wildlife protection, and (3) shift in export products from raw materials to final, highly-processed forest products, such as plywood, 2x4 housing, and door and wood materials for traditional Japanese housing.

5. Relative Price Elasticities

The U.S. timber has been used for construction materials, competing with sugi, the major Japanese timber in the markets. The competitive relation of United States and Japanese timber can be characterized by price elasticity. During recent decades, Japanese timber markets have been highly dependent upon foreign timber because of its low price compared to Japanese timber. Using the annual data for Japanese markets from 1960 to 1973, the cross-price elasticity (relative price elasticity) of U.S. logs (hemlock) with respect to the Japanese log price (sugi) was estimated at 1.83, i.e., very elastic (Yukutake 1977). This implies, for instance, that a 10% increase in the Japanese log price relative to the U.S. log price resulted in an 18.3% increase in demand for U.S. logs. The cross-price elasticity of U.S. logs with respect to the one-period time-lagged price of Japanese logs, became 1.65, which was also very elastic, based on the annual data for Japanese markets from 1960 to 1984 (Yukutake 1989). As for lumber, the lumber price imported from the United States and Canada was also very elastic compared to the price of lumber produced in Japan (including lumber made out of foreign logs). Its cross-price elasticity with a one-period time lag was 1.20. These figures imply that a small fluctuation in the Japanese log and lumber price could cause a large change in the volume of U.S. and Canadian timber demanded in the Japanese markets.

Using the annual data from 1974 to 1994, the cross-price elasticity of U.S. logs (hemlock) with respect to the Japanese log price (sugi), became 0.246, and with one-period time lag the cross-price elasticity of hemlock logs turned out to be 0.456. Using the data for Douglas-fir logs, the cross-price elasticity with respect to the sugi log price was 0.355, and 0.555 with one-period time lag. Decreasing tendency in the cross-price elasticity can be expected. The cross-price elasticity of North American lumber (mainly from the United States and Canada) and Japanese lumber was not statistically satisfactory. The correlation coefficients between North American lumber imports and the relative price of North American lumber (hemlock and Douglas-fir) and sugi lumber, ranged from -0.5 to -0.8. The correlation coefficients between the North American lumber imports and its price ranged from 0.5 to 0.7. This implies that the demand curve for North American lumber has shown a greater shift than the supply curve. This phenomenon is very similar to what occurred for the domestic timber demand and supply before 1960. The supply curve of North American timber cannot move to the right by a large amount owing to increase in demand in the high economic growth period. The North American timber supply could not respond quickly to its demand change in Japanese markets.

6. Timber Price and Production Costs

Figure 5 and 6 show the price trend of sugi, hemlock, and Douglas-fir. Sugi logs were 28% more expensive than hemlock logs in 1978. It became 15% in 1980, and only 1% in 1985. After 1985, however, as the Japanese yen became strong against the U.S. dollar, 100Yen per US$, the price of hemlock logs decreased in Japanese markets. The difference between sugi and hemlock log prices became larger than in 1985. The price of hemlock and Douglas-fir logs increased and became higher than the Miyazaki sugi price in 1989 and the national average sugi price in 1992. This is due to a decrease in harvest in Washington and Oregon States affected by the spotted owl problem since 1987. Note that Miyazaki is one of the main sugi-lumber-producing prefectures and has been leading the markets. After 1992, the price of Douglas-fir remained higher than that of hemlock. In 1995, it became even higher than the national average sugi price. In the case of lumber prices, the national average sugi price was not less than the lumber price of hemlock and Douglas-fir, but the difference between them shrank markedly. The Miyazaki sugi lumber price has been lower than that of lumber imported from the United States and Canada since 1989.

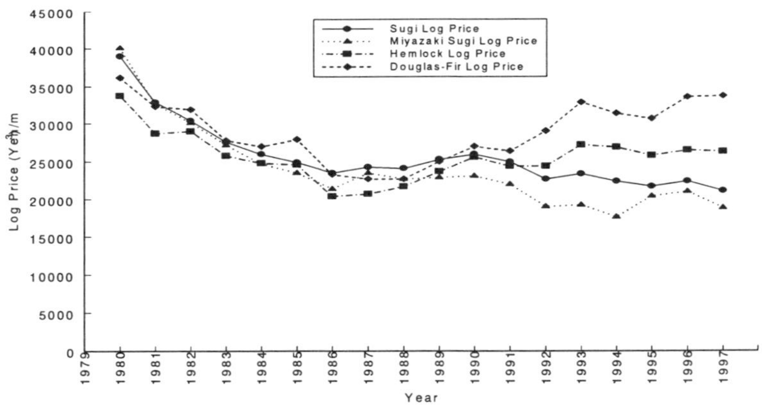

Figure 5. Domestic and imported log price

Reduction in the sugi price has been stimulated by two main factors. One is that as the exchange rate has become favorable to exporters to Japan since 1989, foreign timber became cheap, forcing the domestic timber price lower in order to retain domestic timber production. The other is an increase in merchantable forest resources in Japan. Price-wise, sugi lumber seemed to be competitive against foreign lumber. However, this reduction in the price forced many lumber producers to leave the markets. That is, the price could not cover their production costs. If this is the case, increase in the domestic lumber supply cannot be expected.

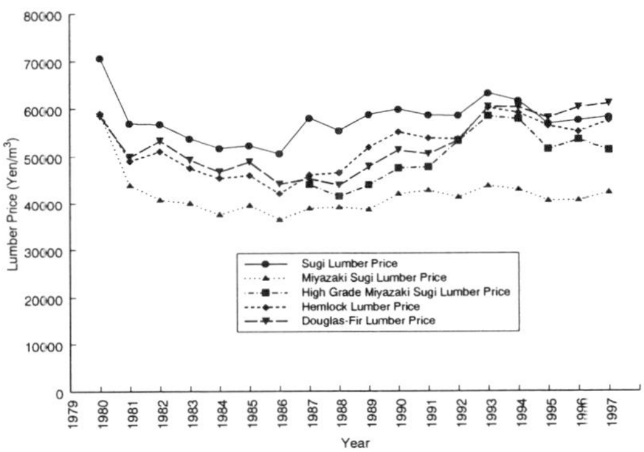

Figure 6. Domestic and imported lumber price

Regarding production costs, the following example from Miyazaki can provide some insights on Japanese forestry. The total silvicultural costs in Miyazaki are approximately 1 to 1.5 million yen per ha on average, while costs in Canada is from 150 to 300 thousand yen per ha, and in New Zealand about 104 thousand yen per ha (Table 1). The costs in Miyazaki are almost five to ten times greater than those in the other countries. This is mainly due to high weeding costs. Weeds are a major silvicultural problem in Japan, and have to be treated regularly. Treatment cost is about 600 thousand yen per ha, which is almost half of the total silvicultural costs. In contrast, weeding cost is almost zero in the other countries.

Other costs in Miyazaki include the costs of harvesting and logging and transportation to a log auction, ranging from 7,000 to 15,000 yen per m^3. On the other hand, the general selling price at a log auction in Miyazaki ranges from 16,000 to 20,000 yen per m^3. Forest stock in a productive site for sugi is approximately 300 m^3 per ha at the 35-year felling period. Thus assuming the average price of sugi logs in the auction market to be 17,000 yen per m^3 and harvesting and logging costs (including the auction market fee) to be

12,000 yen per m^3, a resultant return to the forest owners could become 5,000 yen per m^3 or 1.5 million yen per ha. If the forest owners have to carry out reforestation, the return would not cover the total costs with implementation of weeding and thinning treatments. If a subsidy were provided for reforestation, the forest owners would earn their income from the subsidy not timber production activities.

Table 1. Silvicultural costs (Yen/ha) in 1993

Item	Miyazaki (Japan)	Canada	New Zealand
Site preparation	130,000	30,000	3,000
Plantation	300,000	50,000	21,000
Weeding	600,000		8,400
Pruning	300,000		54,000
Pre-commercial Thinning	30,000	20,000~40,000	13,200
Commercial Thinning		50,000~120,000	
Others			4,500
Total*	1,000,000~1,500,000	150,000~300,000	104,100

Data are based on an interview survey conducted by the authors.
* Value in Total is a rough estimate, not necessarily equal to the sum of each item.

As for the lumber production side, the total costs at sawmills range from 45,500 to 51,500 yen per m^3, while the selling price of lumber is within the range of 41,000 to 50,000 yen per m^3. The price for by-products is about 5,000 yen per m^3. Summing up these, the resultant total profit becomes almost zero. The total costs at sawmills consist of 50% for purchasing raw materials, and about 30% to 40% for processing and related expenses. The rest is for transportation and sales commission (Table 2).

Table 2. Total costs from Miyazaki producing district to Tokyo market (Yen/ha) in 1995

Item	Values
Material Costs	16,000~20,000
Logging Costs	7,000~15,000
Recovery Rate	70%
Processing Costs	6,000~10,000
Other Costs	7,000~10,000
Total Processing Costs	36,000~40,000
Transportation Costs	6,000~7,000
Sales Commission	
Total Costs	45,500~51,500
Selling Price	41,000~50,000

Data are based on an interview survey conducted by the authors.

Transportation cost is also an important factor for lumber production. Transportation costs from the United States and Canada to Japan is approximately 35.00 US$ per m^3, which was half of the transportation cost of 7,000 yen per m^3 from Miyazaki to Tokyo in 1995. Depending upon the exchange rate, shipping products from the United States or Canada to Tokyo could become much cheaper than from Miyazaki to Tokyo.

7. Countries Potentially Exporting Timber to Japan

Currently, because of the wildlife protection movement such as protection of the spotted owl, importing timber from the United States and Canada has become more difficult than ever. This caused a price rise of United States and Canadian timber products in the Japanese markets. Before it became too high, however, timber import from north European countries, Chile, and Russia was stimulated.

As was discussed earlier, the product price, production costs, and transportation costs play an important role in deciding which nation would be able to penetrate into the Japanese markets and occupy more share than the others. Table 3 and Figure 7 show a comparison of the production costs in ABRE (Fegely et al. 1997) for softwood and hardwood pulpwood among countries in South America and Oceania. All in all, Brazil showed the lowest costs, followed by Chile, Australia, and New Zealand.

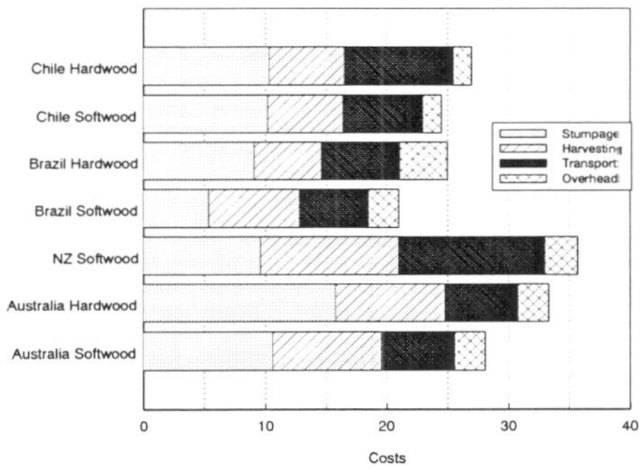

Figure 7. Costs at mill

Table 3. Softwood and hardwood pulpwood costs (US$/m^3) at mill in first quarter of 1996

Country	Stumpage	Harvesting	Transport	Overhead
Australia Softwood	10.6	9.0	6.0	2.5
Australia Hardwood	15.8	9.0	6.0	2.5
NZ Softwood	9.6	11.4	12.0	2.7
Brazil Softwood	5.4	7.4	5.7	2.5
Brazil Hardwood	9.1	5.5	6.5	3.9
Chile Softwood	10.2	6.2	6.6	1.5
Chile Hardwood	10.3	6.2	9.0	1.5

Table 4 shows the costs of softwood production at sawmills in Australia, New Zealand, Chile, and British Columbia in Canada including delivery costs to Korea, not to Japan. This is due to a lack of available Japanese data. Because the distance from these countries to Korea is almost the same as that to Japan, these data can provide enough information on costs of delivering to Japan. Data were converted into the index number by treating Australia as a base country. Compared with the delivery cost from Australia, those from British Columbia and Chile are 10% cheaper, and that from New Zealand is 9% higher. Out of the total costs, the transportation cost was almost one-third while the wood purchase itself was another one-third or more for these four places. The transportation cost from New Zealand made up 30.8% of total costs, from Australia 29.4%, from Chile 28.5%, and from British Columbia 26.6%. Shipping takes about 3 weeks from New Zealand, about 5 weeks from British Columbia, and about 6 weeks from Chile to Korea or Japan.

Table 4. Delivery cost index for softwood sawmill to Korea

Item	Australia	New Zealand	Chile	British Columbia
Wood	37.0	41.8	29.0	24.7
Personnel	6.6	6.3	3.7	13.4
Energy	2.4	1.7	2.5	0.9
Other	7.2	8.3	8.1	6.5
Capital	17.4	19.8	19.2	17.8
Delivery	29.4	30.8	28.5	26.6
Total	100.0	108.7	91.0	89.8

Information Source: Margules Groome Pöyry Ptu Ltd. (1995)

8. Conclusions

The total stock of artificial forests in Japan reached 1 892 million m^3 in 1997, of which sugi and hinoki occupy approximately 80%. Without considering the costs, given the 40-year rotation age, artificial sugi and hinoki forests would be able to cover about 60% of the total domestic lumber demand, 50 million m^3. Increase in the artificial forest stock of 73 million m^3 in 1995 was more than half of the total domestic timber demand, 100 million m^3. If natural forest resources are taken into account, 70% of the total demand would be covered by domestic products.

Compared with United States and Canadian timber, the price of Japanese timber became lower than before and competitive in the market. Problems remain as to the cost of producing lumber. It still needs to be reduced by a large amount in order for Japan to become self-sufficient.

Considering potential timber supply from countries in Oceania and South America, shows that increase in the Japanese harvest level could become even more difficult. These countries are practicing sustainable forest management of eucalyptus (*Eucalyptus spp.*) and radiata pine (*Pinus Radiata*) forests with low production costs and shorter rotation age, less than 10 years for eucalyptus and less than 25 years for radiata pine. In contrast, average rotation age for sugi and hinoki is 40 to 60 years with high production costs. If Japan continues to import a large amount of low-cost timber, including particleboard, MDF, plywood, and laminated wood, maintaining Japanese forests in a sustainable fashion would become almost impossible.

Although Japan has merchantable forest resources for production, economic environments are not favorable to production. The market is largely influenced by inexpensive foreign timber. From a viewpoint of artificial forest environments, this would leave forests unmanaged. Since most artificial forests require proper management, unmanaged forests would create some environmental problems in the near future.

Perfect competition may be the way to go for trading. However, regarding forest resources, because they have not only market values, but also environmental values, cultural values and other type of values, it might be better to let each country consider its own conditions in developing sustainable management policies for conservation and utilization of its own forest resources. Internationally, criteria and indicators for sustainable forest management are being discussed (Funston 1998). An emerging task in Japan should be to establish forest policies that contribute to forest resource utilization that is sustainable throughout the world, and that helps to reduce CO_2 from the atmosphere as well.

8. References

Fegely, A.R., Parsons, M.J. and Pöyry, M. 1997. Forest products trade: The supply challenge from plantations in New Zealand, Chile and Brazil. P. 199-208 *in* Outlook 97, Proc. of the National Agriculture and Resources Outlook Conference, Canberra, 4-6 February, Vol.1, Commodity Market and Resource management, ABRE, Canberra.

Funston, M. 1998. Criteria and indicators for sustainable forest management, http://www.fs.fed.us/land/sustain_dev/:1-3.

Margules Groome Pöyry Ptu Ltd, 1995. The Economic impact of the New South Wales timber industry, B-28.

Nishimura, K. 1998. Structural changes in Japanese timber market and wood industry. Forest Economy (Ringyo Keizai). 597: 1-12.

Rinyacho, 1990. Current states of forest resources. Forestry Agency, Tokyo, 54p. (in Japanese).

Rinyacho, 1995. Current states of forest resources. Forestry Agency, Tokyo, 90p. (in Japanese).

United Nations, 1998. Industrialized countries to cut greenhouse gas emissions by 5.2%, Press Release, http://www.cop3.de/fccc/info/indust.htm: 1-2.

Yukutake, K. 1977. Timber market and demand/supply structure for construction. P. 16-126 *in* Econometric analysis of sawn wood, plywood and paper/pulp. Yuize, Y., and Yukutake, K., (eds). Kohansha.

Yukutake, K. 1989. Simulation analysis of demand/supply relationship of forest. The Current State of Japanese Forestry. Vol. VI: 23-38. The Japanese Forest Economic Society.

Acknowledgements

This research was supported by a Grant-in-Aid for International Scientific Research (No.08041066) from the Ministry of Education, Science, Sports and Culture of Japan.

Japan's Wood Products Import and Forest Sector

- Overview of the changes -

Takashi Kato

Key words: Domestic wood, imported wood, value-added products, structure of industry

Abstract: Over the last few decades, domestic timber has supplied a smaller portion of Japan's wood consumption, which currently stands at approximately 110 million m^3 per year. The country's self-sufficiency in wood products has declined from almost 90% in 1960 to the current rate of slightly more than 20%. Domestic forestry and forest products industries have degraded because of competitiveness of imported products, high costs of harvesting and processing, and shortage of workers. On the other hand, imports of wood, predominantly in the form of logs, chips and other primary manufactured products, have grown from 6.3 million m^3 in 1960 to 89.8 million m^3 in 1996. The major suppliers have been traditionally the Pacific Northwest of the United States, British Columbia of Canada, Russian Far East, southeast Asian countries (mostly Malaysia and Indonesia), and Australia. Currently this supply structure is changing rapidly, as imports of logs from North America and Russia are declining, tropical hardwood sources are cut down, and imports of softwood products from new suppliers like New Zealand, Chile, and Nordic countries are increasing. At the same time, traditional as well as new suppliers are increasingly replacing logs and other primary manufactured products with processed value-added products. The overall trend of wood product imports has led to further declines in domestic wood industries, and this has yielded greater opportunities for foreign producers. In order to restore domestic forestry and forest products industries, the government has continually paid for a variety of programs including promotion of domestic wood consumption, subsidies for regional timber production, development and dissemination of new processing technologies, etc. In the early 1990s, the Forestry Agency announced a new initiative aimed at promotion of forestry activities under the name of "River Basin Forestry Management," which followed introduction of three forestry laws to implement comprehensive measures for pursuing a collective effort and establishing links between forest owners (upstream) and the processing industry (downstream). However, the development of the new initiative will still depend on competitiveness of domestic plantation products against products processed internationally. Promotion programs at the regional level will require a better flow of market information and attainment of greater economic efficiency in harvesting, transport, and manufacturing.

1. Domestic Timber Supply

The forest covers approximately 25 million ha, or about two-thirds of the total land area of Japan, with different forest types, from sub-frigid conifer forest in the north to sub-tropical forest in the south. Slightly over 10 million hectares consists of plantation forests, containing predominantly coniferous species with relatively young age classes.

Private holdings dominate these forests and account for 58% of forest land. Private forest is owned by 2.8 million separate bodies, individuals (mostly farmers), corporations, other organizations, shrines and temples. In terms of number and area, farmers and individuals dominate the ownership of private forest land; approximately 90% of those owners possess fewer than 20 ha, with average holding size 2.6 ha.

The growing stock has increased around 70% during the last 30 years, and the volume is calculated at 3.1 billion m^3, of which 1.6 billion m^3 consists of softwood species in plantation forest. The annual increment of growing stock is estimated at about 70 million m^3. Sugi (Japanese cedar) and hinoki (Japanese cypress) are the main species for plantation forests. During the 1950s and 1960s, extensive reforestation took place under the assumption of intensive management for growing high-quality timber.

The total wood products supply in Japan currently stands at about 110 million m^3 per year. During the last few decades, Japanese domestic timber has remained at a relatively low level, and the increasing demand has been satisfied with imports, traditionally from the United States, Canada, Southeast Asia, and Russia. Domestic timber supply in current years is around 23 million m^3, which accounts for slightly more than 20% of the country's total wood consumption.

As for sawlogs, domestic production supplied around 32%, and for pulpwood domestic production supplied 13% of total consumption in 1995.

Most of the potential for expanding domestic supply lies in the softwood plantations, of which close to 4.8 million ha will enter into harvestable age during the next two decades. The supply potential of this resource, if it were to be managed for maximum biological yield, has been calculated at around 65 million m^3 per year, which is well over twice the current domestic timber harvest.

For many years, the government has predicted an increase in domestic roundwood supply based on 10 million ha of softwood plantations. However, economic factors have worked against any substantial expansion in domestic harvest. The most pressing problem is the high labor intensity of Japanese forestry combined with a shortage of labor and an increase in wages of workers (Japan Forest Technical Association 1992). Predictions of domestic harvest increase are becoming less common, and it could be assumed that domestic timber production will continue to decline for the coming years in spite of increasing mature plantation inventory.

2. Wood Demand

2.1 Trends in total wood demand

Japan is one of the largest consumers of wood in the world. The total demand of wood for industrial use amounted to around 106 - 114 million m^3

in the last decade, peaked at about 114 million m^3 in 1989, to decrease 108 million m^3 in 1994, and peaked again at 112 million m^3 in 1996.

Wood demand is for three major products, 50 million m^3 for sawnwood, 44 million m^3 for pulp and 14 million m^3 for plywood in 1997. For each product category, the majority of raw material requirement was met by imported sources, around 68% for sawnwood, 87% for pulpwood, 99% for plywood in 1997.

2.2 Housing starts

Housing and construction are the main end-use sectors for sawnwood and plywood, using about 80% and 50% of total production, respectively. The condition of the housing sector, therefore, greatly affects wood demand, specifically for sawn wood and plywood. The residential construction market is made up of wooden and non-wooden houses, 754 thousand units and 889 thousand units, respectively, in 1996. After a sharp increase in 1996, housing starts for both types dropped 10% or more in 1997.

Wooden housing construction is composed of three major types of construction: conventional (post and beam) housing, 2 x 4 construction, and prefabricated. The trend for the last decade indicates that the number of housing starts by post and beam method is on the decline (92% in 1986 to 82% in 1996); in contrast, 2 x 4 construction is on the increase (5% in 1986 to 13% in 1996).

Traditionally, most wooden housing was built by carpenters and small companies using traditional post and beam construction methods. The market structure started to change during high economic growth in the 1960s and 1970s, and through an era of low economic growth in the 1980s till the present, as large companies entered into the housing market introducing labor-saving, cost-effective construction methods, and gained a large market share (Theisen and Dirks 1996).

Continual efforts were made by larger companies seeking ways to rationalize housing construction, developing designs and materials that increase performance and reduce labor costs. These less-time-consuming, resource-efficient ways are taking the Japanese housing market towards a more systematic and western style of architecture (Pesonen 1993).

The rationalization process strongly affects the use of materials and delivery systems, and results in increasing use of imported products, preferably value-added products. Engineered wood products are gaining acceptance as builders find uses for them.

Structural laminated wood has been increasingly used in conventional housing construction especially by large companies, because of its quality assurance in size and strength. Also housing construction using pre-cut sawn wood materials has been increasing in recent years, and accounts for more than 30% of post and beam housing construction (Forestry Agency 1997b).

2.3 Paper consumption

Over the past 10 years paper consumption has steadily increased. It reached 30 million metric tons in 1996, a historic record high, that ranks second after the United States. The elasticity of paper consumption to Gross Domestic Product (GDP) is 1.17 for the last decade, indicating paper consumption has increased faster than GDP growth (Yoshimoto 1997).

One distinctive feature of Japan's paper industry is heavy dependence on imported wood chips and waste paper as raw materials. In 1995, about 2[illegible]% of total raw material consumption was supplied by imported wood chips, about 53% by waste paper. For two decades, the use of recycled paper has increased dramatically, which made the paper industry less vulnerable to external control by foreign suppliers. Aside from increasing use of waste paper, another critical factor for the industry maintaining competitiveness has been overseas investments, which are undertaken in order to develop and secure raw material supplies and locating mill sites for production (mostly producing pulp and exporting) for lower input costs (Whitham 1994).

3. Wood Product Import

Japan has long been one of the world's largest consumers of and a leading importer of wood products, having insufficient capacity of domestic timber supply to meet increasing wood demand. Mass supply, uniform quality, and lower price have been the competitive advantage for imported wood. In 1996, Japan imported 90 million m^3 of wood products in log equivalent volume. The share of imports has gradually increased and presently makes up about 80% of total wood products supply. However, trade patterns have reflected shifts from logs or primary products to secondary processed products that include a wide variety of manufactured products. This change reflects a shift in products demanded by the housing market, which is undergoing a continual rationalizing process. Export manufacturers are making great efforts to meet the demand.

North American and Russian softwood logs and lumber, and South Sea hardwood logs have traditionally dominated imports. Now this supply structure is changing, as South Sea logs have been cut down, and imports of logs from North America and Russia are also declining.

In 1997, import of lumber was 12.6 million m^3. Canada is the biggest supplier with 46% of total imports. The balance comes mainly from the United States and Southeast Asian countries.

There have been continual efforts especially by Canadian producers to increase value-added products to be exported to Japan. This has resulted in increased volume of a wide variety of products categorized as custom-ordered and others, to meet specific end-market requirements.

Quite recently, Nordic countries have emerged as important suppliers of softwood logs and lumber. In 1997, timber imports from Nordic countries reached around 109 thousand m^3 for logs and 1,420 thousand m^3 for lumber.

Other newcomers in log exports to Japan are West African countries. Since 1993, Japan has been buying over a half million m^3 of hardwood logs from West Africa for obtaining face and back veneer of plywood.

A sharp rise in the export price of timber from the Pacific Northwest of the United States in 1992 to 1993, largely owing to timber supply restrictions caused by the environmental movement, contributes to these emerging trends.

The short fall in tropical hardwood log imports has been largely replaced by plywood imports. In 1996, import of plywood amounted to 673 million m^2, which is almost equal to domestic production. This trend is partly a consequence of restrictions imposed on log exports and of efforts by exporting countries to promote industrialization, timber processing, and export of value-added products.

Imported pulpwood has long been an important component of fiber supply for Japan's pulp and paper industry. The trend is toward increased reliance on imported pulpwood along with further diversification of its sources.

Most remarkable among the changes in chip import during the last decade has been the rapid increase of hardwood chips, which accounts for two-thirds of the total wood chip imports in recent years. The United States and Canada have traditionally supplied most of Japan's softwood chip imports, but that dominance has been slipping. Taking the place of North America, Australia has emerged as an important source of softwood fiber. In 1997, United States supplied 44% of total softwood chips imported, Australia 28%, and Canada, Brazil, Fiji, Chile, and other countries supplied the rest.

On the other hand, the volume of hardwood chips imported reached over 19.3 million m^3 in 1997. Although Australia has long been the major supplier, its share in total imports has fallen to 25% in 1997, the volume having remained constant. These trends have been the consequences of the effort made by Japanese companies from the late 1980s to diversify their sources of hardwood chips. In 1997, the United States supplied 33% of total hardwood chips imported, Australia 25%, Chile 12%, South Africa 10%, and other Brazil, Thailand, Viet Nam, Indonesia, and other countries supplied the rest.

Although the volume of imported wood-based panels and laminated wood products is not significant compared to total wood product consumption in Japan, the upsurge of engineered wood product imports reflects changes in the product mix demanded by the end-use market. Today's big challenge to the domestic lumber industry is not only competition by lumber from overseas, but also by composite panels and laminated wood products from foreign and domestic sources. Laminated products have provided an alternative for plywood and lumber, substituting for solid wood in many end-use applications. Also finger-jointed lumber is increasingly accepted and used in construction. Big collapses resulting from the Kobe earthquake in 1994, in which North American style wooden buildings fared better than Japanese style construction, may well stimulate engineered wood product sales in Japan.

The major factors driving overall trends in wood product imports toward increasing volume of processed, value-added and wide variety of products are:

- First, efforts by exporters to customize for the Japanese market.
- Second, restriction of log exports and support for exporting processed products by producing countries.
- Third, timber supply restrictions coupled with declining availability of high-quality logs.
- Fourth, decade-long higher appreciation of the Yen exchange rate. Fifth, technological innovations and growing acceptance for new products by end-users.

It can be anticipated as the harvest from plantation and second-growth forests in exporting countries increases, the production of a variety of composite panels and laminated wood products will increase further, and the market for these products will expand substantially.

4. Shifts in the Structure of Industry

Increasing wood product imports, which reflect the shifts from primary products to value-added products, helped shrink domestic production, with consequent decline in the number of sawmills, chip mills and plywood mills. Smaller supplies of tropical hardwood logs is forcing the Japanese plywood industry to decrease its capacity on one hand, and to change its raw material source to softwood logs.

Over the coming years, the predominant factor shaping Japan's forest products industry will be an increase in highly processed, value-added products at the expense of primary and other rough-form wood products.

Except for pulp mills and engineered wood products factories, continuing declines in both number of domestic mills and production capacities are predicted, and this trend translates into shifts in the structure of the industry.

Some of the manufacturers are trying to shift their main production line to value-added products, and others are changing their business to wholesale and imports, collectively changing the structure of the industry.

At present, Japan's forest products industry is adjusting to a rapidly changing business environment.

The sawmill industry, traditionally highly fragmented and small scale in operation, has gone through restructuring. The industry is in transition with substantial consolidation of smaller mills and a steady drop in the number of sawmills. Only larger companies with better productivity and more advanced production systems are surviving. A considerable number of mills are establishing new production lines for value-added products, like kiln drying, planing, and other secondary processing.

Faced with increased competition from imported products and substitute materials, the plywood industry is also in transition with mills closing and a steady drop in production. Problems the industry has encountered are declining raw material sources, increased competition from imported products, higher production cost, etc. In order to survive, the industry has shifted into using new raw materials, softwoods from Russia or radiata pine from New Zealand, and has introduced more efficient processing technology and developed new end uses.

As for the pulp and paper industry, the companies' major concern has been securing raw material sources. Toward the year 2000, the demand for wood chips is expected to rise, and it is anticipated that traditional suppliers will be able to maintain their volumes. In the case of Chile and Brazil where extensive eucalyptus plantations are established, there will be a large surplus of pulpwood, logs, or chips, for export. Establishing plantations overseas to gain access to a new supply has been of major concern for Japanese paper companies. This not only makes the industry possible by increasing fiber supply, but also enables companies to be less susceptible to price control by foreign chip suppliers.

5. Government Support Program for Readjustment

The overall trend in wood product imports has led to further declines in domestic wood industries, and this has yielded greater opportunities to foreign producers. In order to restore domestic forestry and forest product industries, the government has continually paid for a variety of programs including promotion of domestic wood consumption, subsidies for regional timber production, development and dissemination of new processing technologies, etc (Forestry Agency 1997a, Japan FAO Association 1997).

In the early 1990s, the Forestry Agency announced a new initiative aimed at promotion of forestry activities under the name of "River Basin Forestry Management," which followed introduction of three forestry laws to implement comprehensive measures for pursuing a collective effort and establishing links between forest owners (upstream) and the processing industry (downstream). However, the development of the new initiative will still depend on competitiveness of domestic plantation products against products processed internationally. Securing a stable supply of logs to processing mills is considered to be a prerequisite for attaining economies of scale. Promotion programs at the regional level will require a better flow of market information and attainment of greater economic efficiency in harvesting, transport, and manufacturing.

6. References

Forestry Agency. 1997a. Annual report on trends of forestry - fiscal year 1996 (Summary).

Forestry Agency. 1997b. Wood supply/demand and wood products industries 1997. Ministry of Agriculture, Forestry and Fisheries.

Japan Forest Technical Association. 1992. Forestry and forest industries in Japan. Japan Wood Products Information and Research Center.

Japan FAO Association. 1997. Forests and forestry in Japan. Second Edition, Japan FAO Association.

Pesonen, M. 1993. Japanese market for Scandinavian wood products. University of Helsinki, Department of Forest Economics Reports 1.

Theisen, A., and Dirks, J. 1996. A Japanese market profile and sourcebook for Pacific Northwest Value-added products exporters. Center for International Trade in Forest Products, University of Washington, Special Paper 23.

Whitham, J. 1994. The foreign expansion of the Japanese pulp and paper industry. Australian Bureau of Agricultural and Resource Economics, ABARE Conference Paper 94.32.

Yoshimoto, R. 1997. New development and trends in Japanese Fiber, Fiber Asia 1997.

Change in Lumber Market and Behaviors of Lumber Producers in Japan

Junichi Iwamoto

Key words: Lumber market, lumber producer, rural development

Abstract: For the forest industry in Japan, housing is an important field from the viewpoint of marketing. However, the house of traditional Japanese style has gradually lost its share in the wooden house market. Lumber producers, especially in rural forestry areas, have produced lumber mainly for the traditional Japanese style of house construction of and played an important role in rural development. In addition to competition with suppliers of imported lumber and producers of other construction materials, lumber producers in Japan faced problems caused by the change in the lumber market. The purpose of this paper is 1) to describe the change in the lumber market, 2) to examine the effects of the market change on lumber producers and 3) to discuss the role of lumber producers in forestry areas.

1. Introduction

Japan's wood consumption totaled 112.3 million m^3 in 1996. Wood consumption is composed of 44% lumber, 39% pulp and wood chips, 14% plywood, and 3% other. Some 80% of the lumber consumed in Japan is used as construction materials, and some 90% of the lumber for construction is consumed in home construction. Therefore, for lumber producers, home construction is the most important field from the viewpoint of marketing.

There are three types of wooden home construction in Japan. The most popular system is the traditional post-and-beam home-building system. Traditional post-and-beam construction represents 81% of wooden home construction in Japan, 498 thousand units in 1997. However, the market share of this traditional house has steadily decreased. On the other hand, two-by-four construction has shown spectacular growth in recent years. Two-by-four construction represents 13% of wooden home construction, 79 thousand units in 1997. The remaining market share of 6% (34 thousand units) is occupied by houses built of wooden prefabricated components.

2. Background

Traditional Japanese home construction is a kind of post-and-beam construction. Douglas-fir is widely used as beams. On the other hand, one of the most profitable items for a lumber producer processing domestic timber is posts of Japanese cypress (*Chamaechyparis obtusa*). Having a long history of wooden construction, Japanese people have honed a good sense of wooden building materials. One of the examples is posts of

Japanese cypress. In the traditional home-building system, most of the posts are left exposed so that people can enjoy the nice feel of the material. Japanese cypress lumber of high quality has been widely used as "appearance post."

However, in recent years, preferences for building styles are changing, even in the traditional home-building system. The point of change in building style is "westernization." In some western home-building systems, for example, in the two-by-four system, posts are covered with walls. Accordingly, in the Japanese traditional building system, "westernization" means firstly covering posts with walls. The term "structural post" can be defined as the posts intended for applications where appearance is not a primary consideration.

Nowadays, people, especially the younger generation, prefer homes of "westernized" style. This change in preference has affected many lumber producers, especially the sawmills that produce Japanese cypress posts. Exposed posts require a good appearance. From the viewpoint of appearance, knots are very important. Therefore, profitability primarily depends on producing lumber that has fewer knots on the surfaces. This means that production of posts with high-quality appearance promises good profitability to lumber producers. However, the demand for high-quality posts that have good appearance has decreased because of the change in preference of building style.

3. Highlight of the Survey

Owase has been an active forestry region and famous for its high-quality Japanese cypress posts. However, in recent years, market changes described above have affected lumber producers. In order to detect the important factors and find some keys to overcoming the difficulties, we conducted a survey in Owase region (Iwai et al. 1997, Taguchi 1997).

This survey was carried out in two phases. First, in order to grasp the general situation, we had personal interviews with 11 lumber producers in August 1996. Then we conducted a mail survey. The survey questions were: amount of sales, list of products, number of employees, production facilities, timber procurement, features of their products, distribution channels, quality control, and outlook of their business.

The survey was carried out in December 1996 with office support from the municipality of Owase. The municipality of Owase provided a mailing list as well. The survey was sent to 42 lumber producers in Owase forestry region (districts of Owase, Miyama, and Kii-Nagashima). The response rate was 36%, which represents 15 of 42 lumber producers. This survey led us to classify these lumber producers into four main groups.

1) Traditional high-grade post producers

This group has suffered greatly from the market change. They specialized in producing high-grade post. Their sawmills are quite small (operated by 2 or 3 people). Many of them started their business in the late 1950s and had enjoyed good business especially in the 1950s and 1960s, owing to the rapid economic expansion in Japan. Until now, they specialized in producing high-grade appearance posts: 3 meters long and 10.5cm x 10.5cm, or 12cm x 12cm and maintaining their traditional distribution channel. However, they are now losing their demand owing to the market change described above. Many of them are old enough (older than 60) to retire, and some of them intend to close their mills.

2) Traditional premium-grade post producers

The business of this group is stable. Their sawmills are relatively large in the Owase region (operated by 7 or 8 people) and their wood consumption per year is around 3,000 m^3. They specialize in producing appearance posts: 3 meters long and 10.5cm x 10.5cm or 12cm x 12cm. However, the quality of their products are much better than that of "traditional high-grade post producers."

Most of the lumber producers in Japan (including "traditional high-grade post producers" in Owase) purchase the logs at log auction markets. However, having their own workforce for harvesting, lumber producers of this group purchase the logs directly from the forest owners. This is the main reason for stable production of the best-quality posts because they can control the lumber quality, and they can control the timing of the harvest to fit the market condition. Another reason is the balance of supply and demand. The demand for premium-grade posts has not been decreased by recession. From the viewpoint of supply side, it is difficult to find logs of the best quality.

3) Mass structural post producers

Some of the "traditional high-grade post producers" have introduced a new line in recent years. The new line was introduced for efficient mass-production of lower-grade lumber. Their sawmills are relatively large (operated by 7 or 8 people) and their wood consumption per year is around 5 000 m^3. However they have a problem with wood procurement. Most of the forest owners in Japan are Non-industrial Private Owners (NIPF) and the average size of their forest property is only 2.7 ha. So it is difficult to get a continuous supply of enough logs to process by efficient facilities even when they get logs from the local log auction markets.

4) Small-scale market-oriented producers

There is another group of lumber producers who have maintained good business. Their sawmills are small (operated by 3 people) and their wood consumption per year is 1,000 – 2,000 m^3. A significant feature of them is the diversity of the items that they process and the diversity of distribution channels.

They have three major channels. One is the traditional channel to distribute posts. Another channel is local home builders. In Owase, the traditional home built of domestic lumber is quite popular. The residents of this region have much knowledge about wood products and they prefer Japanese cypress lumber as construction materials when they build their homes. The traditional home-building system requires a wide variety of wood products, so they have to process many items at one time. However they say this is a profitable business.

The last channel is newly developed through specific dealers. They supply a wide variety of wood products to some wholesalers. These wholesalers have nation-wide delivery networks. They inform the lumber producers which items are in demand, and lumber producers choose the best combination of items giving careful considerations to the logs they can purchase. Therefore they purchase logs of larger diameter than the logs for post producers so that they can hold a wide selection. This results in avoiding competition with post producers in wood procurement.

4. Conclusion

Lumber producers in Japan are at a crossroads owing to the market change described above. They are struggling with this change and trying to find new ways to compete. This survey revealed some of the new challenges that need to be overcome.

What surprised us is that “small-scale market-oriented producers” have continued to have good business. When discussing the strategy of the wood product industry, many people support efficient mass-production, and actually the number of mass-producers are increasing. However, from the viewpoint of rural development, this strategy is not good for several reasons. For one thing, mass-production will result in the deterioration of local forest resources. In addition, there will be fewer surviving producers. Finally, the mass-producers could be only the servant of the housing industry. There will be a pressure to keep wood products prices low, and this will reduce the share for local people (forest owners, forestry workers, lumber producers, lumber dealers).

On the other hand, the strategy of “small-scale market-oriented producers” is important. Their strategy fits the forest ownership structure and traditional home-building system in Japan. They will be a good core when people establish a new network of lumber production and distribution that benefits the local economy as a whole.

5. References

Iwai, Y., Fujiwara, M., Matsushita, K., Iwamoto, J., et al. 1997. Marketing research report of cypress lumber produced in Owase. (in Japanese).

Taguchi, K. 1997. The sawmill industry in forestry areas in Japan. P. 162-165 *in* Proceedings of IUFRO Symposium in Kyoto 1997.

The Forest Planning System of Japan

Hirokazu Yamamoto

Key words: Forest planning, forest resources, public function, timber product, preferential treatment, certification system

Abstract: The objective of the Forest Planning System is to secure a stable supply of forest products and enhance the public functions of forest resources. This system works as a means to increase forest resources, on the premise that the forest is a public good. Forest resources are not only timber resource, but also water conservation, land conservation, and environmental conservation. The system includes not only forestry planning, but also environmental conservation and landscape planning. The basic principle of the system is guidance for personal forest owners to meet the goal of the forest plan. An advantageous tax system and subsidy system support this guidance based on the premise that forestry is a profitable business. If forestry does not profit, the Forest Planning System does not go well. A social system that evaluates public functions is required. I propose a new type of certification system, that connects to the Forest Planning System. In this system, forest owners and local self-governing bodies make an agreement in each district. This agreement promises that forest owners receive a subsidy for well-managed forests, as certified by an international disinterested party, and that local government supports the planning and practices.

1. The Forest Planning System in Japan

This paper introduces the Forest Planning System of Japan and clarifies its problems. The Japanese Forest Planning System is the most complete and detailed forest planning system in the world. It is based of the Forest Law, which was amended in 1951. The Japanese Government is responsible for the system. Accordingly this system is a top-down system. The primary objective of the Forest Planning System is to secure a stable supply of forest products and enhance public functions of forest resources. These forest resources are timber production, water conservation, land conservation, and environmental conservation. Forests are closely related to the national economy and society. Although forests may be renewable resources, it takes a long time to recover their functions if they are damaged. Because most forest is located in steep mountainous regions in Japan, a neglected forest invites natural disasters like landslides and floods, which affect the national economy. Accordingly adjustment between the nature of forests as private property and their nature as public property is required.

In order to preserve and cultivate forests, increase forest productivity, maintain the land, and develop the national economy, the government is required to clarify the necessary administrative guidance for forests, and provide such guidance to forest owners. The Forest Planning System has been instituted to cover the national, prefectural, city, town, and village levels and the forest owners themselves. The Forest Planning System consists of four kinds of forest plans, the Nation-Wide Forest Plan, the Regional Forest Plan, the Municipal Forest Improvement Plan and the Forest Management Plan (Forestry Agency 1997, Figure 1).

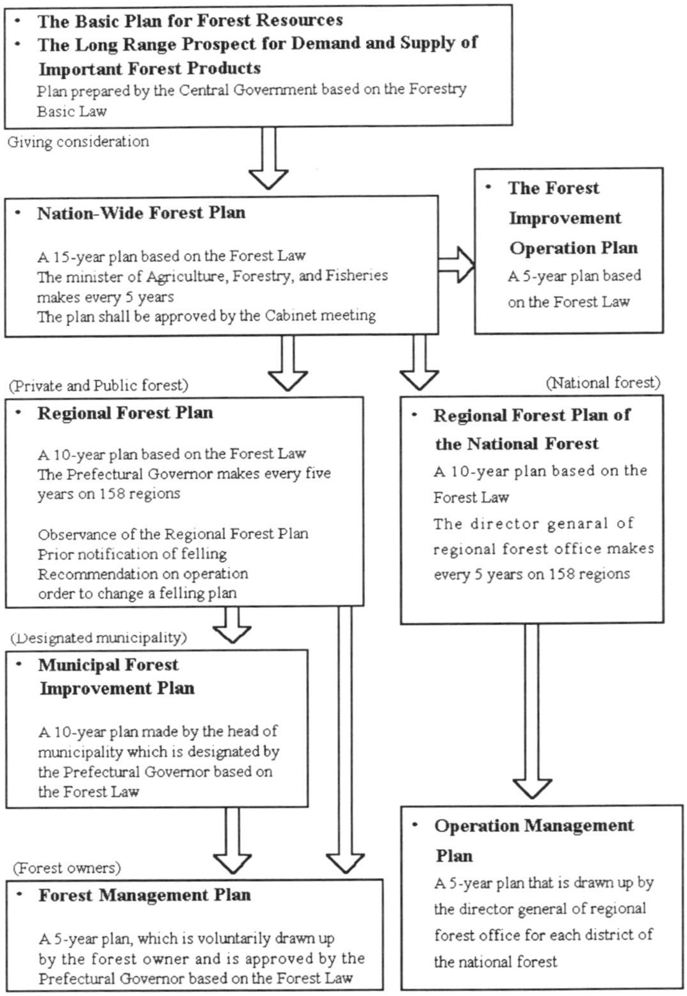

Figure 1. Schematic presentation of the Forest Planning System

1.1 The Nation-Wide Forest Plan

The Minister of Agriculture, Forestry, and Fisheries revises the Nation Wide Forest Plan every 5 years. This is a 15-year plan that covers all the forests in the nation in accordance with "The Basic Plan for Forest Resources and the Long-Range Prospect for Demand and Supply of Important Forest Products," which determines the national fundamental long-term policy toward forest resources and products supply and demand based on the Forestry Basic Law.

1.1.1 The Basic Plan for Forest Rresources and the Long-Range Prospect for Demand and Supply of Important Forest Products

The Basic Plan for Forest Resources was drawn up by the Cabinet in 1996. It described the concept of forest resources and a target for strengthening forest resources. The purpose of this plan is to promote coordination of forest management and forest industry with the country's long-term plans, which include the National Land-Usage Plan. The Basic Plan for Forest Resources classifies the whole forest into three groups according to the target of forest management and shows the policy of forest improvement. Approximately 50% of the whole forest is called "the forests to respect the land conservation and water conservation function," 30% of the whole forest is called "the forests to respect the continuing use of forest resources," and the rest is called "the forests to respect the coexistence of humans and nature." This means that the plan gives first priority to the function of protecting against natural disasters and second priority to the people's benefit. In addition, the Basic Plan for Forest Resources was formulated to emphasize the relation with The Long Range Prospect for Demand and Supply of Important Forest Products, which considers the supply of national timber products in light of supply and demand for future timber. This plan gives a forecast of the supply-demand situation in 2015 by classifying the demand as timber, wood panel material, pulpwood and other timber; and by classifying the supply as domestic or imported (Tsumoto 1997).

1.1.2 Planning matters of the Nation-Wide Forest Plan

The Nation-Wide Forest Plan is a national plan to cover fundamental matters relating to forest management. The plan provides guidelines for the Regional Forest Plan and the Regional Forest Plan of the National Forest as well as guidance to forest owners for forest consolidation. Under the Nation-Wide Forest Plan, the whole country is divided into 44 broad blocks, mainly on the basis of river basins with consideration also given to geographical features and other conditions. The main planning items follow.

a) According to each forest function, targets for forest consolidation and adjustment.
   Examples: area of tended multi-storied forest is 2,380,000 ha, forest accumulation is 165 m^3 per ha, rate of construction of forest road is 66%.
b) Felling of standing trees, reforestation and tending, construction of forest roads, conservation of forest land, and protection of facilities.

Examples: felling timber volume is 40 million m^3 per year, afforestation area is 67,000 ha per year, construction of forest road is 3,800 km per year.

c) Consolidation of forests where specific forest management should be performed.

Examples: management of multi-storied forests and other forest management especially for maintaining and promoting public functions.

Furthermore, planning matters are separately defined for each block. The Nation-Wide Forest Plan should give thoughtful consideration for the creation and conservation of satisfactory natural environments and to other conditions essential for the promotion of public functions of forests.

1.1.3 The Forest Improvement Operation Plan

The Forest Law stipulates that the purpose of the Forest Improvement Operation Plan is to attain the target of forest consolidation indicated in the Nation-Wide Forest Plan. This implies forest consolidation as a means to cope with a recent economic situation surrounding forests and the forestry industry. Such a situation makes it difficult to attain the targets for forest consolidation projected in the Nation-Wide Forest Plan. This forest consolidation plan should therefore be formulated closely with the related Nation-Wide Forest Plan. Furthermore, the applicable projects in this forest consolidation plan are specified as basic projects for forest consolidation, including afforestation, thinning, cultivation, forest road construction, and improvements.

1.2. The Regional Forest Plan

The Nation-Wide Forest Plan clarifies the nation's basic policy for the forests throughout the country. To realize this policy smoothly and appropriately, it is necessary to establish administrative targets suitable for the character of the local forests and to define concrete guidelines and regulations for forestry. Under the provisions of the Forest Law prefectural governors are required to draw up a regional forest plan in accordance with the Nation-Wide Forest Plan and map out a plan for private forests in each forest planning district every 5 years for the next 10 years. Prefectural regions are currently divided by river basins into 158 forest planning districts. The regional forest plan is to be formulated according to the Nation-Wide Forest Plan. This stipulation is provided to prevent the plan from running counter to the aim of the Nation-Wide Forest Plan and to promote a relation to the social and economic conditions of the nation. The regional forest plan is required to flexibly cope with the conditions of the region to which it applies.

1.2.1 Planning matters of the regional forest plan

The main planning items of the regional forest plan follow.

a) Forest land area covered

b) According to each forest function, the location, size, and consolidation target of forests and other fundamental matters respecting the consolidation and adjustment of the forest
c) The cut volume, standard cutting age, and cutting method
Standard cutting age is defined as the most productive cutting age. Standard cutting age affects forest management and regional forestry directly.
d) Reforestation areas, planted species, and standard planting methods
e) Volume of thinned standing trees and standard methods of thinning
f) Land area of specific management forest and management method
g) Collaboration in forest management
h) Protection of surface soil and roots for forest conservation
i) Maintenance of protection forests

1.2.2 Operation of the regional forest plan

Regional forest plans are applied to private forests in various areas with different patterns of ownership, to enhance the preservation of forests, to increase the productivity of forests, and to contribute to national land conservation. In order to realize the regional forest plan, the Forest Law provides the following regulations.

a) Observance of the regional forest plan: Forest owners must conform to the regional forest plan regarding operation use and profits. This provision does not have any legal force. It expects that targets will be accomplished through common knowledge of the relevant plan owing to administrative action, administrative guidance, and the voluntary observance of the plan by forest owners.
b) Prior notification of felling: Prior to felling, forest owners must submit notifications that mention the location, size, and felling methods to the prefectural governor. It is important to grasp the actual state of cutting in order to understand the forest resource.
c) Recommendation on forest operation: The prefectural governor can advise forest owners to observe the regional forest plan if they are considered not to be doing so.
d) Order to change felling plan: The prefectural governor can order forest owners to change their felling plan if the felling notification is considered not to conform to the regional forest plan.

1.2.3 The Regional Forest Plans of the National Forest

The National Forests have been administered in accordance with an original regional management plan pertaining to the Nation-Wide Forest Plan. However, in order to realize the regional forestry plan, both national forests and private forests are required to incorporate it. Therefore, the Regional Forest Plans of the National Forests were legally connected with the regional forest plan for private forests when the Forest Law was amended in 1991. The affected forests are dealt with in the same manner as private forests are dealt with in the district forest plan. In addition, the planning items are the same as those of the regional forest plan in order to clarify targets of forest maintenance in each district.

To establish district forest plans, directors of related regional forest offices and chiefs of related branch offices are required to ask for the opinions of relevant prefectural governors and relevant mayors and heads of municipalities.

1.3 The Municipal Forest Improvement Plan

The prefectural governor can designate a municipality that meets the conditions of the regional forest plan as a Forest Improvement Municipality and cooperate with said municipality to promote systematic and rational forest improvement suitable to actual local circumstances under the leadership of the city, town, or village. The purpose of this system is for the municipalities most closely related to the regions to carry out plans for thinning and tending of forests according to the actual local situation. However the decline in forestry profitability and the reduction of the forestry labor force have caused forestry activities to become sluggish. The result is an increased need for cooperation in regional forest maintenance guarantees of work for people engaged in forestry and with management for promotion of mechanization.

The designated cities, towns, and villages are required to formulate municipal forest consolidation plans. Approximately 2,000 cities, towns, and villages have been designated as forest maintenance municipalities, and they are required every 5 years to map out 10-year plans for private forests. Plans include thinning, tending, and other maintenance activities taking into account geographical features and other conditions.

1.3.1 Planning matters of Municipal Forest Improvement Plan

The main planning items are:

a) Thinning, tending, and the other fundamental matters respecting forest consolidation.
b) Promotion of cooperation in forest management.
c) Training and provision for people engaged in forestry.
e) Promotion of installation of machines necessary for modernizing forest management.
f) Consolidation of facilities necessary for the spur road network and other forest management.

Recommendations by mayors and heads of municipalities on forest maintenance and mediation recommendations, and judgement by prefectural governors are provided to ensure the execution of needed thinning and tending under the municipal forest consolidation plans.

1.4 The Forest Management Plan for forest owners

The forest management planning system is generally divided into two types of plans. The first is the general Forest Management Plan, the purpose of which is to continue the production of timber. The second is the designated forest management plan, the purpose of which is to fully demonstrate the public functions of forests.

1.4.1 Planning matters of the Forest Management Plan

The purpose of this system is to promote rational and planned forest management based on the free will of forest owners and to contribute to the attainment of the objectives of the Nation-Wide Forest Plan and the regional forest plan, and thereby continuously cultivate forests, enhance forestry productivity, and increase the income of forest owners. The functions provided by forests are important for people's lives. Because it is important to carry out forest management properly, these plans are used as standards for the forest management performed by forest owners. The Forest Law provides several types of systems to carry out the plans. However, under the current planning system, it is expected that the plans will be carried out voluntarily by forest owners, and the system does not force them to do so by authoritative means. From this viewpoint, the most effective method is for the forest owners themselves to voluntarily make the Forest Management Plans according to the Nation-Wide Forest Plan and the regional forest plan, and then carry out the pertinent management. This planning system has been instituted taking these considerations into account. The Forest Law stipulates that "forest owners can formulate their own forest plans for each forthcoming five year period and can submit them to the relevant Prefectural Governors who have jurisdiction over the locations of the forests in order to obtain approval that the relevant forest management plans are appropriate." The total approved forest area is 12,833,000 ha, 74% of private forest land. In the whole country, 64,200 forest management plans are established by 4 million forest owners. The average forest area of a forest management plan is 200 ha.

Furthermore, the law provides that forest owners should determine a long-term policy and formulate plans according to that policy. The word "long-term" means a duration of twenty years or more, and the plan must clarify the ideas of the forest owner about improvements in species of trees or forest phase, forest management, and so forth. There are two types of Forest Management Plans. Personal Forest Management Plan is formulated to cover all the forests of an applicable forest owner. Community Forest Management Plan is undertaken by a number of owners for a forest exceeding 30 ha. Now, 87% of the plans are community plans.

The main planning items are:

a) The area of the forests under study by location, artificial or other, the species of trees, forest situation, the age of the forest, and the lumber volume of the trees.
b) For logging: time, area, standing tree volume, and method by location.
c) For reforestation: time, size, species of trees, and method by location.
d) The tending area by species of tree.

The approval conditions are:

a) Plans to meet the standards for modernization of forest operations, systematic improvement of the species of trees or the forest situation. Appropriate stand age and harvest schedule, and sustainable yield.
b) It is suitable for the regional forest plan

The approved forest owner must abide by the forest management plan and must offer notification regarding cutting or reforestation to the prefectural governor. If the approved forest owner does not abide by the project, the prefectural governor can revoke the approval of the plan.

1.4.2 The Designated Forest Management Plan for forest owners

To meet the expectations of the people regarding display of the public functions of forests, it is important to maintain forests and to ensure the constancy of the various forest functions. This requires that forests be kept and cultivated in sound form without reducing their functions and that the management of multi-storied forests be promoted in order to prevent forests from being laid bare or to shorten the bare-state period. A specific forest management plan is formulated to promote the management of multi-storied forests and the effective and efficient long-term management of cut forests. For artificial forests, on the other hand, it is necessary to equalize tree ages.

The Designated Forest Management Plan requires long-term execution of the designated forest management for the applicable forest. The long-term policy here is that the basic operational methods be performed for a period of 40 years or more and standing trees be cut every 5 years for each operational method in afforestation areas and tending areas.

1.5 Preferential treatment and incentive measures

The government offers preferential treatment and incentive measures to forest owners who have obtained approval for their forest management plans or designated forest management plans and have carried out forest management accordingly. The preferential treatment and incentive measures relate to taxation, finance, and subsidies. This is because forest owners who have carried out management according to a forest management plan or a designated forest management plan are actually cooperating in the forest policy that the government has formulated. The government offers measures necessary for further promotion of cooperation in the expectation that as many forest owners as possible will cooperate in the accomplishment of the nation's policy.

The taxation system includes income tax, corporation tax, inheritance tax, and a special land tax. In the income tax, 20% of forest income is exempted. In the corporation tax, 30% of planting expenses may transfer into their loss account as special depreciation. In the inheritance tax, payment of tax is deferred in connection with the cutting plan. Forest land covered by a forest management plan is exempted from the special land tax.

In the assistance system, afforestation covered by a community forest management plan or a personal forest management plan may receive preferential treatment and incentives at a higher assistance rate than that for general afforestation. Forest disaster relief projects are applied to forests under forest management plans approved under the Violent Disaster Law. One important matter for regions covered by urgent construction projects for forest roads is that the assistance rate for areas in an approved forest management plan is about 70% or more. Under the Loan System of the Funds from Agriculture, Forestry, and Fishery Finance Corporation, persons who have an approved forest management plan can be issued a special loan for forest land acquisition at an advantageous condition. Also a special arrangement for reforestation funds may be made for a planned forest at lower interest rate.

2. Effects of the Forest Planning System

A Forest Planning System covering the whole country was introduced in 1939, but it did not work well because of the Second World War (Tezuka 1986). The existing system was established by the Forest Law amended in 1951. In the early 1950s, the main purpose of the Forest Planning System was to restore the forests damaged by over-cutting during the World War. In order to control the cutting, a permission system of cutting was started for the restricted forests and young plantations. After the late 1950s, the objective of the system changed into increasing the forest resources. The logging restriction on broad-leaved trees was relaxed, and expansive afforestation was accelerated in order to promote productivity of the forestland. Nearly half of the artificial private forests were planted during the 15 years prior to 1970. After 1970, the main social issue has been the environmental destruction caused by economic growth. So the Forest Land Development Permission System was started, and forest plans that respect the public functions were established. The Forest Planning System halted disorderly forestland development. Thus, the Forest Planning System has responded to social requests. Especially, this system played an important role in increasing forest resources by establishing plantations. Recently, demands from forest resources have shifted from timber products to land conservation and water resources as people's affluence increases. Consequently many kinds of forests, like multi-storied forests and broad-leaved forests, are required. However, it is hard to respond quickly to such varied requests because there are problems to solve regarding forest owners' intentions and stand density.

3. Problems of the Forest Planning System

The Forest Planning System works as a guideline and as a standard to steadily increase forest resources, which are timber production, water conservation, land conservation, and environmental conservation. There is a limit to interference with individual forest management because most forests that display public functions, are private property. Society requires various kinds of forest and a rich stock of forests; on the other hand, forest owners seek efficient timber production. If a forest treatment, that the forest owner desires satisfies social requests, the Forest Planning System is not necessary. The basic principle of the Forest Planning System is to guide private forest owners to the goal of the forest plan. An advantageous tax system and subsidy system support this guidance on the premise that the forestry is a profitable business. If forestry does not profit, the Forest Planning System does not work well in Japan. Recently, requirements for public functions of the forest are increasing, but the economic condition of Japanese forestry is becoming more serious, and forestry is not a profitable business. Accordingly, the Forest Planning System does not work to promote the functions of forest resources in Japanese society. The rate of accomplishment for the initial plan is 72% of planned harvest and 56% of planned planting. This system works well to limit cutting, but does not work to accelerate harvesting efficiently because this system assigns forest owners a duty of no action rather than a duty of action (Fujisawa 1996). It seems that the failure of the system does not affect society now, because it takes a long time for problems to occur. However, we cannot neglect plantations and declining mountain villages. We have to restore the supply of domestic timber resources in order to achieve the sustainable forest management in our society.

Originally the purpose of the Forest Planning System was resource and land-use management. However, the Forestry Basic Law of 1964 was intended to improve the forestry industry. Since then there have been dual roles for forestry policy. At first the purpose of the system was only forest resource improvement, but now the system has to include processing and distribution of timber products because most forests have matured. This makes the Forest Planning System ambiguous and difficult to understand for forest owners.

4. Proposals for Future Forest Planning

Generally in economic principles, present value has more priority than future value. However in forest management, we should give priority to future value, because it takes a long time to grow a forest. The problem is that forest owners do not have bright prospects under the present condition of forestry, and there is no system that considers social desire concerning public functions. A social system that estimates the public functions of forests reasonably and promotes those functions is required.

I propose a new type of certification system that connects to the Forest Planning System. In this system, forest owners and local self-governing bodies make an agreement in each district. This agreement promises that the forest owners receive a subsidy for well-managed forests, as certified by an international disinterested party, and that the local government supports the planning and practices. In this system the Forest Planning System supports forest owners in attaining certification. This management system should be carried out in a large-scale community with voluntary cooperation and responsibility of owners because an individual private forest is not large enough to attain certification. The government should not manage this system directly in order to make the owners' responsibility clear. If this system enhances the profit of forest owners, the Forest Planning System will go well.

Because the Forest Planning System is a top-down system from the central government to prefectural government to forest owner, similar forest plans are apt to be established all over the country without regard for the regional situation and forest owner's intentions. To improve this system, cities, towns, and villages have to be responsible for forming their own forest plans. It is necessary for local self-governing bodies to expand the financial basis and employ capable persons. Non-governmental businesses in forest management and forest practice are required to carry out their own responsibilities.

5. References

Forest Agency. 1997. Outline of the forest planning system. Forestry Agency. 26p.

Fujisawa, H. 1996. Gendai sinrin keikakuron. Nihon ringyo chosakai. 270p. (in Japanese).

Tezuka, H. 1986. History of the forest planning system of Japan. Shinrin Keikaku Kenkyukai kaihou. No. 300: 2-7. (in Japanese).

Tsumoto, Y. 1997. The basic plan for forest resources and the long term demand and supply forest for important forest product: Concept and outline of the amendment. Shinrin Keikaku Kenkyukai Kaihou. No. 375: 3-15. (in Japanese).

FORESTRY SCIENCES

1. P. Baas (ed.): *New Perspectives in Wood Anatomy.* Published on the Occasion of the 50th Anniversary of the International Association of Wood Anatomists. 1982 ISBN 90-247-2526-7
2. C.F.L. Prins (ed.): *Production, Marketing and Use of Finger-Jointed Sawnwood.* Proceedings of an International Seminar Organized by the Timber Committee of the UNECE (Halmar, Norway, 1980). 1982 ISBN 90-247-2569-0
3. R.A.A. Oldeman (ed.): *Tropical Hardwood Utilization.* Practice and Prospects. 1982 ISBN 90-247-2581-X
4. P. den Ouden (in collaboration with B.K. Boom): *Manual of Cultivated Conifers.* Hardy in the Cold- and Warm-Temperate Zone. 3rd ed., 1982 ISBN Hb 90-247-2148-2; Pb 90-247-2644-1
5. J.M. Bonga and D.J. Durzan (eds.): *Tissue Culture in Forestry.* 1982 ISBN 90-247-2660-3
6. T. Satoo: *Forest Biomass.* Rev. ed. by H.A.I. Madgwick. 1982 ISBN 90-247-2710-3
7. Tran Van Nao (ed.): *Forest Fire Prevention and Control.* Proceedings of an International Seminar Organized by the Timber Committee of the UNECE (Warsaw, Poland, 1981). 1982 ISBN 90-247-3050-3
8. J.J. Douglas: *A Re-Appraisal of Forestry Development in Developing Countries.* 1983 ISBN 90-247-2830-4
9. J.C. Gordon and C.T. Wheeler (eds.): *Biological Nitrogen Fixation in Forest Ecosystems.* Foundations and Applications. 1983 ISBN 90-247-2849-5
10. M. NÇmeth: *Virus, Mycoplasma and Rickettsia Diseases of Fruit Trees.* Rev. (English) ed., 1986 ISBN 90-247-2868-1
11. M.L. Duryea and T.D. Landis (eds.): *Forest Nursery Manual.* Production of Bareroot Seedlings. 1984; 2nd printing 1987 ISBN Hb 90-247-2913-0; Pb 90-247-2914-9
12. F.C. Hummel: *Forest Policy.* A Contribution to Resource Development. 1984 ISBN 90-247-2883-5
13. P.D. Manion (ed.): *Scleroderris Canker of Conifers.* Proceedings of an International Symposium on Scleroderris Canker of Conifers (Syracuse, USA, 1983). 1984 ISBN 90-247-2912-2
14. M.L. Duryea and G.N. Brown (eds.): *Seedling Physiology and Reforestation Success.* Proceedings of the Physiology Working Group, Technical Session, Society of American Foresters National Convention (Portland, Oregon, USA, 1983). 1984 ISBN 90-247-2949-1
15. K.A.G. Staaf and N.A. Wiksten (eds.): *Tree Harvesting Techniques.* 1984 ISBN 90-247-2994-7
16. J.D. Boyd: *Biophysical Control of Microfibril Orientation in Plant Cell Walls.* Aquatic and Terrestrial Plants Including Trees. 1985 ISBN 90-247-3101-1
17. W.P.K. Findlay (ed.): *Preservation of Timber in the Tropics.* 1985 ISBN 90-247-3112-7
18. I. Samset: *Winch and Cable Systems.* 1985 ISBN 90-247-3205-0
19. R.A. Leary: *Interaction Theory in Forest Ecology and Management.* 1985 ISBN 90-247-3220-4
20. S.P. Gessel (ed.): *Forest Site and Productivity.* 1986 ISBN 90-247-3284-0
21. T.C. Hennessey, P.M. Dougherty, S.V. Kossuth and J.D. Johnson (eds.): *Stress Physiology and Forest Productivity.* Proceedings of the Physiology Working Group, Technical Session, Society of American Foresters National Convention (Fort Collins, Colorado, USA, 1985). 1986 ISBN 90-247-3359-6

FORESTRY SCIENCES

22. K.R. Shepherd: *Plantation Silviculture.* 1986 ISBN 90-247-3379-0
23. S. Sohlberg and V.E. Sokolov (eds.): *Practical Application of Remote Sensing in Forestry.* Proceedings of a Seminar on the Practical Application of Remote Sensing in Forestry (Jnping, Sweden, 1985). 1986 ISBN 90-247-3392-8
24. J.M. Bonga and D.J. Durzan (eds.): *Cell and Tissue Culure in Forestry.* Volume 1: General Principles and Biotechnology. 1987 ISBN 90-247-3330-4
25. J.M. Bonga and D.J. Durzan (eds.): *Cell and Tissue Culure in Forestry.* Volume 2: Specific Principles and Methods: Growth and Development. 1987 ISBN 90-247-3331-2
26. J.M. Bonga and D.J. Durzan (eds.): *Cell and Tissue Culure in Forestry.* Volume 3: Case Histories: Gymnosperms, Angiosperms and Palms. 1987 ISBN 90-247-3332-0
Set ISBN (Volumes 24–26) 90-247-3333-9
27. E.G. Richards (ed.): *Forestry and the Forest Industries: Past and Future.* Major Developments in the Forest and Forest Industries Sector Since 1947 in Europe, the USSR and North America. In Commemoration of the 40th Anniversary of the Timber Committee of the UNECE. 1987 ISBN 90-247-3592-0
28. S.V. Kossuth and S.D. Ross (eds.): *Hormonal Control of Tree Growth.* Proceedings of the Physiology Working Group, Technical Session, Society of American Foresters National Convention (Birmingham, Alabama, USA, 1986). 1987 ISBN 90-247-3621-8
29. U. Sundberg and C.R. Silversides: *Operational Efficiency in Forestry.* Volume 1: Analysis. 1988 ISBN 90-247-3683-8
30. M.R. Ahuja (ed.): *Somatic Cell Genetics of Woody Plants.* Proceedings of the IUFRO Working Party S2.04-07 Somatic Cell Genetics (Grosshansdorf, Germany, 1987). 1988 ISBN 90-247-3728-1
31. P.K.R. Nair (ed.): *Agroforestry Systems in the Tropics.* 1989 ISBN 90-247-3790-7
32. C.R. Silversides and U. Sundberg: *Operational Efficiency in Forestry.* Volume 2: Practice. 1989 ISBN 0-7923-0063-7
Set ISBN (Volumes 29 and 32) 90-247-3684-6
33. T.L. White and G.R. Hodge (eds.): *Predicting Breeding Values with Applications in Forest Tree Improvement.* 1989 ISBN 0-7923-0460-8
34. H.J. Welch: *The Conifer Manual.* Volume 1. 1991 ISBN 0-7923-0616-3
35. P.K.R. Nair, H.L. Gholz, M.L. Duryea (eds.): *Agroforestry Education and Training.* Present and Future. 1990 ISBN 0-7923-0864-6
36. M.L. Duryea and P.M. Dougherty (eds.): *Forest Regeneration Manual.* 1991 ISBN 0-7923-0960-X
37. J.J.A. Janssen: *Mechanical Properties of Bamboo.* 1991 ISBN 0-7923-1260-0
38. J.M. Bonga and P. Von Aderkas: *In Vitro Culture of Trees.* 1992 ISBN 0-7923-1540-5
39. L. Fins, S.T. Friedman and J.V. Brotschol (eds.): *Handbook of Quantitative Forest Genetics.* 1992 ISBN 0-7923-1568-5
40. M.J. Kelty, B.C. Larson and C.D. Oliver (eds.): *The Ecology and Silviculture of Mixed-Species Forests.* A Festschrift for David M. Smith. 1992 ISBN 0-7923-1643-6
41. M.R. Ahuja (ed.): *Micropropagation of Woody Plants.* 1992 ISBN 0-7923-1807-2
42. W.T. Adams, S.H. Strauss, D.L. Copes and A.R. Griffin (eds.): *Population Genetics of Forest Trees.* Proceedings of an International Symposium (Corvallis, Oregon, USA, 1990). 1992 ISBN 0-7923-1857-9

FORESTRY SCIENCES

43. R.T. Prinsley (ed.): *The Role of Trees in Sustainable Agriculture.* 1993 ISBN 0-7923-2030-1
44. S.M. Jain, P.K. Gupta and R.J. Newton (eds.): *Somatic Embryogenesis in Woody Plants*, Volume 3: Gymnosperms. 1995 ISBN 0-7923-2938-4
45. S.M. Jain, P.K. Gupta and R.J. Newton (eds.): *Somatic Embryogenesis in Woody Plants*, Volume 1: History, Molecular and Biochemical Aspects, and Applications. 1995 ISBN 0-7923-3035-8
46. S.M. Jain, P.K Gupta and R.J. Newton (eds.): *Somatic Embryogenesis in Woody Plants*, Volume 2: Angiosperms. 1995 ISBN 0-7923-3070-6
    Set ISBN (Volumes 44–46) 0-7923-2939-2
47. F.L. Sinclair (ed.): *Agroforestry: Science, Policy and Practice.* Selected Papers from the Agroforestry Sessions of the IUFRO 20th World Congress (Tampere, Finland, 6–12 August 1995). 1995 ISBN 0-7923-3696-8
48. J.H. Goldammer and V.V. Furyaev (eds.): *Fire in Ecosystems of Boreal Eurasia.* 1996 ISBN 0-7923-4137-6
49. M.R. Ahuja, W. Boerjan and D.B. Neale (eds.): *Somatic Cell Genetics and Molecular Genetics of Trees.* 1996 ISBN 0-7923-4179-1
50. H.L. Gholz, K. Nakane and H. Shimoda (eds.): *The Use of Remote Sensing in the Modeling of Forest Productivity.* 1996 ISBN 0-7923-4278-X
51. P. Bachmann, M. Köhl and R. Päivinen (eds.): *Assessment of Biodiversity for Improved Forest Planning.* Proceedings of the Conference on Assessment of Biodiversity for Improved Planning (Monte Verità, Switzerland, 1996). 1998 ISBN 0-7923-4872-9
52. G.M.J. Mohren, K. Kramer and S. Sabaté (eds.): *Impacts of Global Change on Tree Physiology and Forest Ecosystems.* Proceedings of the International Conference on Impacts of Global Change on Tree Physiology and Forest Ecosystems (Wageningen, The Netherlands, 26–29 November 1996). 1997 ISBN 0-7923-4921-0
53. P.K.R. Nair and C.R. Latt (eds.): *Directions in Tropical Agroforestry Research.* Selected Papers from the Symposium on Tropical Agroforestry organized in connection with the annual meeting of USA (Indianapolis, U.S.A., 5 November 1996). 1998 ISBN 0-7923-5035-9
54. K. Sassa (ed.): *Environmental Forest Science.* Proceedings of the IUFRO Division 8 Conference Environmental Forest Science. (Kyoto, Japan, 19–23 October 1998). 1998 ISBN 0-7923-5280-7
55. S.M. Jain, P.K Gupta and R.J. Newton (eds.): *Somatic Embryogenesis in Woody Plants*, Volume 4. 1999 ISBN 0-7923-5340-4
56. J.R. Boyle, J.K. Winjum, K. Kavanagh, E.J. Jensen (eds.): *Planted Forests: Contribution to the Quest for Sustainable Societies.* 1999 ISBN 0-7923-5468-0
57. K. von Gadow and G. Hui: *Modelling Forest Development.* 1999 ISBN 0-7923-5488-5
58. J. Abildtrup, F. Helles, P. Holten-Andersen, J.F. Larsen and B.J. Thorsen (eds.): *Modern Time Series Analysis in Forest Products Markets.* 1999 ISBN 0-7923-5524-5
59. S.M. Jain, P.K. Gupta and R.J. Newton (eds.): *Somatic Embryogenesis in Woody Plants*, Volume 5. 1999 ISBN 0-7923-5553-9
60. D. Auclair and C. Dupraz (eds.): *Agroforestry for Sustainable Land-Use.* Fundamental Research and Modelling with Emphasis on Temperate and Mediterranean Applications. 1999 ISBN 0-7923-5799-X

FORESTRY SCIENCES

61. F. Helles, P. Holten-Andersen and L. Wichmann (eds.): *Multiple Use of Forests and Other Natural Resources*. Aspects of Theory and Application. 1999 ISBN 0-7923-5859-7
62. A. Yoshimoto and K. Yukutake (eds.): *Global Concerns for Forest Resource Utilization*. Sustainable Use and Management. 1999 ISBN 0-7923-5968-2
63. C. Mátyás (ed.): *Forest Genetics and Sustainability*. 1999 ISBN 0-7923-6011-7
64. S.M. Jain and S.C. Minocha (eds.): *Molecular Biology of Woody Plants*, Volume 1. 1999 ISBN 0-7923-6012-5

KLUWER ACADEMIC PUBLISHERS – DORDRECHT / BOSTON / LONDON